Progress in Inflammation Research

Series Editor

Prof. Dr. Michael J. Parnham
PLIVA
Research Institute
Prilaz baruna Filipovica 25
10000 Zagreb
Croatia

Forthcoming titles:
The Hereditary Basis of Allergic Diseases, J. Holloway, S. Holgate (Editors), 2002
Bone Morphogenetic Proteins: From Laboratory to Clinical Practice, S. Vukicevic, K.T. Sampath (Editors), 2002
Inflammatory Processes and Cancer, D.W. Morgan, J. Winkler, U. Forssman (Editors), 2002
Anti-Inflammatory or Anti-Rheumatic Drugs, R.O. Day, D.E. Furst, P.L. van Riel (Editors), 2002
Heart Failure and Inflammation, G.Z. Feuerstein (Editor), 2002

(Already published titles see last page.)

Mechanisms and Mediators of Neuropathic Pain

Annika B. Malmberg
Sandra R. Chaplan

Editors

Springer Basel AG

Editors

Annika B. Malmberg (PhD)
NeurogesX, Inc.
969C Industrial Road
San Carlos, CA 94070
USA

Sandra R. Chaplan (MD)
RW Johnson Pharmaceutical Institute
3210 Merryfield Row
San Diego, CA 92121
USA

A CIP catalogue record for this book is available from the Library of Congress, Washington D.C., USA

Deutsche Bibliothek Cataloging-in-Publication Data
Mechanisms and mediators of neuropathic pain / Annika B. Malmberg ; Sandra R. Chaplan, ed.. - Basel ; Boston ; Berlin : Birkhäuser, 2002
(Progress in inflammation research)
ISBN 978-3-0348-9448-7 ISBN 978-3-0348-8129-6 (eBook)
DOI 10.1007/978-3-0348-8129-6

The publisher and editor can give no guarantee for the information on drug dosage and administration contained in this publication. The respective user must check its accuracy by consulting other sources of reference in each individual case.

The use of registered names, trademarks etc. in this publication, even if not identified as such, does not imply that they are exempt from the relevant protective laws and regulations or free for general use.

ISBN 978-3-0348-9448-7

Originally published by Birkhäuser Verlag in 2002
Softcover reprint of the hardcover 1st edition 2002
Member of the BertelsmannSpringer Publishing Group
Printed on acid-free paper produced from chlorine-free pulp. TCF ∞
Cover design: Markus Etterich, Basel
Cover illustration: Neuropeptide Y-immunoreactive neurons in the mouse spinal cord. Photo contributed by Nigel A. Calcutt.

ISBN 978-3-0348-9448-7

9 8 7 6 5 4 3 2 1

www.birkhauser-science.com

Contents

List of contributors

Ralf Baron, Klinik für Neurologie, Christian-Albrechts-Universität Kiel, Niemannsweg 147, 24105 Kiel, Germany; e-mail: r.baron@neurologie.uni-kiel.de

Joel A. Black, Department of Neurology and PVA/EPVA Neuroscience Research Center, Yale University School of Medicine, New Haven, CT 06510; and Rehabilitation Research Center, VA Connecticut, West Haven, CT 06516, USA; e-mail: Joel.Black@yale.edu

Hartmut Buerkle, Klinik und Poliklinik für Anästhesiologie und operative Intensivmedizin des Universitätsklinikums Münster, Albert-Schweitzer-Str. 33, 48149 Münster, Germany; e-mail: buerkle@anit.uni-muenster.de

Nigel A. Calcutt, Department of Pathology, 0612, University of California San Diego, La Jolla, CA 92093-0612, USA; e-mail: ncalcutt@ucsd.edu

Susan M. Carlton, Department of Anatomy and Neurosciences, Marine Biomedical Institute, University of Texas Medical Branch, Galveston, TX 77555-1069, USA; e-mail: smcarlto@utmb.edu

Victoria Chapman, School of Biomedical Sciences, University of Nottingham, E-Floor Medical School, Queen's Medical Centre, Nottingham, NG7 2UH, UK; e-mail: Victoria.Chapman@nottingham.ac.uk

Jin Mo Chung, Marine Biomedical Institute, University of Texas Medical Branch, 301 University Boulevard, Galveston, Texas 77555-1069, USA; e-mail: jmchung@utmb.edu

Kyungsoon Chung, Marine Biomedical Institute, University of Texas Medical Branch, 301 University Boulevard, Galveston, Texas 77555-1069, USA; e-mail: Kchung@utmb.edu

Richard E. Coggeshall, Department of Anatomy and Neurosciences, Marine Biomedical Institute, University of Texas Medical Branch, Galveston, TX 77555-1069, USA; e-mail: recogges@utmb.edu

Theodore R. Cummins, Department of Neurology and PVA/EPVA Neuroscience Research Center, Yale University School of Medicine, New Haven, CT 06510; and Rehabilitation Research Center, VA Connecticut, West Haven, CT 06516, USA; e-mail: Theodore.Cummins@yale.edu

Sulayman D. Dib-Hajj, Department of Neurology and PVA/EPVA Neuroscience Research Center, Yale University School of Medicine, New Haven, CT 06510; and Rehabilitation Research Center, VA Connecticut, West Haven, CT 06516, USA; e-mail: sulayman.dib-hajj@yale.edu

Anthony H. Dickenson, Department of Pharmacology, University College London, Gower Street, London WC1 E 6BT, UK

Stuart A. Dunbar, Dept. of Anesthesiology, Tufts University School of Medicine, Baystate Medical Center, 759 Chestnut Street, Springfield, MA 01199, USA; e-mail: sdunbar@library.bhs.org

Per Kristian Eide, Department of Neurosurgery, The National Hospital, University of Oslo, Sognvannsveien 20, 0027 Oslo, Norway; e-mail: per.kristian.eide@rikshospitalet.no

Jason D. Freshwater, Department of Pathology, 0612, University of California San Diego, La Jolla, CA 92093-0612, USA; e-mail: jfreshwater@ucsd.edu

Richard H. Gracely, Clinical Measurement and Mechanisms Unit, Pain and Neurosensory Mechanisms Branch, National Institute of Dental and Craniofacial Research, National Institutes of Health, Bethesda, MD 20892, USA; e-mail: Richard.Gracely@nih.gov

Kelly A. Gratto, Cameco MS/Neuroscience Research Center, Saskatoon City Hospital, Room 5800, 701 Queen Street, Saskatoon, Saskatchewan, Canada S7K 0M7

Laurie A. Karchewski, Neural Plasticity Research Group, Dept. of Anesthesia and Critical Care, Massachusetts General Hospital and Harvard Medical School, 149 13th St., Room 4309, Charlestown, MA 02129, USA

Martin Michaelis, Aventis Pharma Deutschland, Industriepark Hoechst, Building H821, 65926 Frankfurt/Main, Germany; e-mail: Martin.Michaelis@aventis.com

Karen M. Park, Department of Anesthesiology and Critical Care Medicine, George Washington Medical Center, Washington, D.C. 20037, USA

Srinivasa N. Raja, Department of Anesthesiology and Critical Care Medicine, The Johns Hopkins Medical Institutions, 600 N. Wolfe St., Osler 292, Baltimore, MD 21287, USA; e-mail: sraja@jhmi.edu

Linda S. Sorkin, Anesthesia Research Labs, University of California, San Diego, La Jolla, CA 92103-0818, USA; e-mail: lsorkin@ucsd.edu

Charles P. Taylor, CNS Pharmacology, Pfizer Global Research & Development, 2800 Plymouth Rd., Ann Arbor, MI 48105, USA; e-mail: charles.taylor@pfizer.com

Valerie M.K. Verge, Cameco MS/Neuroscience Research Center, Saskatoon City Hospital, Room 5800, 701 Queen Street, Saskatoon, Saskatchewan, Canada S7K 0M7; e-mail: Verge@sask.usask.ca

Mark S. Wallace, Department of Anesthesiology, University of California, San Diego, School of Medicine, 9500 Gilman Drive #0924, La Jolla, CA 92093-0924, USA; e-mail: mswallace@ucsd.edu

Stephen G. Waxman, Department of Neurology and PVA/EPVA Neuroscience Research Center, Yale University School of Medicine, New Haven, CT 06510; and Rehabilitation Research Center, VA Connecticut, West Haven, CT 06516, USA; e-mail: stephen.waxman@yale.edu

Tracy D. Wilson-Gerwing, Cameco MS/Neuroscience Research Center, Saskatoon City Hospital, Room 5800, 701 Queen Street, Saskatoon, Saskatchewan, Canada S7K 0M7

Preface

The first formal description of neuropathic pain as a distinct clinical entity occurred during the American Civil War. In 1864 army surgeons Silas Weir Mitchell, G.R. Moorhouse and William W. Keen reported on their treatment of wounded soldiers with unusual pain syndromes in a book entitled *Gunshot Wounds and Other Injuries of Nerves* [1]. The large numbers of such injuries, the contemporary practice of housing patients with similar disorders in dedicated facilities, and Dr. Weir Mitchell's background in neurology and research methodology (he had studied with Claude Bernard) surely helped to create favourable circumstances for important insights into the unique consequences of nerve damage. The features that brought these neuropathic syndromes to particular attention were their severity, their lack of response to analgesic therapies, and the permanence of the pain state despite wound healing. The unusual characteristics of the pain syndromes that these authors documented included spontaneous burning and lancinating pain, pain responses to non-noxious stimuli (allodynia) and exaggerated pain responses to noxious stimuli (hyperalgesia).

More than a century later, neuropathic conditions are now recognised to result from a wide range of etiologies, including but not limited to trauma, infection, metabolic derangement, autoimmune diseases, malignancies, and drug/radiation exposure. Until very recently, treatment options remained unsatisfactory: a lifetime of marginally effective opiate maintenance, or progressive neurodestructive surgeries, the latter often resulting in cumulative neuropathic syndromes.

While the specific "causes" of neuropathic pain remain the subject of active research, recent significant advances in pain management have emerged both from clinical recognition of the unique pharmacology of the post nerve injury state, and from improved preclinical understanding of the underlying mechanisms. Novel pharmacological agents are bringing new hope into the clinic as well as a surge of interest in accurate and specific diagnostic tools, controlled trials, and quantitative clinical research methodology.

Major improvements in our understanding have come from the development of animal models, most of which involve partial peripheral nerve injury. Models in which some primary afferent neurons are preserved allow behavioural analysis of changes in sensory processing as well as tissue-based studies. These models have shown that nerve injury initiates a cascade of events resulting in altered neurochemistry and molecular biology of peripheral neurons, dorsal root ganglion cells, and the dorsal horn of the spinal cord. Moreover, nerve injury produces anatomical changes with functional consequences; for example, sprouting of postganglionic noradrenergic axons occurs around DRG neurons, and non-nociceptive large myelinated primary afferents extend their central terminals into areas of nociceptive processing in the spinal dorsal horn.

This book aims to summarise the current understanding of the pathophysiological processes in the peripheral and central nervous system that contribute to neuropathic pain. We have divided the book into four sections: (I) Peripheral consequences of nerve injury, (II) Spinal cord plasticity following peripheral nerve injury, (III) Neuropathic pain: preclinical and clinical paradigms, and (IV) New treatment strategies for neuropathic pain. We hope that this book provides an insight into current understandings of some of the mechanisms and modulators of neuropathic pain. We recommend this book for clinicians and scientists interested in the molecular machinery involved in nerve injury induced pain states.

Annika B. Malmberg
Sandra R. Chaplan

1 Mitchell SW, Moorhouse GR, Keen WW (1864) *Gunshot Wounds and Other Injuries of Nerves*. Lippincott, Philadelphia

Peripheral consequences of nerve injury

Electrophysiological characteristics of injured peripheral nerves

Martin Michaelis

Aventis Pharma Deutschland, Industriepark Hoechst, Building H821, 65926 Frankfurt/Main, Germany

Introduction

Injury of a peripheral nerve interrupts the centripetal action potential conduction in axons of many sensory neurons. Since it is the peripheral, receptive ending that is the only part of the sensory neuron specialised for action potential generation under normal conditions, one would expect proximal parts of sensory neurons to become and remain silent following injury. Consequently, patients with peripheral nerve lesions should complain of sensory deficits.

Paradoxically, the opposite often happens and many patients with peripheral nerve lesions suffer from positive symptoms like paresthesias, pain, hyperalgesia or allodynia [1]. The explanation is that proximal parts of injured sensory neurons start to produce ectopic action potential activity. This chapter reviews experimental evidence for nerve injury induced hyperexcitability in primary afferent neurons along with considerations about fundamental pathophysiological mechanisms. Correlations of original sensory function and forms of injury induced hyperexcitability are particularly emphasised.

Injury discharge

In many affected axons, discharges were elicited at the time when the injury hits the nerve. The onset of such injury discharge generally consisted of a brief burst of action potentials [2]. With respect to the duration, there were significant differences between functional types of afferent nerve fibres: While injury discharge faded within seconds in most myelinated axons of rapidly adapting cutaneous afferents [2, 3], it lasted up to 15 min in about 20% of unmyelinated axons of cutaneous afferents many of them being nociceptors [3]. This prolonged injury discharge in unmyelinated axons exhibited an irregular pattern at a mean frequency of 2 Hz 15 seconds after injury and at less than 0.5 Hz after 4 min. Some slowly adapting cutaneous

Mechanisms and Mediators of Neuropathic Pain, edited by Annika B. Malmberg and Sandra R. Chaplan

and muscle spindle afferents exhibited injury discharges of higher frequency (5–10 Hz) that lasted even longer than 30 min. [4].

The major factor responsible for the generation of injury discharges is membrane depolarisation close to the lesion site. Injured axons remain depolarised for at least a few minutes until the cut axon end reseals and membrane potential starts to recover [5].

Injury discharge in sensory axons is conducted centrally to the CNS. Here, it has two consequences: First, it will directly cause paresthesias and pain that last as long as the injury discharge itself; secondly, injury discharge may trigger longer lasting events (e.g. central sensitisation). Support for the latter comes from studies that investigated pain related behaviour in animals that had received experimental nerve lesions before testing: When injury discharge was prevented by local anesthetic block of the sciatic nerve prior to injury, resulting hyperalgesia and autotomy were modest [6–8]. *Vice versa*, intensifying the injury discharge resulted in an increased degree of autotomy [6]. These results supported the assumption that preventing intense sensory input from gaining access to the CNS in patients might help to prevent or reduce postoperative pain. However, it is still a matter of debate under which conditions such pre-emptive analgesia significantly reduces pain [9–11].

Evoked ectopic activity

When the peripheral axon of a sensory neuron is injured, the cell soma together with the axon proximal to the lesion often survive while the distal part including the receptive ending which is no longer connected to the cell soma dies (Wallerian degeneration). In surviving sensory neurons, fine sprouts start to grow out from the proximal axon end within a few hours after injury [12]. Some sprouts finally find their way and reinnervate target tissue. When regenerative outgrowth is blocked, sprouts form a tangled mass, a neuroma, that may enclose all surviving axons, e.g. in a nerve end neuroma following nerve ligation, or just a few axons in microneuromata that can be found distally to the prior lesion site along the regenerated part of the nerve. The tips of regenerating afferent sprouts, no matter whether growing towards target tissue or kept in a neuroma are particularly sensitive and are an important site of ectopic activity generation.

Ectopic mechanosensitivity

An impressive feature of tips of outgrowing afferent axons is their delicate sensitivity for mechanical stimuli. This ectopic mechanosensitivity has been demonstrated in a number of studies on different mammalian species, including primates, by recording activity proximal to the lesion site from afferent axons that responded to gentle mechanical stimulation of nerve end neuromata [13–23]. Likewise, mechanical stimulation of sites where a peripheral nerve or dorsal root ganglion has been chronically compressed, excited many lesioned afferents [24–27]. Although most studies concentrated on ectopic mechanosensitivity in myelinated A fibres, unmyelinated C fibres exhibited this pathophysiological property too [16–18, 21]. In this context it is important to note that almost no ectopic mechanosensitivity was found when intact nerves were stimulated with similar intensities (Fig. 1) [19, 21]. This clearly shows that the ectopic mechanosensitivity is a consequence of prior nerve injury and not a trivial result due to the experimental procedure. Ectopic mechanosensitivity is restricted to injured axons of afferent neurons and does not occur in axotomised motoneurons [19, 20].

Two patterns of mechanically evoked discharge have been observed: Some axons responded with brief bursts of action potentials at the moment of stimulus onset. Others exhibited slowly adapting discharges for as long as a constant force was applied [21, 25]. These two patterns probably reflect differences in underlying encoding mechanisms (see below).

The temporal onset of ectopic mechanosensitivity following nerve transection was different in cutaneous rat A and C fibre axons: Almost no myelinated axon responded for eight hours after nerve lesion. Then, the first mechanosensitive A fibres appeared and subsequently their incidence started to rise comprising about 30% of all A fibres 24 h after lesion (Fig. 1) [19, 21, 28]. In contrast, 5–8% of C fibre axons responded to the same mechanical stimuli already 30 min after injury. Moreover, their prevalence did not exhibit a significant change with time (Fig. 1) [21].

There is no uniform information about the incidence of ectopic mechanosensitivity in chronic neuromata. Some reports state that the incidence of mechanosensitive afferents reached a peak about one month after injury and decreased thereafter [14, 17, 23, 26] whilst others found most mechanosensitive axons later than two months after lesion but not earlier [16]. Still others did not find a significant correlation between incidence of ectopic mechanosensitivity and time after injury [18].

The clinical relevance of the aforementioned animal data has been demonstrated by microneurographic recordings from peripheral nerve fibres in awake patients [29]. These patients suffered from paresthesias which could be evoked by different manoeuvres. In all patients, mechanically evoked ectopic activity correlated in intensity and time course with the reported symptoms.

Ectotopic thermosensitivity

The effect of thermal stimuli on activity in lesioned afferents has been investigated less intensive. One recent systematic study on ectopic thermosensitivity in axotomised cutaneous afferents [30] demonstrated that 4–24 h after sural nerve tran-

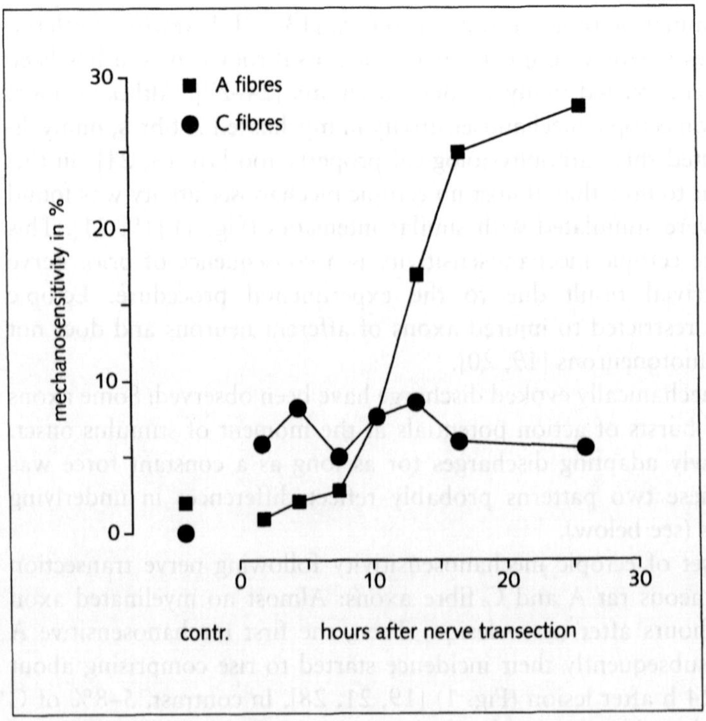

Figure 1
Prevalence of sural nerve axons that were excited by mechanical stimuli applied either to the intact nerve (contr.) or to the lesion site at different times after nerve injury.
The sural nerve was ligated and cut close to the ankle. Repetitive brief probing stimuli using a small blunt glass rod were applied to the nerve 1–2 mm proximal to the ligature. In control experiments, the intact sural nerve was stimulated close to the ankle by mechanical stimuli of virtually the same intensity as in nerve lesion experiments. Number of axons excited by such stimuli are given in percent of axons responding to electrical stimulation of the sural nerve separately for myelinated A fibres (cond. vel. >2 m/s) and unmyelinated C fibres (cond. vel. ≤ 2 m/s). Data obtained from [21], with permission.

section in rats, some 10% of unmyelinated axons responded in a graded fashion to thermal stimulation of the lesion site, whilst all A fibres were unresponsive. Most C fibres were either cold- or warm-sensitive; just a few were both cold- and warm-sensitive. Response characteristics differed in that cold fibre responses were significantly more phasic than warm fibre responses [30]. The same type of cold and warm stimuli excited some unlesioned C fibres, but the incidence of ectopic thermosensitivity was significantly lower in intact than in lesioned nerves [30].

In chronically lesioned nerves, some myelinated afferent fibres also exhibited ectopic thermosensitivity [31, 32]: A few days after crushing the plantar nerve in cats, some sprouting A fibres were excited by cooling the lesion site but not by heating [31] whereas A fibres in a rat sciatic nerve end neuroma were exclusively excited by heating but not by cooling [32]. Unmyelinated chronic neuroma afferents are preferentially excited by cooling and suppressed by heating [32].

Ectopic cold sensitivity of neuroma afferents may be a relevant mechanism in the generation of cold intolerance and cold hyperalgesia in patients with peripheral nerve lesions.

Ectopic chemosensitivity

Mediators and metabolic factors of inflammation

Every nerve lesion is accompanied by a local inflammatory reaction resulting in enhanced concentrations of inflammatory mediators in the micromilieu surrounding tips of outgrowing axon. Bradykinin, prostaglandins or histamine locally applied to experimental neuromata for 2–5 min either separately or in combination excited some afferents preferentially those with thinly myelinated and unmyelinated axons [18, 33–35]. Neuroma superfusion with acid solutions that mimic lowered pH in inflamed tissue likewise excited some axotomised C fibres [35]. The excitatory responses often consisted of a burst of action potentials lasting for a few seconds. Sometimes, the response outlasted the application time for up to 5 min. Responses to a second stimulus were usually much smaller indicating strong tachyphylaxis. Excitatory responses to inflammatory mediators could be demonstrated as early as 2 h after nerve lesion [35] whereas intact nerve trunks were unresponsive [31, 35]. This ectopic chemosensitivity remained for months in chronic neuromata [18]. On average, about 10% of unmyelinated axons responded to at least one of the mediators when tested hours after nerve lesion [35]; the incidence tended to be lower after days or weeks without a clearcut time-dependency [18].

In addition to immediate excitatory effects, inflammatory mediators also enhanced the sensitivity of axotomised nerve fibres to mechanical and thermal stimuli. When a mixture of several mediators was applied to the nerve lesion site, von Frey thresholds declined and response magnitudes increased in most ectopically mechanosensitive C fibres, and in some A fibres, too. Moreover, a few previously silent fibres became mechanosensitive after mediator application [36]. The sensitisation lasted for 10-40 minutes following a brief application of two minutes [36]. Bradykinin alone enhanced the ectopic mechanosensitivity in some axotomised C fibres, albeit to a lesser extent than the mixture of mediators [37]. Moreover, bradykinin significantly enhanced the C fibre responses evoked by local warming of the nerve stump [37] and lowered their activation threshold, in some fibres even below 37 °C [38].

The aforementioned studies have shown that inflammatory mediators excite axotomised neurons or enhance their ectopic excitability and that these effects are relatively short lasting when the mediators have been applied for a few minutes. But there is also evidence for a chronic correlation of inflammation and neuropathic pain. In the chronic constriction injury model of neuropathic pain, inflammation around the injury has been proven to be fundamental for the development of hyperalgesia [39, 40]: reduction of the perineural inflammation blocked the development of pain related behaviour, while on the other hand pain related behaviour was augmented when the inflammation was enhanced [40].

Thus, certain inflammatory factors may critically contribute to the generation of neuropathic pain states. Promising candidate molecules in addition to the aforementioned classical mediators are pro-inflammatory cytokines, e.g. tumour necrosis factor α (see the chapter by L. Sorkin), and neurotrophins [41, 42].

Catecholamines

Catecholamines are sympathetic transmitters that do not excite receptive endings of intact cutaneous afferents [43]. Accordingly, catecholamines are not algesic in healthy humans [44]. However, induced by peripheral nerve lesions, many neuroma afferents some of them probably being nociceptors get excitable by sympathetic transmitters [13, 14, 16, 18, 45–48]. In line with this experimental evidence is the finding that in humans with painful chronic nerve end neuromata, perineuromal injections of epinephrine exacerbated their pain [49].

Axon tips are not the only sites of afferent neurons that may become sensitive to catecholamines after nerve lesion. Moreover, if a nerve is only partially lesioned, receptive endings of some spared intact afferents acquire novel ectopic adrenosensitivity [50-53]. In addition, somata of some sensory neurons in DRGs that contain axotomised sensory neurons start to respond to sympathetic transmitters [54–59] (see also below).

Ectopic catecholamine sensitivity of afferent neurons following peripheral nerve injury may be important for the generation of sympathetically maintained pain in some neuropathic pain patients [60].

Mechanisms

Anterograde axoplasmic transport and ectopic mechanosensitivity

The sensory properties of intact primary afferent neurons are essentially based on ensembles of specific receptors and ion channels expressed in the axon membrane of receptive endings. Novel membrane components that are permanently being synthesised in afferent cell somata get to the receptive ending *via* anterograde axonal transport where the axon membrane is constantly being replaced by new components arriving from the cell soma. Injury of a peripheral nerve suddenly blocks the anterograde transport at the lesion site. Since the anterograde transport continues after axotomy, all trafficking proteins gradually accumulate just proximal to the lesion site as has been demonstrated for selected neuronal proteins (e.g. sodium channels; rat [61, 62]; human [63, 64]). Based on this consideration, it has been hypothesised that membrane components which were originally thought for the receptive ending, not only accumulate but are in addition functionally incorporated into axon membrane proximal to a nerve lesion site. This erroneous membrane remodelling is thought to be fundamental for the generation of ectopic excitability [65].

There is convincing experimental evidence that membrane remodelling is indeed essential for the development of ectopic mechanosensitivity in axotomized A fibres: First, the incidence of mechanosensitive A fibres characteristically increased with time after nerve transection (Fig. 1) [19, 21] indicating that transported organelles are gradually incorporated into the membrane as they accumulate. Second, briefly after pharmacological blockade of anterograde axonal transport at mid nerve, myelinated axons became mechanosensitive at the site of the block [66]. And finally, comparing proximal and distal nerve end after nerve transection, it was always exclusively the proximal end where anterogradely transported material accumulates but never the distal nerve end that became mechanosensitive [19].

In lesioned unmyelinated fibres, however, novel incorporation of mechanotransducers into axon membrane does not seem to be comparably important for the development of ectopic mechanosensitivity. This is concluded from experimental data showing that some 5% of axotomised C fibres exhibited ectopic mechanosensitivity as early as 30 min after nerve transection and that the incidence of ectopic mechanosensitivity in C fibres did not change unequivocally with time after lesion (Fig. 1) [21]. A likely explanation for this fast appearance of ectopic C fibre mechanosensitivity is that already intact C fibre axons have the capability of transducing mechanical forces into action potential activity but that under normal conditions axonal generator potentials do not reach threshold. Axotomy now unstabilise the axon membrane potential close to the lesion site with the consequence that mechanically evoked generator potentials now reach threshold so that ectopic mechanosensitivity becomes apparent.

A general hypothesis

The final result of the pathophysiological processes after nerve lesion is likely the same in A and C fibres, in spite of different underlying mechanisms: The axon tip of each injured afferent neuron has the tendency to develop the same types of sensitivities that the receptive ending of this neuron had before injury.

Although direct evidence for this general hypothesis is missing, there is accumulating supportive evidence. First, in acute skin nerve neuromata, many axotomised afferents with thickly myelinated axons were mechanosensitive but not thermosensitive [21, 30] very much like intact myelinated skin afferents with axons of same calibre [67]. Secondly, responses to vibratory stimuli applied to skin nerve neuromata revealed that according to the frequency range of maximum sensitivity, the A fibres could be classified into three groups that were reminiscent of three well known classes of cutaneous afferents [19]. Thirdly, most axotomized muscle afferents exhibited slowly adapting discharges to sustained mechanical stimuli [20, 68], whereas among lesioned skin afferents rapidly adapting responses were more frequent [20, 36]. Similarly, most intact skin afferents are rapidly adapting and most intact muscle afferents are slowly adapting mechanoreceptors [69, 70]. Moreover, in three species of electric fish producing electric fields of characteristically different frequencies, many axotomised electrosensory neurons exhibited preserved frequency differences in their ectopic spontaneous firing [71]. A remarkable corollary of these results is that characteristic patterns of response firing in myelinated afferents apparently expresses inherent properties of axon endings independent of specific receptor organs.

A recent study has determined the distribution of sensory properties among axotomized cutaneous C fibres and this distribution was found to be very similar to the distribution of receptor types among intact skin afferents [72]. This corroborates the general hypothesis that axotomised afferents tend to exhibit the same sensory properties as their former receptive endings.

Ectopic catecholamine sensitivity

The novel ectopic sensitivity to catecholamines in some lesioned afferents is not a re-expression of a former property, since excitatory sensitivity to catecholamines does not belong to the normal sensory repertoire of intact afferents (see above).

Two different mechanisms should be considered for the generation of ectopic catecholamine sensitivity. First, functional α-adrenoceptors are expressed in lesioned afferent neurons and their activation is excitatory [73–75]. Such direct excitatory catecholamine effect is not normally found, although some intact primary afferents express adrenoceptors at their central terminals in the spinal cord but their activation inhibits transmitter release [76]. Second, peripheral nerve lesions enhance the sensitivity of axotomised sensory neurons to hypoxia. Thus, vasoconstriction evoked by catecholamine release from sympathetic neurons (or by other factors) may lower oxygen pressure below the critical level so that hypoxia-sensitive axotomised afferent neurons are excited [46, 59, 77].

Ongoing ectopic activity

After peripheral nerve injuries, many lesioned afferent neurons start to produce ongoing action potential activity in the absence of any obvious stimulus ("spontaneous" activity). Ectopic sites where ongoing activity may originate are tips of regenerating axons and dorsal root ganglia (DRG) that contain somata of lesioned afferents.

Different discharge patterns have been observed in lesioned afferents: rhythmic discharges with constant interspike intervals that appear either in bursts or tonic, and non-rhythmic irregular discharges.

Ongoing activity originating from axon tips

Transection of the sciatic nerve has often been used as an experimental model to study ectopic ongoing activity in neuroma afferents, since the sciatic nerve is the largest peripheral nerve and it is easy accessible. In sciatic nerve A fibres, most discharge was rhythmic [14, 22, 46, 47, 78]. The incidence of fibres exhibiting ongoing activity was almost zero during the first hours, reached a peak of about 25% 14 days after lesion [14, 46, 78] declining thereafter to some 5%. Mean burst frequency was about 20 Hz [14, 47, 78] and did not change with time after lesion [78].

Differences between skin and muscle afferents in their ability to generate ectopic ongoing activity after axotomy were investigated after transection of nerves supplying exclusively either skin or skeletal muscle. By far most firing myelinated neuroma fibres originally supplied muscle [20, 68, 79] whereas most or almost all myelinated skin neuroma fibres were silent [16–21, 68, 79]. As in sciatic nerve neuroma afferents, the most common discharge pattern was rhythmic with a mean burst frequency of about 20 Hz [20, 68, 79]. In addition, low frequency irregular activity was found in 3–8% of all C fibre afferents in skin neuromata in rats [18, 21] and in 10–13% of the C fibres in cat and monkey [16, 17].

Almost all spontaneously active axotomised A fibres were particularly sensitive to mechanical stimulation of the nerve lesion site. Therefore it has been suggested that in these A fibres ongoing discharges are the expression of extremely low threshold for mechanical excitation [20, 25]. This is very likely not the case in most C fibres briefly after skin nerve transection because only a minority of C fibres with ongoing activity were additionally mechanosensitive. However, most spontaneously firing C fibres were additionally thermosensitive [72] so that in these C fibres spontaneous activity is more likely a consequence of low thermal threshold.

Ongoing activity originating from DRG

Cell somata of afferent neurons in DRGs containing axotomised sensory neurons contribute to the spontaneous ectopic barrage. Systematic studies in rats revealed that after sciatic nerve lesion a maximum of 10% of all axotomized DRG neurons with myelinated axons projecting into the sciatic nerve produce ongoing activity [15, 80–82]. Ectopic discharges originated in DRG also after chronic constriction injury [56, 57, 83–85] and after spinal nerve lesion [86–88]. There was a substantial rise in the incidence of DRG neurons generating ongoing activity 16–24 h following spinal nerve injury [87, 88] but not before three days after sciatic nerve injury [88]. Thus, ongoing activity in the DRG ocurred early when the nerve lesion was close to the DRG, and later after a remote lesion.

Ongoing activity in DRG neurons with unmyelinated axons was only rarely encountered after sciatic nerve lesion [82] and it was totally absent after spinal nerve lesion [86, 88]. The discharge of DRG origin was often rhythmic during the first days after injury, later the pattern was preferentially irregular at a mean frequency of <5 Hz [56, 57, 82, 83, 88, 89].

Whether or not a DRG neuron was susceptible of developing spontaneous activity of DRG origin after peripheral nerve lesion, essentially depended on the target tissue of the neuron prior to axotomy. Whilst in the rat about 20% of all neurons with myelinated axons projecting in the lesioned gastrocnemius-soleus (GS) nerve fired spontaneously, all DRG neurons projecting into the cut sural nerve remained silent [89]. Thus, somata of muscle afferents but not of cutaneous afferents produce ongoing discharges after peripheral nerve lesion. In the cat, similar results have been obtained showing that there was no spontaneous activity of DRG origin in lesioned skin afferents [16] but in some injured muscle afferents [90]. Therefore it is reasonable to assume that ongoing activity of DRG origin that has been observed after lesion of mixed nerves like the sciatic nerve [55, 80–82] or spinal nerve L5 [86–88] very likely also appeared exclusively in muscle afferents. A similar discrepancy has been reported for muscle and skin neuroma afferents to produce ongoing discharges (see above).

Almost all muscle afferents with spontaneous activity had medium or low conduction velocity [89] a third of them being nociceptors [91]. Ectopic ongoing activity in such thinly myelinated muscle afferents may evoke ongoing muscle pain or abnormal proprioceptive sensations and, moreover, pain referred to the skin or other deep somatic structures [92, 93]. A cardinal symptom of neuropathic pain is tactile allodynia which is based on a state of central sensitisation. Activity in muscle afferents can produce a much stronger central sensitisation than activity in skin afferents [94]. It is thought that input from C fibre but not A fibre muscle afferents induces central sensitisation. However, nerve lesions trigger BDNF (brain derived neurotrophic factor) expression in many large diameter DRG neurons [95, 96]. Since there is evidence that BDNF is released from DRG neurons in the spinal cord [97] where it enhances the excitability of central nociceptive neurons [98], such a phenotypic switch of myelinated muscle afferents for BDNF could explain how ectopic activity in these afferents induces central sensitisation, even in the absence of C fibre activity [88].

Mechanisms

Ongoing activity from DRG: indications for a paracrine signal

It has been suggested that a paracrine signal is liberated in DRG in consequence of a peripheral nerve injury and that it is this paracrine signal that triggers the ectopic impulse generation in muscle afferents. Two lines of experimental evidence speak in favour of this hypothesis. First, the number of lesioned DRG neurons determine the firing rate of muscle DRG neurons: When in rats only two small branches of the sciatic nerve had been sectioned, about 20% of DRG neurons projecting into the GS nerve exhibited ongoing discharges at a mean discharge rate of 3 Hz but when all major branches of the sciatic nerve had been cut, mean discharge rate of GS neurons was 6 Hz (Fig. 2) [89]. Secondly, GS neurons themselves do not have to be lesioned themselves but lesion of neighbouring afferents sharing the same DRG is sufficient to make GS neurons firing: In rats with intact GS nerve but transected major branches of sciatic nerve, almost 20% of unlesioned GS neurons exhibited ongoing discharges. The percentage of GS neurons with ongoing activity and their mean discharge rate were similar in those rats with all peripheral sciatic nerve branches including the GS nerve being cut (Fig. 2) [89]. GS neurons in intact DRG were, however, silent. Thus, a nerve lesion is pre-requisite to the development of ectopic discharges. The signal that induces ectopic ongoing activity following nerve injury is yet unknown. Speculation is permitted that it may be a neurotrophin or cytokine which significantly determine injury induced neuronal hyperexcitability under a variety of conditions [99].

Ectopic impulse generation depends on membrane potential oscillations

Recently it has been demonstrated that ectopic action potential generation in DRG essentially depends on certain intrinsic resonant properties of the cell membrane of these neurons [100]. Some DRG neurons exhibited subthreshold oscillations in their membrane potential and nerve injury enhanced the incidence of DRG neurons exhibiting oscillations. A TTX-sensitive Na^{+} conductance contributed to the oscillations. Oscillations gave rise to action potentials when they reached threshold (Fig. 3A). Most important, the presence of oscillations were a necessary condition for all patterns of ongoing discharges (Fig. 3A, B); neurons without membrane potential oscillations were incapable of sustained discharge [100, 101].

It is worthwhile to assume that the aforementioned paracrine signal that is presumably liberated in DRG following peripheral nerve injury, enhances the incidence of membrane potential oscillations. Moreover, it is likely that membrane potential oscillations are also pre-requisite for generation of ectopic action potential generation at outgrowing axons tips.

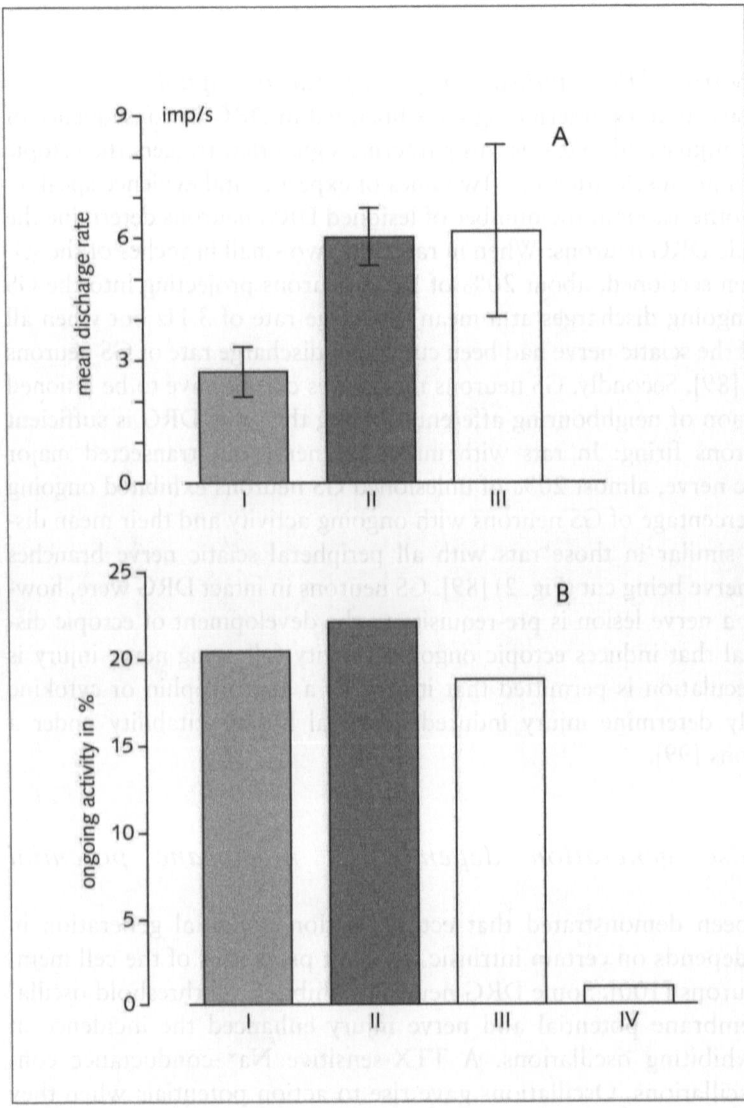

Figure 2
Mean discharge rate (± SEM) (A) and incidence (B) of dorsal root ganglion neurons with myelinated axons projecting into the gastrocnemius-soleus (GS) nerve exhibiting ongoing activity. Three experimental groups of rats have been used: I, GS and sural nerve cut (small lesion); II, tibial, peroneal, sural and GS nerve cut (large lesion); III, GS nerve intact, tibial, peroneal, sural nerve cut (large lesion). Electrophysiological experiments were performed 6–12 days after nerve injury. Rats without prior nerve lesion served as control group (IV). Data obtained from [102], with permission.

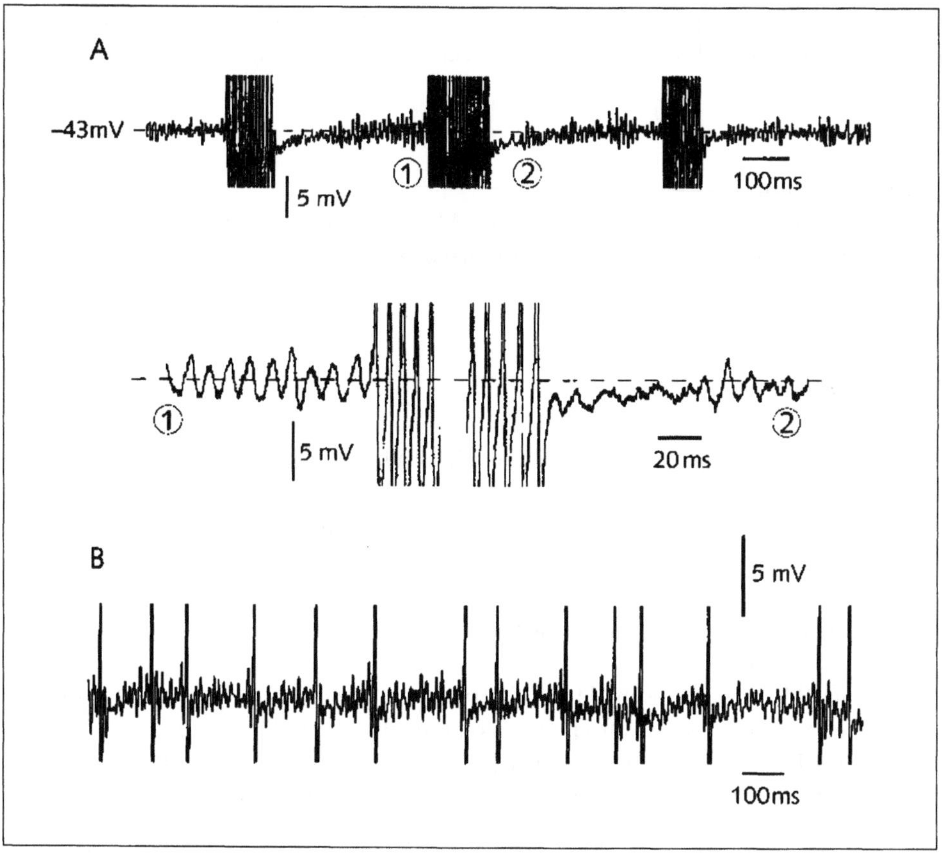

Figure 3
Subthreshold membrane potential oscillations trigger action potentials in DRG neurons. (A) Spontaneous repetitive bursts of action potentials in a DRG A_0 neuron from an immature rat. Segments of the record at time points 1 and 2 are shown below on a faster timebase to illustrate the triggering of spike bursts by membrane potential oscillations. (B) DRG A_0 neuron from an adult rat exhibiting oscillations with intermittent spiking seven days after cutting the spinal nerve L5.
Spikes are truncated throughout. Modified from [102] (A) and [101] (B), with permission.

Conclusion

Experimental as well as clinical studies have proven that sensory neurons start to generate ectopic activity following nerve lesions. Moreover, this ectopic activity develops as a result of highly specific pathophysiological processes. Since ectopic activity evokes neuropathic pain and related symptoms, increasing knowledge of these mechanisms will help to improve selective pharmacological treatment of neuropathic pain.

Acknowledgements
The expert help with the illustrations of Ms Eike Tallone is gratefully acknowledged. The author's work has been supported by the Deutsche Forschungsgemeinschaft.

References

1 Bennett GJ (1994) Neuropathic pain. In: PD Wall, R Melzack (eds): *Textbook of pain.* Churchill Livingstone, Edinburgh, 201–224
2 Wall PD, Waxman S, Basbaum AI (1974) Ongoing activity in peripheral nerve: injury discharge. *Exp Neurol* 45: 576–589
3 Blenk K-H, Jänig W, Michaelis M, Vogel C (1996) Prolonged injury discharge in unmyelinated nerve fibres following transection of the sural nerve in rats. *Neurosci Lett* 215: 185–188
4 Chung JM, Leem JW, Kim SH (1992) Somatic afferent fibers which continuously discharge after being isolated from their receptors. *Brain Res* 599: 29–33
5 Spira ME, Benbassat D, Dormann A (1993) Resealing of the proximal and distal cut ends of transected axons: electrophysiological and ultrastructural analysis. *J Neurobiol* 24: 300–316
6 Seltzer Z, Beilin B, Ginzburg R, Paran Y, Shimko T (1991) The role of injury discharge in the induction of neuropathic pain behavior in rats. *Pain* 46: 327–336
7 Dougherty PM, Garrison CJ, Carlton SM (1992) Differential influence of local anesthetic upon two models of experimentally induced peripheral mononeuropathy in the rat. *Brain Res* 570: 109–115
8 Yamamoto T, Shimoyama N, Mizuguchi T (1993) Role of the injury discharge in the development of thermal hyperesthesia after sciatic nerve constriction injury in the rat. *Anesthesiology* 79: 993–1002
9 Bach S, Noreng MF, Tjellden NU (1988) Phantom limb pain in amputees during the first 12 month following limb amputation, after preoperative lumbar epidural blockade. *Pain* 33: 297–301
10 Nikolajsen L, Ilkjaer S, Christensen JH, Kroner K, Jensen TS (1997) Randomised trial of epidural bupivacaine and morphine in prevention of stump and phantom pain in lower-limb amputation *Lancet* 350: 1353–1357
11 Warncke T, Stubhaug A, Jorum E (2000) Preinjury treatment with morphine or ketamine inhibits the development of experimentally induced secondary hyperalgesia in man. *Pain* 86: 293–303

12 Fawcett JW, Keynes RJ (1990) Peripheral nerve regeneration. *Annu Rev Neurosci* 13: 43-60

13 Wall PD, Gutnick M (1974) Properties of afferent nerve impulses originating from a neuroma. *Nature* 248: 740–743

14 Scadding JW (1981) Development of ongoing activity, mechanosensitivity, and adrenaline sensitivity in severed peripheral nerve axons. *Exp Neurol* 73: 345–364

15 Burchiel KJ (1984) Effects of electrical and mechanical stimulation on two foci of spontaneous activity which develop in primary afferent neurons after peripheral axotomy. *Pain* 18: 249–265

16 Blumberg H, Jänig W (1984) Discharge pattern of afferent fibers from a neuroma. *Pain* 20: 335–353

17 Meyer RA, Raja SN, Campbell JN, Mackinnon SE, Dellon AL (1985) Neural activity originating from a neuroma in the baboon. *Brain Res* 325: 255–260

18 Welk E, Leah JD, Zimmermann M (1990) Characteristics of A- and C-fibers ending in a sensory nerve neuroma in the rat. *J Neurophysiol* 63: 759–766

19 Koschorke GM, Meyer RA, Tillman DB, Campbell JN (1991) Ectopic excitability of injured nerves in monkey: entrained responses to vibratory stimuli. *J Neurophysiol* 65: 693–701

20 Proske U, Iggo A, Luff AR (1995) Mechanical sensitivity of regenerating myelinated skin and muscle afferents in the cat. *Exp Brain Res* 104: 89–98

21 Michaelis M, Blenk K-H, Jänig W, Vogel C (1995) Development of spontaneous activity and mechanosensitivity in axotomized afferent nerve fibers during the first hours after nerve transection in rats. *J Neurophysiol* 74: 1020–1027

22 Babbedge RC, Soper AJ, Gentry CT, Hood VC, Campbell EA, Urban L (1996) *In vitro* characterization of a peripheral afferent pathway of the rat after chronic sciatic nerve section. *J Neurophysiol* 76: 3169–3177

23 Bongenhielm U, Robinson PP (1996) Spontaneous and mechanically evoked afferent activity originating from myelinated fibres in ferret inferior alveolar nerve neuromas. *Pain* 67: 399–406

24 Tal M, Eliav E (1996) Abnormal discharge originates at the site of nerve injury in experimental constriction neuropathy (CCI) in the rat. *Pain* 64: 511–518

25 Chen Y, Devor M (1998) Ectopic mechanosensitivity in injured sensory axons arises from the site of spontaneous electrogenesis. *Eur J Pain* 2: 165–178

26 Bongenhielm U, Robinson PP (1998) Afferent activity from myelinated inferior alveolar nerve fibers in ferrets after constriction or section and regeneration. *Pain* 74: 123–132

27 Hu SJ, Xing JL (1998) An experimental model for chronic compression of dorsal root ganglion produced by intervertebral foramen stenosis in the rat. *Pain* 77: 15–23

28 Koschorke GM, Meyer RA, Campbell JN (1994) Cellular components necessary for mechanoelectrical transduction are conveyed to primary afferent terminals by fast axonal transport. *Brain Res* 641: 99–104

29 Nordin M, Nyström B, Wallin U, Hagbarth KE (1984) Ectopic sensory discharges and

paresthesiae in patients with disorders of peripheral nerves, dorsal roots and dorsal columns. *Pain* 20: 231–245

30 Blenk K-H, Michaelis M, Vogel C, Jänig W (1996) Thermosensitivity of acutely axotomized sensory nerve fibers. *J Neurophysiol* 76: 743–752

31 Zimmermann M, Sanders K (1982) Responses of nerve axons and receptor endings to heat, ischemia and algesic substances. Abnormal excitability of regenerating nerve endings. In: WJ Culp, J Ochoa (eds): *Abnormal nerves and muscles as impulse generators.* Oxford University Press, Oxford, 513–532

32 Matzner O, Devor M (1987) Contrasting thermal sensitivity of spontaneously active A- and C-fibers in experimental nerve-end neuromas. *Pain* 30: 373–384

33 Hartung M, Leah J, Zimmermann M (1989) The excitation of cutaneous nerve endings in a neuroma by capsaicin. *Brain Res* 499: 363–366

34 Devor M, White DM, Goetzl EJ, Levine JD (1992) Eicosanoids, but not tachykinins, excite C-fiber endings in rat sciatic nerve-end neuromas. *NeuroRep* 3: 21–24

35 Michaelis M, Vogel C, Blenk K-H, Jänig W (1997) Algesics excite axotomized afferent nerve fibres within the first hours following nerve transection in rats. *Pain* 72: 347–354

36 Michaelis M, Vogel C, Blenk K-H, Arnarson A, Jänig W (1998) Inflammatory mediators sensitize acutely axotomized nerve fibers to mechanical stimulation in the rat. *J Neurosci* 18: 7581–7587

37 Arnarson A, Michaelis M, Jänig W (1999) Sensitizing effect of bradykinin on ectopic mechanical and thermal sensitivity in axotomized afferent nerve fibers. *Pflügers Arch Suppl* 437: R129 (Abstract)

38 Becker S, Reeh PW, Michaelis M (1999) Thermosensitivity of axotomized cutaneous afferents and the sensitizing effect of bradykinin, *in vitro*. In: *Abstracts of the 9th World Congress on Pain,* 31 (Abstract)

39 Maves TJ, Pechman PS, Gebhart GF, Meller ST (1993) Possible chemical contribution from chromic gut sutures produces disorders of pain sensation like those seen in man. *Pain* 54: 57–69

40 Clatworthy AL, Illich PA, Castro GA, Walters ET (1995) Role of periaxonal inflammation in the development of thermal hyperalgesia and guarding behavior in a rat model of neuropathic pain. *Neurosci Lett* 184: 5–8

41 Ren K, Thomas DA, Dubner R (1995) Nerve growth factor alleviates a painful peripheral neuropathy in rats. *Brain Res* 699: 286–292

42 Siuciak JA, Lewis D, Lindsay R (1996) Blockade of NGF hyperalgesia with trkA-IgG receptor body. *Pain* 8: 120 (Abstract)

43 Jänig W, Koltzenburg M (1992) Possible ways of sympathetic afferent interaction. In: W Jänig, RF Schmidt (eds): *Reflex sympathetic dystrophy. Pathophysiological mechanisms and clinical implications.* VCH Verlagsgesellschaft, Weinheim, New York, 213–243

44 Torebjörk E, Wahren LK, Wallin G, Hallin R, Koltzenburg M (1995) Noradrenaline-evoked pain in neuralgia. *Pain* 63: 11–20

45 Devor M, Jänig W (1981) Activation of myelinated afferents ending in a neuroma by stimulation of the sympathetic supply in the rat. *Neurosci Lett* 24: 43–47

46 Burchiel KJ (1984) Spontaneous impulse generation in normal and denervated dorsal root ganglia: sensitivity to alpha-adrenergic stimulation and hypoxia. *Exp Neurol* 85: 257–272
47 Chen Y, Michaelis M, Jänig W, Devor M (1996) Adrenoceptor subtype mediating sympathetic-sensory coupling in injured sensory neurons. *J Neurophysiol* 76: 3721–3730
48 Rubin G, Kaspi T, Rappaport ZH, Cohen S, Ravikovitch M, Lomazov P, Devor M (1997) Adrenosensitivity of injured afferent neurons does not require the presence of postganglionic sympathetic terminals. *Pain* 72: 183–191
49 Chabal C, Jacobson L, Russell LC, Burchiel KJ (1992) Pain response to perineuromal injection of normal saline, epinephrine, and lidocaine in humans. *Pain* 49: 9–12
50 Sato J, Perl ER (1991) Adrenergic excitation of cutaneous pain receptors induced by peripheral nerve injury. *Science* 251: 1608–1610
51 Bossut DF, Perl ER (1995) Effects of nerve injury on sympathetic excitation of Aδ mechanical nociceptors. *J Neurophysiol* 73: 1721–1723
52 O'Halloran KD, Perl ER (1997) Effects of partial nerve injury on the responses of C-fiber polymodal nociceptors to adrenergic agonists. *Brain Res* 759: 233–240
53 Ali Z, Ringkamp M, Hartke TV, Chien HF, Flavahan NA, Campbell JN, Meyer RA (1999) Uninjured C-fiber nociceptors develop spontaneous activity and alpha-adrenergic sensitivity following L6 spinal nerve ligation in monkey. *J Neurophysiol* 81: 455–466
54 McLachlan EM, Jänig W, Devor M, Michaelis M (1993) Peripheral nerve injury triggers noradrenergic sprouting within dorsal root ganglia. *Nature* 363: 543–546
55 Devor M, Jänig W, Michaelis M (1994) Modulation of activity in dorsal root ganglion (DRG) neurons by sympathetic activation in nerve-injured rats. *J Neurophysiol* 71: 38–47
56 Xie Y, Zhang J, Petersen M, LaMotte RH (1995) Functional changes in dorsal root ganglion cells after chronic nerve constriction in the rat. *J Neurophysiol* 73: 1811–1820
57 Zhang JM, Song XJ, LaMotte RH (1997) An *in vitro* study of ectopic discharge generation and adrenergic sensitivity in the intact, nerve-injured rat dorsal root ganglion. *Pain* 72: 51–57
58 Liu X, Chung K, Chung JM (1999) Ectopic discharges and adrenergic sensitivity of sensory neurons after spinal nerve injury. *Brain Res* 849: 244–247
59 Häbler H-J, Eschenfelder S, Brinker H, Grunow B, Jänig W, Liu X-G (2000) Neurogenic vasoconstriction in the dorsal root ganglion may play a crucial role in sympathetic-afferent coupling after peripheral nerve injury. In: M Devor, M Rowbotham, Z Wiesenfeld-Hallin (eds): *Proceedings of the 9th world congress on pain.* IASP Press, Seattle, 661–667
60 Michaelis M (2000) Coupling of sympathetic and somatosensory neurons following nerve injury: mechanisms and potential significance for the generation of pain. In: M Devor, M Rowbotham, Z Wiesenfeld-Hallin (eds): *Proceedings of the 9th world congress on pain.* IASP Press, Seattle, 645–656
61 Devor M, Govrin-Lippmann R, Angelides K (1993) Na channel immunolocalization in

peripheral mammalian axons and changes following nerve injury and neuroma formation. *J Neuroscience* 13: 1976–1992

62 Novakovic SD, Tzoumaka E, McGivern JG, Haraguchi M, Sangameswaran L, Gogas KR, Eglen RM, Hunter JC (1998) Distribution of the tetrodotoxin-resistant sodium channel PN3 in rat sensory neurons in normal and neuropathic conditions. *J Neurosci* 18: 2174–2187

63 England JD, Happel LT, Kline DG, Gamboni F, Thouron CL, Liu ZP, Levinson SR (1996) Sodium channel accumulation in humans with painful neuromas. *Neurology* 47: 272–276

64 Coward K, Plumpton C, Facer P, Birch R, Carlstedt T, Tate S, Bountra C, Anand P (2000) Immunolocalization of SNS/PN3 and NaN/SNS2 sodium channels in human pain states. *Pain* 85: 41–50

65 Devor M (1983) Nerve pathophysiology and mechanisms of pain in causalgia. *J Auton Nerv Syst* 7: 371–384

66 Proske U, Luff AR (1998) Mechanical sensitivity of muscle afferents in a nerve treated with colchicine. *Exp Brain Res* 119: 391–398

67 Leem JW, Willis WD, Chung JM (1993) Cutaneous sensory receptors in the rat foot. *J Neurophysiol* 69: 1684–1699

68 Johnson RD, Munson JB (1991) Regenerating sprouts of axotomized cat muscle afferents express characteristic firing patterns to mechanical stimulation. *J Neurophysiol* 66: 2155–2158

69 Lewin GR, McMahon SB (1991) Physiological properties of primary sensory neurons appropriately and inappropriately innervating skeletal muscle in adult rats. *J Neurophysiol* 66: 1218–1231

70 Lewin GR, McMahon SB (1991) Physiological properties of primary sensory neurons appropriately and inappropriately innervating skin in the adult rat. *J Neurophysiol* 66: 1205–1217

71 Devor M, Keller CH, Ellisman MH (1990) Spontaneous discharge of afferents in a neuroma reflects original receptor tuning. *Brain Res* 517: 245–250

72 Michaelis M, Blenk KH, Vogel C, Jänig W (1999) Distribution of sensory properties among axotomized cutaneous C-fibres in adult rats. *Neuroscience* 94: 7–10

73 Campbell JN, Meyer RA, Raja SN (1992) Is nociceptor activation by alpha-1 adrenoreceptors the culprit in sympathetically maintained pain? *APS J* 1: 3–11

74 Perl ER (1994) A re-evaluation of mechanisms leading to sympathetically related pain. In: HL Fields, JC Liebeskind (eds): *Pharmacological approaches to the treatment of chronic pain: new concepts and critical issues. Progress in Pain Research and Management*, Vol. 1. IASP Press, Seattle, 129–150

75 Perl ER (1999) Causalgia, pathological pain, and adrenergic receptors. *Proc Natl Acad Sci USA* 96: 7664–7667

76 Yaksh TL (1985) Pharmacology of spinal adrenergic systems which modulate spinal nociceptive processing. *Pharmacol Biochem Behav* 22: 845–858

77 Korenman EM, Devor M (1981) Ectopic adrenergic sensitivity in damaged peripheral nerve axons in the rat. *Exp Neurol* 72: 63–81
78 Govrin-Lippmann R, Devor M (1978) Ongoing activity in severed nerves: source and variation with time. *Brain Res* 159: 406–410
79 Tal M, Wall PD, Devor M (1999) Myelinated afferent fiber types that become spontaneously active and mechanosensitive following nerve transection in the rat. *Brain Res* 824: 218–223
80 Wall PD, Devor M (1983) Sensory afferent impulses originate from dorsal root ganglia as well as from the periphery in normal and nerve injured rats. *Pain* 17: 321–339
81 Devor M, Wall PD (1990) Cross-excitation in dorsal root ganglia of nerve-injured and intact rats. *J Neurophysiol* 64: 1733–1746
82 Michaelis M, Devor M, Jänig W (1996) Sympathetic modulation of activity in dorsal root ganglion neurons changes over time following peripheral nerve injury. *J Neurophysiol* 76: 753–763
83 Kajander KC, Wakisaka S, Bennett GJ (1992) Spontaneous discharge originates in the dorsal root ganglion at the onset of a painful peripheral neuropathy in the rat. *Neurosci Lett* 138: 225–228
84 Petersen M, Zhang J, Zhang J-M, LaMotte RH (1996) Abnormal spontaneous activity and responses to norepinephrine in dissociated dorsal root ganglion cells after chronic nerve constriction. *Pain* 67: 391–397
85 Study RE, Kral MG (1996) Spontaneous action potential activity in isolated dorsal root ganglion neurons from rats with a painful neuropathy. *Pain* 65: 235–242
86 Liu X, Eschenfelder S, Blenk K-H, Jänig W, Häbler H (2000) Spontaneous activity of axotomized afferent neurons after L5 spinal nerve injury in rats. *Pain* 84: 309–318
87 Han HC, Lee DH, Chung JM (2000) Characteristics of ectopic discharges in rat neuropathic pain model. *Pain* 84: 253–261
88 Liu CN, Wall PD, Ben-Dor E, Michaelis M, Amir R, Devor M (2000) Tactile allodynia in the absence of C-fiber activation: altered firing properties of DRG neurons following spinal nerve injury. *Pain* 85: 503–521
89 Michaelis M, Liu X, Jänig W (2000) Axotomized and intact muscle afferents but no skin afferents develop ongoing discharges of dorsal root ganglion origin after peripheral nerve lesion. *J Neurosci* 20: 2742–2748
90 DeSantis M, Duckworth JW (1982) Properties of primary afferent neurons from muscle which are spontaneously active after a lesion of their peripheral processes. *Exp Neurol* 75: 261–274
91 Mense S, Meyer H (1985) Different types of slowly conducting afferent units in cat skeletal muscle and tendon. *J Physiol Lond* 363: 403–417
92 Torebjörk HE, Ochoa JL, Schady W (1984) Referred pain from intraneural stimulation of muscle fascicles in the median nerve. *Pain* 18: 145–156
93 Laursen RJ, Graven-Nielsen T, Jensen TS, Arendt-Nielsen L (1999) The effect of compression and regional anaesthetic block on referred pain intensity in humans. *Pain* 80: 257–263

94 Wall PD, Woolf CJ (1984) Muscle but not cutaneous C-afferent input produces prolonged increases in excitability of the flexion reflex in the rat. *J Physiol Lond* 356: 443–458
95 Michael GJ, Averill S, Shortland PJ, Yan Q, Priestley JV (1999) Axotomy results in major changes in BDNF expression by dorsal root ganglion cells: BDNF expression in large trkB and trkC cells, in pericellular baskets, and in projections to deep dorsal horn and dorsal column nuclei. *Eur J Neurosci* 11: 3539–3551
96 Zhou XF, Chie ET, Deng YS, Zhong JH, Xue Q, Rush RA, Xian CJ (1999) Injured primary sensory neurons switch phenotype for brain-derived neurotrophic factor in the rat. *Neuroscience* 92: 841–853
97 Mannion RJ, Costigan M, Decosterd I, Amaya F, Ma QP, Holstege JC, Ji RR, Acheson A, Lindsay RM, Wilkinson GA et al (1999) Neurotrophins: peripherally and centrally acting modulators of tactile stimulus-induced inflammatory pain hypersensitivity. *Proc Natl Acad Sci USA* 96: 9385–9390
98 Kerr BJ, Bradbury EJ, Bennett DL, Trivedi PM, Dassan P, French J, Shelton DB, McMahon SB, Thompson SW (1999) Brain-derived neurotrophic factor modulates nociceptive sensory inputs and NMDA-evoked responses in the rat spinal cord. *J Neurosci* 19: 5138–5148
99 McMahon SB, Bennett DLH (1999) Trophic factors and pain. In: PD Wall, R Melzack (eds): *Textbook of pain*. Churchill Livingstone, Edinburgh, 105–128
100 Amir R, Michaelis M, Devor M (1999) Membrane potential oscillations in dorsal root ganglion neurons: role in normal electrogenesis and neuropathic pain. *J Neurosci* 19: 8589–8596
101 Liu C-N, Michaelis M, Amir R, Devor M (2000) Spinal nerve injury enhances subthreshold membrane potential oscillations in DRG neurons: relation to neuropathic pain. *J Neurophysiol* 84: 205–215
102 Cesare P, McNaughton P (1996) A novel heat-activated current in nociceptive neurons and its sensitization by bradykinin. *Proc Natl Acad Sci USA* 93: 15435–1543

Sodium channels and the molecular basis for pain

Joel A. Black, Theodore R. Cummins, Sulayman D. Dib-Hajj and Stephen G. Waxman

Department of Neurology and PVA/EPVA Neuroscience Research Center, Yale University School of Medicine, New Haven, CT 06510, USA; and Rehabilitation Research Center, VA Connecticut, West Haven, CT 06516, USA

Introduction

Primary sensory neurons (dorsal root ganglion (DRG) and trigeminal neurons) initiate the nociceptive (pain) response. These neurons extend a single process that bifurcates, sending an axon to peripheral targets (i.e., skin, muscle, viscera) and also centrally to enter the ascending pathways that carry information to the brain. The thinly-myelinated (Aδ-type) and unmyelinated (C-type) nociceptive DRG and trigeminal neurons normally have a high threshold for firing and are relatively quiescent unless activated by noxious stimuli. When stimulated, these neurons produce a highly-modulated series of action potentials that convey graded information about pain and temperature sensation to the brain. Following peripheral nerve or tissue injury, however, primary sensory neurons can become hyperexcitable, and can give rise to spontaneous action potential or abnormal high-frequency activities which may contribute to chronic neuropathic and inflammatory pain [1–4].

Action potential electrogenesis in most mammalian neurons depends on the depolarizing activity of voltage-gated sodium channels. Sodium channels within neurons are composed of a single α-subunit, which forms the ion selective and voltage-sensing pore, and auxilliary β-subunits, which have modulatory roles in channel gating and assembly and targeting actions [5, 6]. Sodium channel α-subunits are members of a gene family; there are at least eight different voltage-gated sodium channels present within the nervous system. These channels share a common motif, but are encoded by different genes which endow them with different amino acid sequences. Distinct voltage-dependence, kinetic and pharmacological properties are associated with the differing sodium channel isoforms. Moreover, these channels are expressed in a spatially- and temporally-specific manner within the nervous system [7], which conveys unique functional properties on subpopulations of neurons during development and in different regions of the nervous system.

Molecular and electrophysiological methods have begun to identify a complex repetoire of sodium channels in adult DRG neurons, and to enumerate injury-induced alterations in the expression and distribution of these channels. Notably,

Mechanisms and Mediators of Neuropathic Pain, edited by Annika B. Malmberg and Sandra R. Chaplan

three sensory neuron-specific sodium channels – PN1, SNS and NaN – are preferentially expressed in DRG and trigeminal neurons, in addition to several channels with more widespread expression throughout the nervous system. Hyperexcitability of injured DRG neurons results, at least in part, from changes in sodium channel expression, including the down-regulation of several sodium channel genes and the up-regulation of at least one previously silent sodium channel gene. This chapter reviews recent advances in the understanding of sodium channel expression in small (<25 μm diam.) DRG neurons, which are predominantly nociceptive, in relation to models of neuropathic and inflammatory pain, and discusses the implications of these advances for the development of new treatment strategies for pain management.

Multiple sodium channels are expressed in DRG neurons

DRG neurons are unique in expressing several distinct sodium currents, which can be differentiated on the basis of different voltage-dependence and kinetic properties, and also by pharmacological differences, including varying degrees of sensitivity to the neurotoxin, tetrodotoxin (TTX) [8–11]. Some individual DRG neurons, in fact, produce multiple discrete sodium currents (e.g. Fig. 1), implying that these neurons can co-express several types of sodium channels [12–14]. Virtually all small DRG neurons express measurable levels of classically fast, TTX-sensitive sodium current and approximately 85% of these neurons also express TTX-resistant sodium currents [12].

Reverse transcription polymerase chain reaction (RT-PCR), RNAase protection and *in situ* hybridization techniques have yielded detailed information about which sodium channel genes are transcriptionally active in DRG neurons. These studies show that at least six transcripts encoding different sodium channels are expressed in normal adult rat DRG neurons [15, 16] (Fig. 2). Sodium channel types I [17] and Na6 [18] (also termed PN4 [19]) produce TTX-sensitive sodium currents and are moderately expressed in DRG neurons; these channels are also expressed at high levels by several neuronal cell types within the CNS. In addition to the sodium channels that are found within the CNS, DRG neurons preferentially express at least three sodium channel mRNAs which are not present at significant levels in CNS neurons: (1) PN1 [20] is present in virtually all DRG neurons and encodes a TTX-sensitive channel. There is strong evidence that PN1 sodium channels, in DRG neurons, respond to small, slow depolarizations close to the resting potential and activate to produce inward (depolarizing) currents, which would poise these channels to act as boosters of sensory generator potentials [21]. (2) SNS [22] (also termed PN3 [23]) is expressed preferentially in small and medium-diameter DRG neurons [15, 22–25]. These channels produce a sodium current that is relatively resistant to TTX (K_i ~30–60 μM), activates between –40 and –30 mV and is slowly-inactivat-

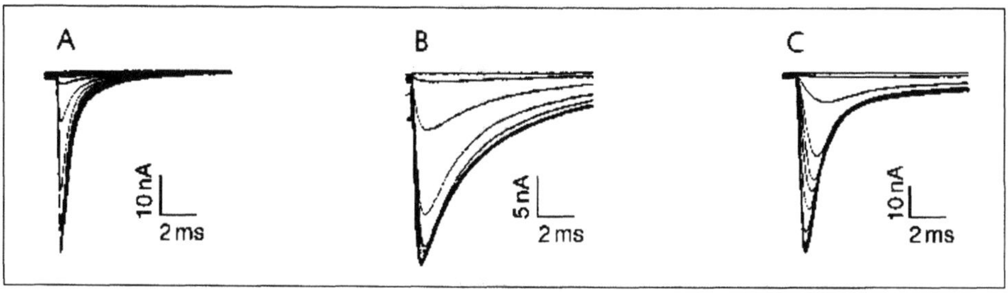

Figure 1
Voltage-gated sodium channels recorded by whole-cell patch-clamp in DRG neurons. (A) Fast, TTX-sensitive sodium current is recorded in a muscle afferent DRG neuron. (B) A small IB4$^+$ neuron exhibits only slow, TTX-resistant sodium current. (C) Both fast, TTX-sensitive and slow, TTX-resistant sodium currents are recorded in a small IB4$^-$ neuron. Modified and reproduced with permission from [86].

ing [22–24]. (3) NaN [16, 26] (also termed SNS2 [24]) is also expressed preferentially in small DRG neurons [16, 24, 25], where there is substantial co-expression with SNS [24, 27]. NaN is predicted to encode a TTX-resistant sodium channel, based on the presence of a serine residue at position 355 [16]. In a study on SNS knockout mice in which functional SNS channels are not expressed [28], Cummins et al. [13] observed a TTX-resistant (K_i = 39 ± 9 μM) persistent sodium current which is attributable to NaN channels (Fig. 3). A similar TTX-resistant persistent current can be recorded from rat [13] and human [14] DRG neurons. The persistent sodium current exhibits a hyperpolarized voltage-dependence of activation (threshold ~–70 mV; midpoint of activation = –41 mV in mouse DRG neurons) and steady-state inactivation (midpoint = –44 mV). There is substantial overlap between activation and steady-state inactivation of this channel (Fig. 3E), indicating that persistent sodium currents should be active near the resting membrane potential [13]. Persistent sodium currents have been implicated in setting membrane resting potential [29], subthreshold membrane potential oscillations [30], amplification of depolarizing inputs [31] and impulse initiation [32]. As a result of its hyperpolarized voltage-dependence of activation, NaN appears to participate in setting resting potential and thus regulating excitability in small DRG neurons.

The expression of two TTX-resistant sodium channels, SNS and NaN, within small DRG neurons, provides a molecular basis for the electrophysiological observations [8–10, 33, 34] that these cells express prominent TTX-resistant sodium currents, in addition to the more conventional TTX-sensitive sodium currents that are present in most other neurons. In addition, electrophysiological studies suggest that TTX-resistant sodium channels participate in action potential conduction within nociceptive sensory neurons and their axons [35-37], and SNS and NaN channel

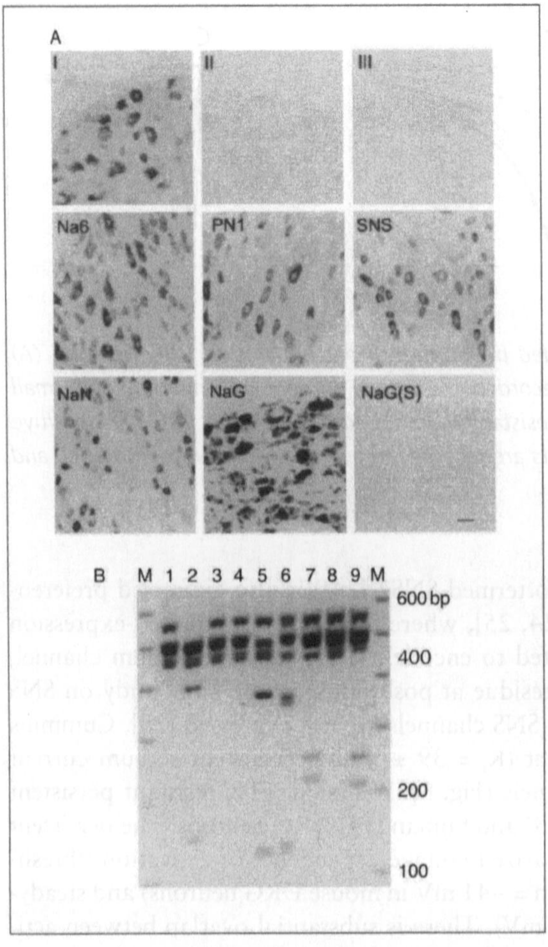

Figure 2
(A) Multiple sodium channels in adult DRG neurons. Sodium channel α-subunit mRNAs visualized in rat DRG neurons by in situ *hybridization with subtype-specific antisense riboprobes. Transcipts for six different sodium channels (I, Na6, PN1, SNS, NaN and NaG) are present at moderate-to-high levels in DRG neurons. Sodium channels II and III are not detectable in adult rat DRG neurons. Hybridization with sense riboprobes, e.g., for NaG (S), does not result in signal. Scale bar, 50 μm.*
(B) Restriction mapping (sodium channel domain 1) demonstrates a similar distribution of sodium channel mRNAs in DRG to that shown in (A). "M" lanes contain 100-bp ladder marker. Lane 1 contains the RT-PCR amplification product from DRG cDNA. Lanes 2–9 show results with EcoRV, EcoN1, AvaI, Sph1, BamH1, Af/III, XbaI, *and* EcoRI, *which are specific to subunits I, II, III, Na6, PN1, SNS, NaG, and NaN, respectively. Modified and reproduced with permission from [15] and [16], © 1996 National Academy of Sciences, USA.*

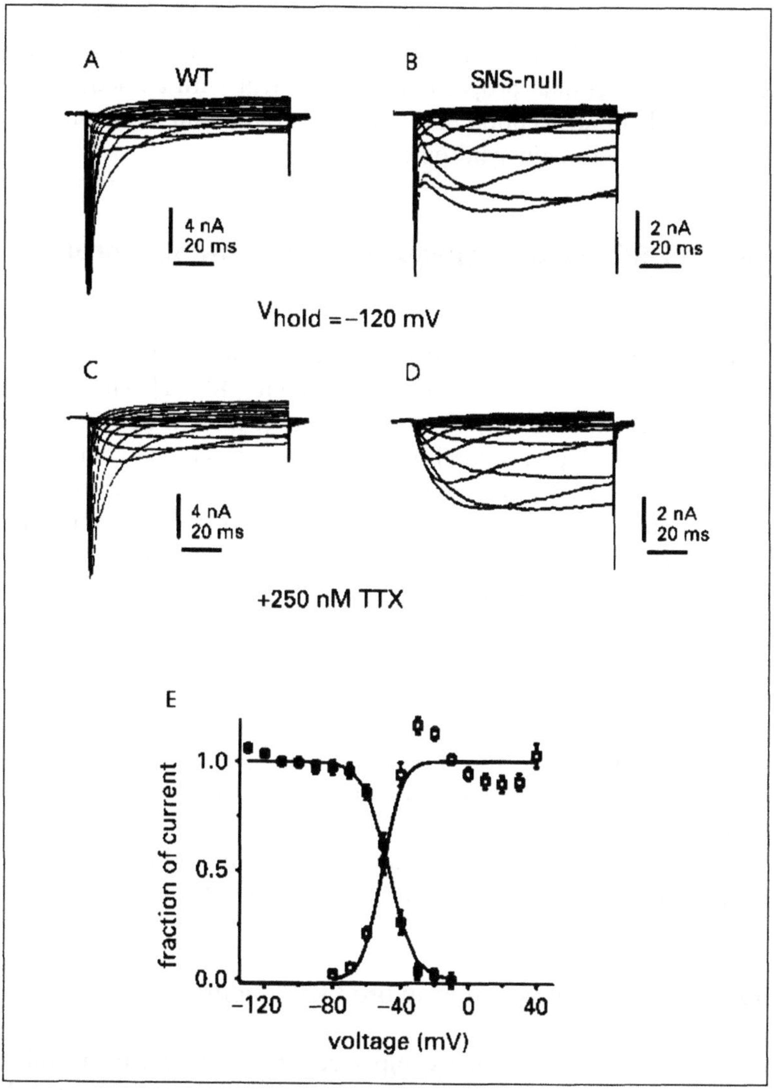

Figure 3

Multiple sodium currents in small DRG neurons. (A, B) Representative recordings from a holding potential of –120 mV in wild type (WT) and SNS-null mice. (C, D) Addition of 250 nM TTX blocks the fast sodium current. WT DRG neurons have both a slowly-inactivating and persistent TTX-resistant sodium current, while SNS-null DRG neurons have only the persistent TTX-resistant sodium current. (E) Activation (unfilled squares) and steady-state inactivation (filled squares) curves for the TTX- resistant current in SNS-null neurons show significant overlap. Steady-state inactivation was measured with 500 ms prepulses. Modified and reproduced with permission from [13], © 1999 Society for Neuroscience.

immunostaining have been demonstrated in unmyelinated fibers [38, 39]. The highly specific expression of SNS and NaN channels in C- and Aδ-type DRG neurons, coupled with their unique functional properties, are consistent with an active participation of these channels in nociception and implies that aberrant expression of these channels may contribute to pain syndromes.

Sodium channels and DRG neurons hyperexcitability following axonal injury

Early electrophysiological studies utilizing microelectrodes showed that motor neurons can become hyperexcitable following axonal transection [40, 41]. These studies suggested that sodium channel density was increased over the cell body and the dendrites following axotomy. Hyperexcitability of sensory neurons following axotomy has also been shown [4, 42, 43]. Pharmacological studies and ion substitution experiments have confirmed a role of sodium channels in neuronal hyperexcitability [44, 45]. Moreover, immunocytochemical observations have demonstrated abnormal accumulations of sodium channels at the distal tips of injured axons [46-48], which have been suggested as sites of ectopic electrogenesis contributing to neuropathetic pain. More recently, subtype-specific antibodies have been utilized to show that type III [49], a TTX-sensitive sodium channel [50] generally not expressed in normal adult tissue [7], and SNS [51] sodium channels accumulate in axons at the site of nerve injury. Taken together with observations of the partial efficacy of sodium channel-blocking agents in experimental neuropathic pain and in humans with chronic neuropathic pain (see, for example, [52–55]), these results support the conclusion that activation of sodium channels within injured DRG neurons can contribute to sensory neuron hyperexcitability associated with chronic pain.

Abnormal sodium channel expression in injured DRG neurons

Peripheral nerve injury in humans is often accompanied by neuropathic pain, which may be manifested as spontaneous pain, thermal and mechanical hyperalgesia and/or allodynia. Early studies demonstrated a significant up-regulation in the expression of the previously silent type III sodium channel gene in DRG neurons following sciatic nerve injury [56, 57] (Fig. 4A, B, C). More recently, the up-regulation of type III mRNA in noceptive-type DRG neurons has been shown to be accompanied by robust type III immunoreactivity in these injured neurons [49] (Fig. 4D, E). In parallel with the up-regulation of type III mRNA and protein in injured DRG neurons, axonal injury also triggers the emergence of a rapidly-repriming (i.e. recovers rapidly from inactivation) TTX-sensitive sodium current in these neurons [12]. The change in repriming kinetics in axotomized DRG neurons is shown in Fig. 5. The

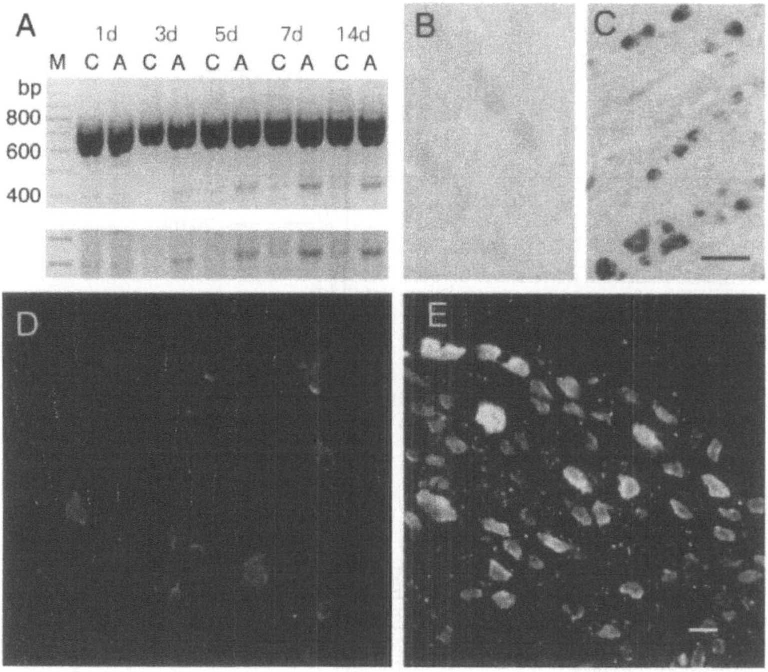

Figure 4
Sodium channel III is upregulated in DRG neurons following peripheral nerve axotomy. (A) RT-PCR shows products of co-amplification of type III and β-actin transcripts in control (C) and axotomized (A) DRG (days post-axotomy indicated above gels), with computer enhanced images of amplification products shown below gels. (B, C) The micrographs show in situ *hybridizations in control (B) and 5–7 days post-axotomy (C) DRG. Scale bar, 50 µm. (D, E) Type III immunostaining is generally not detectable within control DRG (D), while peripherally-axotomized DRG neurons exhibit increased type III immunostaining (E). Scale bar, 50 µm. Modified and reproduced with permission from [49] and [57], © 1996, National Academy of Sciences, USA.*

time constant for recovery of TTX-sensitive sodium currents from inactivation is accelerated by up to four-fold following axotomy throughout a range of test voltages. Cummins and Waxman [12] proposed that the type III sodium channel produces the rapidly-repriming sodium current. This hypothesis is supported by the following observations: (1) rapidly-repriming TTX-sensitive current and expression of type III sodium channel protein display parallel patterns of up-regulation following axotomy of the peripherally-directed (sciatic nerve) axons of DRG neurons; however, transection of the centrally-directed (dorsal root) axons of these cells does not

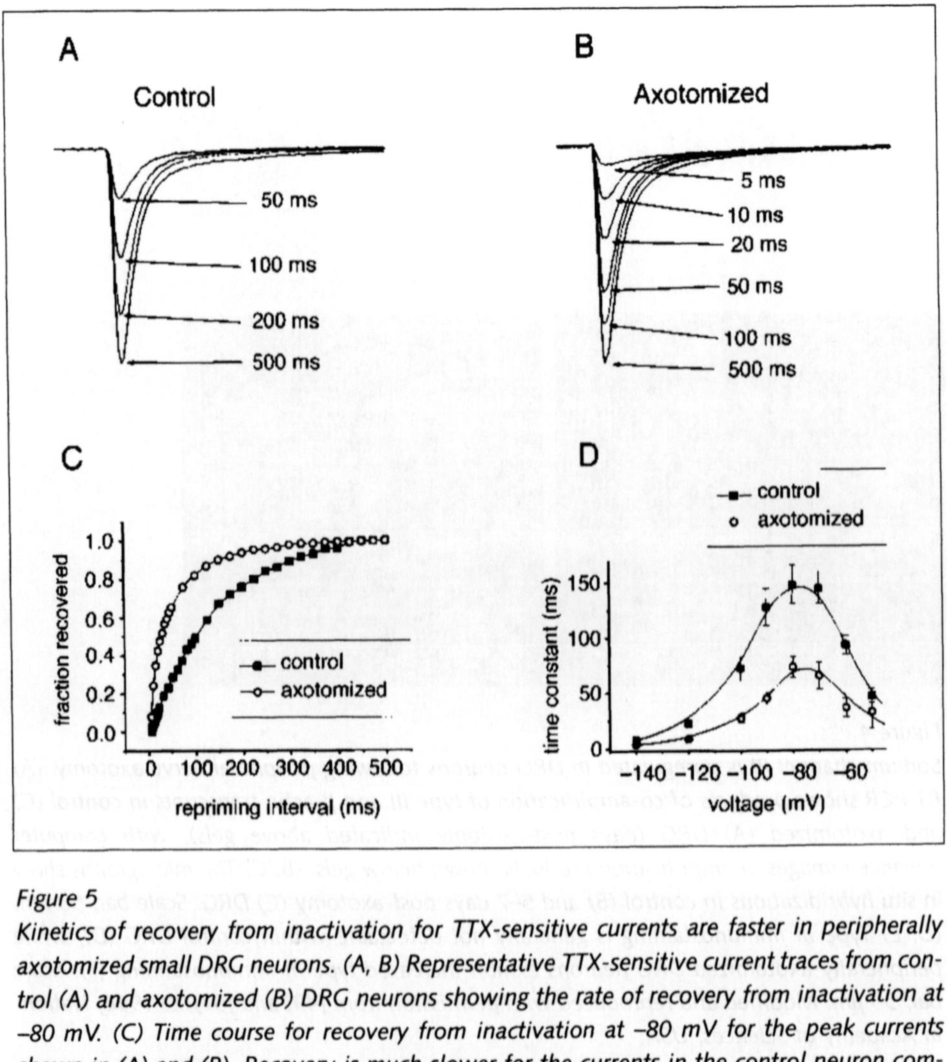

Figure 5
Kinetics of recovery from inactivation for TTX-sensitive currents are faster in peripherally axotomized small DRG neurons. (A, B) Representative TTX-sensitive current traces from control (A) and axotomized (B) DRG neurons showing the rate of recovery from inactivation at –80 mV. (C) Time course for recovery from inactivation at –80 mV for the peak currents shown in (A) and (B). Recovery is much slower for the currents in the control neuron compared to axotomized neuron. (D) Time constants for recovery from inactivation for control (closed squares) and axotomized (open circles) plotted as a function of voltage. Time constants for axotomized neurons are smaller than for control neurons at all voltages tested. Modified and reproduced with permission from [49].

elicit either the emergence of a rapidly-repriming TTX sensitive current or of type III immunostaining [49]; (2) rapid-repriming has been observed in a mammalian expression system (HEK 293 cells) when these cells are transfected with a type III sodium channel construct (Cummins, Dib-Hajj and Waxman, unpublished observa-

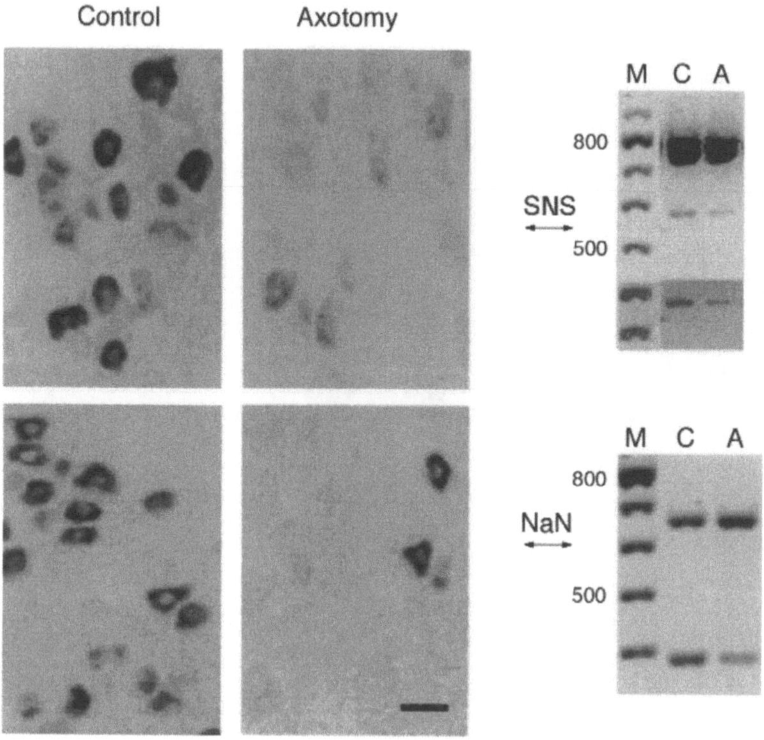

Figure 6
Transcripts for TTX-resistant sodium channels are downregulated following axotomy. Micrographs (left side) show in situ *hybridizations for detection of SNS and NaN transcripts in control and peripherally-axotomized DRG. Both SNS and NaN mRNAs are down-regulated in axotomized neurons. Scale bar, 50 µm. RT-PCR (right side) show products of co-amplification of SNS (572 bp) with β-actin and NaN (392 bp) with GAPDH in control (C) and axotomized (A) DRG. Both SNS and NaN transcripts show decreased expression in axotomized compared to control DRG. Modified and reproduced with permission from [16], © 1998 National Academy of Sciences, USA, and [57] © 1996 National Academy of Sciences, USA.*

tions); (3) abnormal accumulations of type III sodium channel protein are present within the distal stump of transected sciatic nerves [49], where hyperexcitability has been demonstrated [58–60].

In contrast to the upregulation of type III sodium channels, there is a downregulation of SNS [57, 61] and NaN [16, 24] transcripts in DRG neurons following transection of the sciatic nerve (Fig. 6). The decrease in SNS and NaN transcript levels following peripheral axotomy is paralleled by an attenuation of SNS and NaN

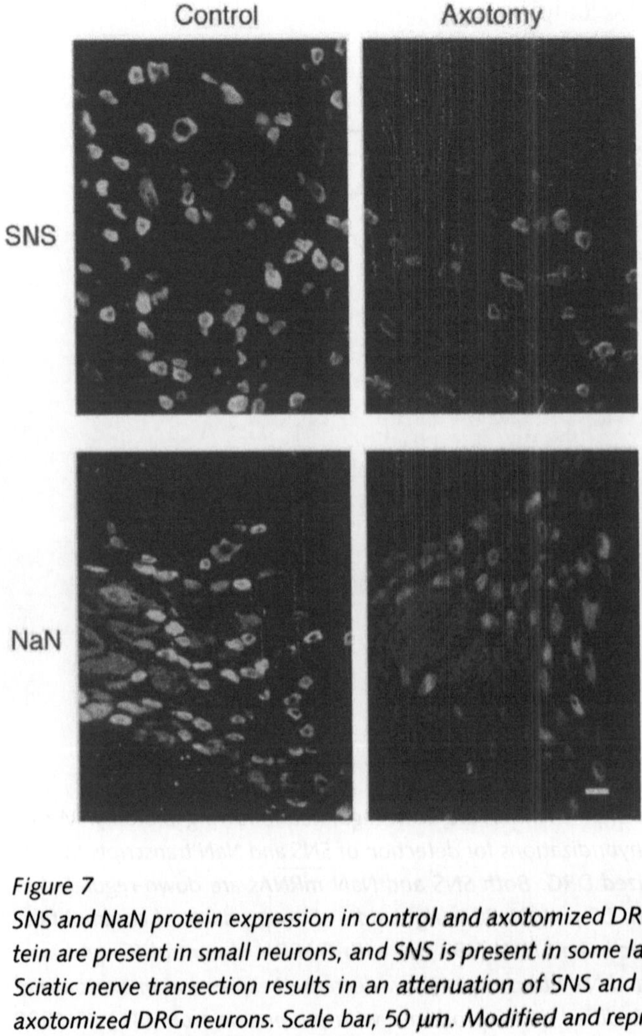

Figure 7
SNS and NaN protein expression in control and axotomized DRG neurons. SNS and NaN protein are present in small neurons, and SNS is present in some larger neurons, in control DRG. Sciatic nerve transection results in an attenuation of SNS and NaN immunoreactivity in the axotomized DRG neurons. Scale bar, 50 μm. Modified and reproduced with permission from [25] © 2000 Society for Neuroscience.

immunoreactivity in axotomized DRG neurons (Fig. 7). Interestingly, non-axotomized neurons within ipsilateral DRG (only ~70% of L4/L5 DRG neurons are axotomized by a midthigh sciatic nerve transection [62]) exhibit a slight decrease in SNS and NaN immunoreactivity compared to contralateral control neurons [25]. Thus, it is possible that there is a limited *indirect* effect (i.e. paracrine) of axotomy on the expression of SNS and NaN in these neurons.

The downregulation of SNS and NaN transcripts and protein in DRG neurons following sciatic nerve transection suggests that a reduction in TTX-resistant sodi-

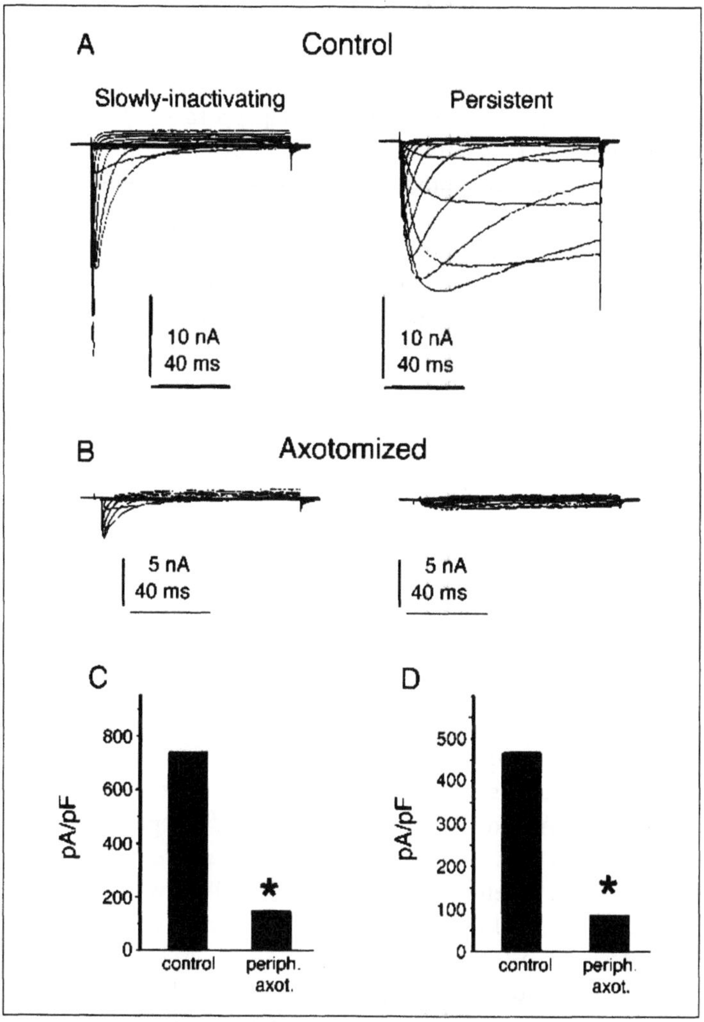

Figure 8
TTX-resistant sodium currents in small DRG neurons are reduced following peripheral axotomy. (A) Slowly-inactivating and persistent TTX-resistant currents recorded from control neurons. Predominantly slowly-inactivating currents are recorded if the neurons are held at –60 mV and a 500 ms step to –120 mV precedes the test pulse. Persistent sodium current is determined by subtracting the slowly inactivating current from total TTX-resistant current, which is measured from a holding potential of –120 mV. (B) Following peripheral axotomy, both slowly-inactivating and persistent currents are attenuated in small DRG neurons. (C, D) Quantification of the mean peak amplitudes of slowly-inactivating (C) and persistent (D) sodium currents in control and axotomized small DRG neurons. Modified and reproduced with permission from [25], © 2000 Society for Neuroscience.

um currents would be expected in these cells following axotomy. Patch-clamp studies have demonstrated a significant reduction of *total* TTX-resistant sodium currents in DRG neurons following peripheral axotomy [12, 63], which persists for at least 60 days following injury [12]. More recently, utilizing protocols to separate slowly-inactivating and persistent TTX-resistant sodium currents, it has been demonstrated that *both* currents are attenuated in DRG neurons following peripheral axotomy (Fig. 8).

In contrast to peripheral axotomy, transection of the central projections (dorsal root) of DRG neurons is not accompanied by changes in transcript or protein levels of either SNS or NaN, or of slowly-inactivating or persistent TTX-resistant sodium currents (Fig. 9). These observations support the suggestion that axotomy per se does not induce the alterations in sodium channel expression, but rather that the nature of the DRG axonal target (i.e. peripheral *versus* central) is an important determinant in events that modulate gene expression.

Several other experimental models of neuropathic pain have also been utilized to examine nerve injury-induced alterations in sodium channel expression, including chronic constriction injury (CCI) of the sciatic nerve [51, 61, 64] and diabetic neuropathy [61]. In general, similar, but less extensive, attenuation of SNS and NaN transcripts, and also TTX-resistant sodium currents, are reported following CCI compared to control DRG neurons [61, 64, 65]. Figure 10 illustrates the reduction of both SNS and NaN mRNAs, and an attenuation of *total* TTX-resistant sodium currents (protocols to separate slowly-inactivating and persistent TTX-resistant currents had not been established at the time of this study), in DRG neurons following CCI compared to control DRG neurons. CCI-induced injury is also associated with an upregulation of type III mRNA and the emergence of rapidly-repriming TTX-sensitive currents in DRG neurons (Fig. 11). These results are consistent with previous observations of ≥80% loss of myelinated fibers and 60–80% loss of unmyelinated fibers in the CCI model [66], and mirror results obtained with complete transection of the sciatic nerve [12, 25, 57, 63].

There are a number of reasons for predicting that the observed changes in the expression of multiple sodium channels following peripheral nerve injury might cause DRG neurons to fire spontaneously, or at inappropriately high frequencies. First, increased sodium conductance due to increased numbers of channels, *per se*, would be expected to lower the threshold for action potential generation [67, 68]. Second, shifting ratios of sodium channel isoforms, with differing overlapping of activation and steady-state inactivation curves, may permit membrane potential oscillations and thereby produce abnormal action potential activity [69]. Third, because of the rapidly-repriming of TTX-sensitive sodium current in DRG neurons following peripheral axotomy, a reduced refractory period would be expected in injured DRG neurons, so that these cells would sustain higher firing frequencies [12]. Fourth, some sodium channel subtypes in DRG neurons appear to be inactivated close to resting potential (see for example [8, 10]); if the persistent sodium

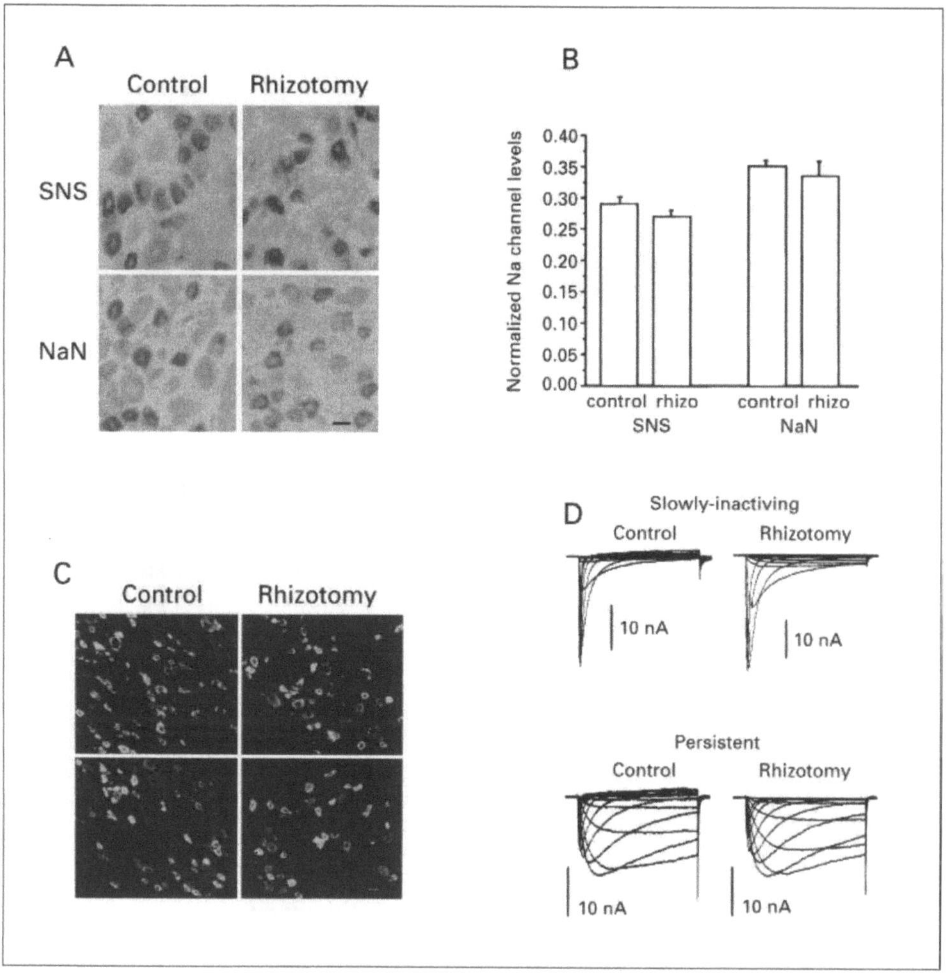

Figure 9
SNS and NaN sodium channels and their TTX-resistant currents are not decreased following dorsal rhizotomy. (A) In situ *hybridization of sections of control and centrally-axotomized DRG show similar levels of SNS mRNA in control and rhizotomized DRG, and also similar levels of NaN mRNA in control and rhizotomized DRG. Scale bar, 25 µm. (B) Quantitative RT-PCR analyses of SNS and NaN transcript levels show no significant differences in the transcript levels of the respective sodium channels between control and rhizotomized DRG. (C) SNS and NaN protein expression in control and rhizotomized DRG show similar levels of immunostaining for the respective channels between the two conditions. Scale bar, 25 µm. (D) Dorsal rhizotomy does not attenuate either slowly-inactivating or persistent TTX-resistant currents compared to control neurons. Modified and reproduced with permission from [25], © 2000 Society for Neuroscience.*

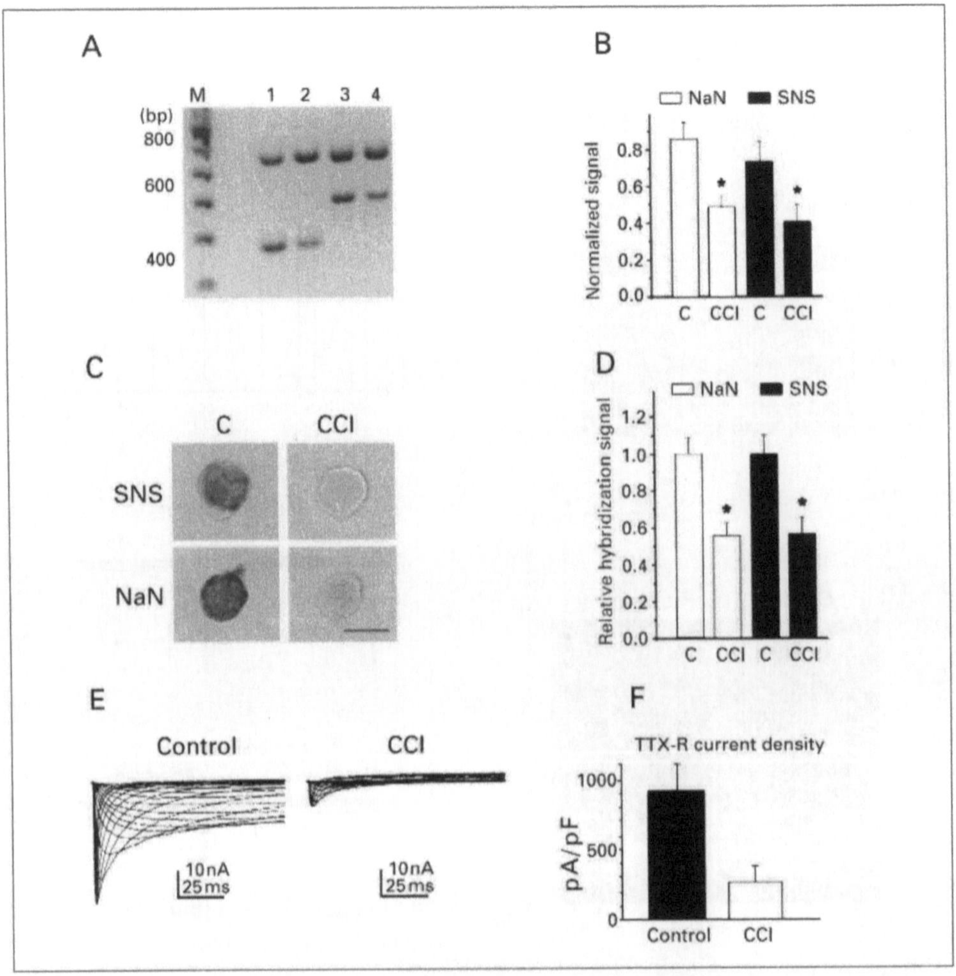

Figure 10

Chronic constriction injury (CCI) of the sciatic nerve downregulates SNS and NaN sodium channels and TTX-resistant sodium current. (A) Quantification of RT-PCR products from control and CCI DRG 14 days post surgery. Representative gel analysis of the co-amplified NaN (392 bp), SNS (512 bp) and GAPDH (666 bp). "M" indicates 100 bp standard. Lanes 1 and 2 GAPDH/NaN from control and CCI DRG, respectively; lanes 3 and 4 GAPDH/SNS from control and CCI DRG, respectively. (B) Mean levels (± S.D.) of NaN and SNS amplification products normalized to GAPDH. There is a significant reduction in both SNS and NaN transcripts in CCI DRG compared to control DRG. (C) In situ *hybridization of small neurons from control and CCI DRG. Both SNS and NaN transcripts are downregulated in DRG neurons following CCI compared to control neurons. Scale bar, 20 µm. (D) Quantification of relative SNS and NaN hybridization signals in control and CCI DRG neurons. SNS and NaN lev-*

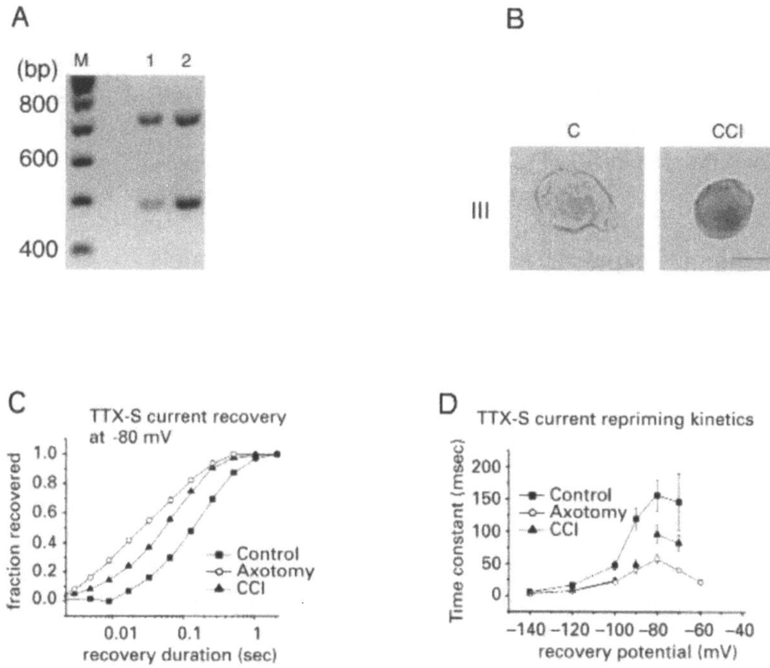

Figure 11
Type III sodium channel transcripts and rapidly-repriming TTX-sensitive sodium current are upregulated in CCI neurons. (A) Representative RT-PCR gel analysis of co-amplified type III (412 bp) and GAPDH (666 bp) products. "M" indicates 100 bp standard. Lanes 1 and 2 GAPDH/type III from control and CCI DRG, respectively. (B) In situ *hybridization of small neurons from control and CCI DRG. Type III transcripts are upregulated in CCI neurons. Scale bar, 20 µm. (C) Recovery from inactivation is accelerated after CCI. Repriming kinetics of CCI neurons are intermediate between values obtained for control and peripherally-axotomized neurons. Modified and reproduced with permission from [64].*

---

els are significantly lower in small CCI neurons compared to control neurons. (E) Representative TTX-resistant sodium currents in control and CCI small neurons. (F) Mean TTX-resistant peak current densities from control and CCI neurons. The TTX-resistant current density is significantly smaller in CCI neurons. Modified and reproduced with permission from [64].

channel NaN contributes to resting potential in DRG neurons or their axons, in a manner similar to the persistent current demonstrated in axons within the optic nerve [29], reduction of TTX-resistant currents following axotomy could lead to a hyper-polarizing shift in resting potential in DRG neurons. A hyper-polarizing shift might increase the availability of TTX-sensitive sodium current by relieving resting inactivation; this would, in turn, increase excitability by lowering the threshold for action potential generation [12].

Sodium channel expression in inflammatory pain models

Inflammatory pain results, in part, from sensitization of the primary nociceptive afferents, which may involve abnormal expression of sodium channels in these neurons. Molecules released during inflammatory events (e.g., prostaglandins, adenosine and serotonin) can modulate TTX-resistant sodium currents in DRG neurons *in vitro* [70], possibly *via* protein kinase A and/or protein kinase C initiated-cascades [71, 72]. Several models of *in vivo* inflammation have also examined sodium channel expression following this type of injury. Tanaka et al. [73] studied sodium channel expression in L4/5 DRG neurons in rats four days following injection of the inflammatory agent carrageenan into the hind paw. In these experiments, there was an increase in SNS mRNA expression in small DRG neurons projecting to the inflamed limb, compared with DRG neurons from the contralateral side or naive (uninjected) controls (Fig. 12). In addition, there was a significant increase in the mean peak amplitude of TTX-resistant sodium current in small ipsilateral DRG neurons (31.7 ± 3.3 nA) compared to contralateral neurons (20.1 ± 2.1 nA) four days post-injection [73]. In contrast to these observations, injection of complete Freund's adjuvant into the hindpaw has been reported not to significantly alter SNS mRNA levels at three days [61] following administration.

Additional studies have demonstrated an increase in NaN mRNA within ipsilateral DRG 7 days following injection of complete Freund's adjuvant into the plantar surface of the hindpaw [24]. Utilizing a pan-specific sodium channel antibody, increased sodium channel immunoreactivity has also been reported in DRG neurons following injection of complete Freund's adjuvant into their peripheral projection fields, which can persist for at least 2 months [74]. The changes in sodium channel expression in the DRG neurons affected by the complete Freund's adjuvant roughly parallel the behavioral changes measured in this model of inflammatory pain.

Neurotrophic factors and sodium channel expression in DRG neurons

Although a detailed understanding of the mechanisms responsible for alterations in sodium channel expression in DRG neurons during the pathogenesis of neuropath-

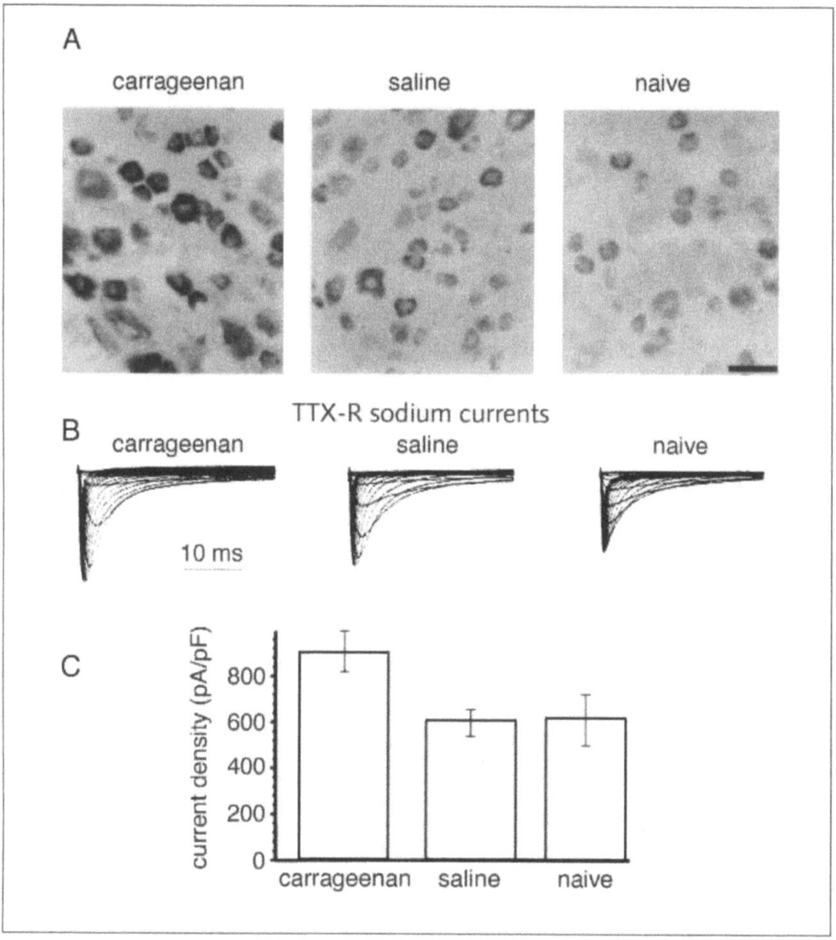

Figure 12

Carrageenan upregulates SNS mRNA and TTX-resistant sodium current. (A) Four days following ipsilateral injection of carrageenan into the hindpaw SNS transcript levels in DRG neurons are upregulated compared to contralateral saline-injected or naive DRG. Scale bar, 50 μm. (B) Families of TTX-resistant current traces from representative carrageenan, saline or naive small DRG neurons. (C) Mean TTX-resistant current density of carrageenan-injected DRG is significantly greater than saline-injected or naive DRG. Modified and reproduced with permission from [73].

ic and inflammatory pain is lacking, altered levels of neurotrophic factors have been implicated as contributors to these changes. Neurotrophic factors, including the neurotrophins, transforming growth factor β-related factors, epidermal growth fac-

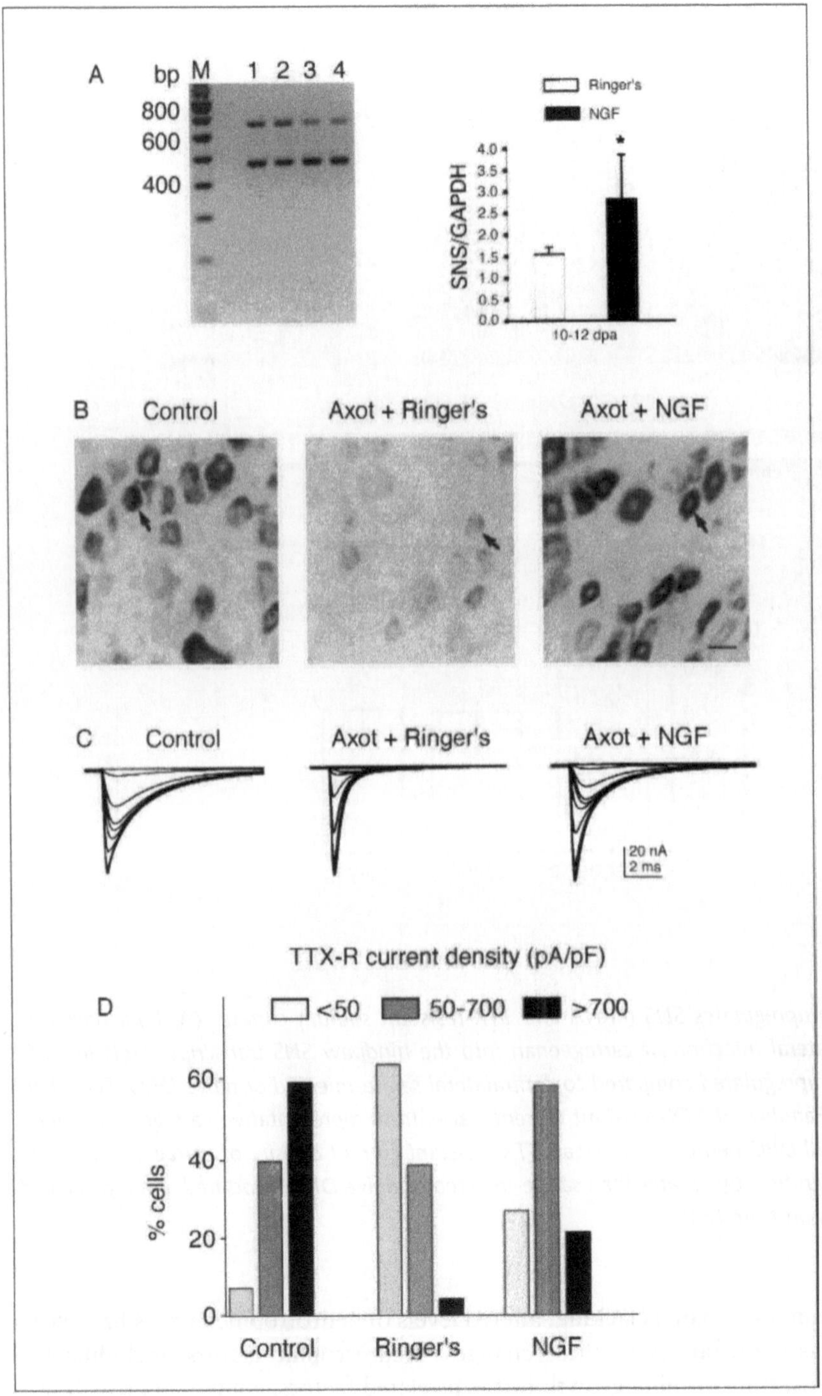
A
bp M 1 2 3 4
800
600
400
Ringer's
NGF
SNS/GAPDH
4.0
3.5
3.0
2.5
2.0
1.5
1.0
0.5
0.0
10-12 dpa
B Control
Axot + Ringer's
Axot + NGF
C Control
Axot + Ringer's
Axot + NGF
20 nA
2 ms
TTX-R current density (pA/pF)
D
<50
50-700
>700
% cells
60
40
20
0
Control
Ringer's
NGF

tor and fibroblast growth factor, have been shown to exert essential roles in the development and maintenance of neuronal phenotypes. Nerve growth factor (NGF), a widely-studied neurotrophin, is produced by Schwann cells, keratinocytes and fibroblasts within peripheral target tissues, and was suggested as a retrograde trophic signal nearly 20 years ago [75]. It is also well-known that inflammation stimulates NGF production in immune cells, and increased NGF concentrations have been observed within tissues exposed to inflammatory agents [76, 77]. In contrast to enhanced tissue levels of NGF during inflammation, nerve injury may prevent target-derived factors from being transported from the periphery to cell bodies, thus affecting synthetic pathways.

Early observations established that NGF can affect sodium channel expression in DRG neurons *in vitro* [78, 79]. These results suggested that the effects of peripheral nerve injury on sodium channel expression in DRG neurons [56, 57] may result, at least in part, from a loss of access to a peripheral supply of NGF. Black et al. [80] studying an *in vitro* model that mimics axotomy, demonstrated that NGF, delivered directly to DRG cell bodies *in vitro*, downregulates the expression of type III mRNA and upregulates the expression of SNS mRNA in small DRG neurons. Support for a modulatory role for NGF in the expression of SNS in DRG neurons *in vivo* is provided by several recent studies. Dib-Hajj et al. [81] demonstrated that delivery of exogenous NGF to the transected sciatic nerve, *via* an osmotic pump attached to the nerve stump, resulted in a partial rescue of SNS mRNA levels and TTX-resistant sodium current in small DRG neurons (Fig. 13). Moreover, there is also evidence indicating that NGF is required for maintaining SNS expression in uninjured adult

Figure 13

NGF rescues downregulation of SNS mRNA and TTX-resistant sodium current following axotomy. (A) On the left side, representative gel analysis of the coamplified SNS and GAPDH products. "M" 100 bp standard. Lanes 1 and 2, Ringer's-treated axotomized DRG. Lanes 3 and 4, NGF-treated axotomized DRG. On the right side, quantification of SNS amplification products normalized to GAPDH for Ringer's- and NGF-treated axotomized DRG. There is a significant upregulation of SNS in the NGF-treated DRG compared to Ringer's-treated. (B) In situ *hybridization of sections from control, Ringer's-treated and NGF-treated DRG. Axotomy-induced downregulation of SNS is partially reversed by NGF treatment. Scale bar, 25 µm. (C) TTX-resistant sodium currents recorded from representative control, Ringer's- and NGF-treated axotomized small DRG neurons. (D) Small neurons from control, Ringer's- and NGF-treated DRG were assigned to 1 of 3 groups based on their TTX-resistant current density. Substantially fewer neurons with high levels of TTX-resistant current density were present in Ringer's-treated compared to control DRG. NGF-treatment partially rescues the percentage of neurons expressing high levels of TTX-resistant current. Modified and reproduced with permission from [81].*

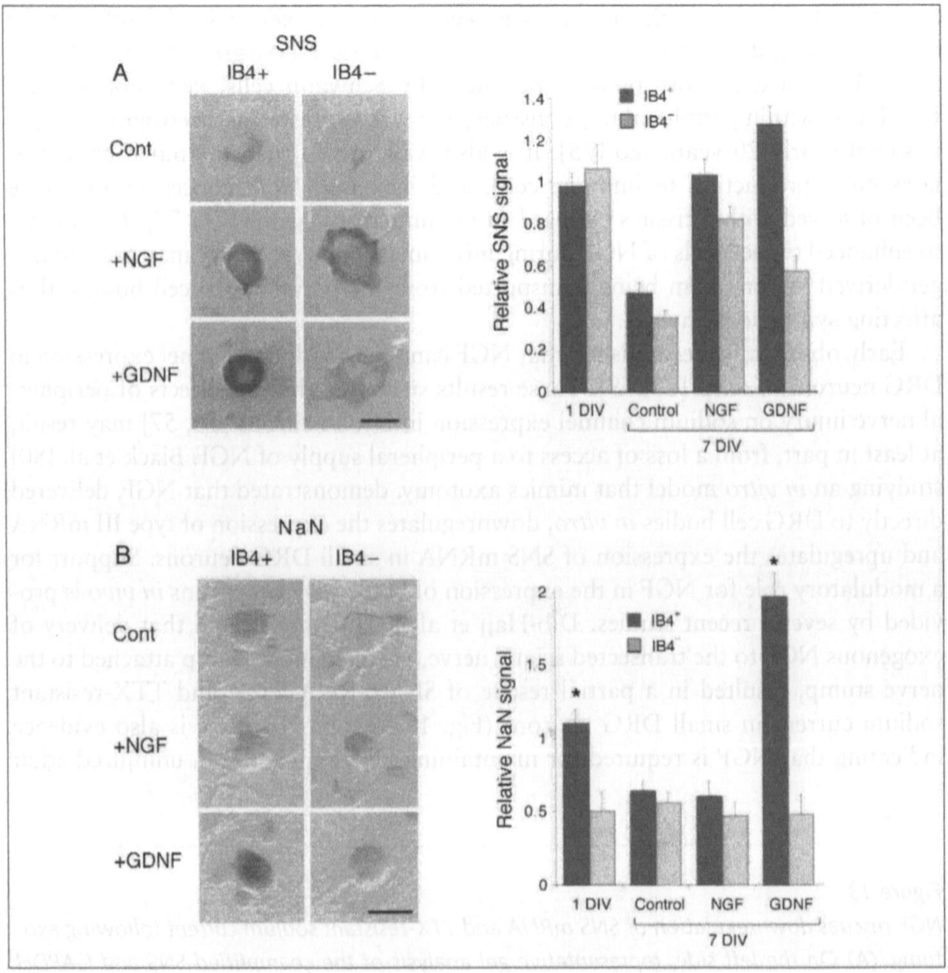

Figure 14

Effects of NGF and GDNF on expression of SNS and NaN transcripts. (A) In situ hybridization for SNS mRNA in representative IB4⁺ and IB4⁻ DRG neurons 7 days in culture without (control) or with NGF or GDNF treatment. Scale bar, 25 μm. (B) Quantification of SNS hybridization signal in IB4⁺ and IB4⁻ DRG neurons in control, NGF and GDNF conditions. Both NGF and GDNF produced significant differences in the respective subpopulations compared to 7 days in culture control neurons. (C) In situ hybridization for NaN mRNA in representative IB4⁺ and IB4⁻ DRG neurons 7 days in culture without (control) or with NGF or GDNF treatment. Scale bar, 25 μm. D. Quantification of NaN hybridization signal in IB4⁺ and IB4⁻ DRG neurons in control, NGF and GDNF conditions. Only GDNF treatment of IB4⁺ neurons resulted in a significant upregulation of NaN compared to control neurons. Modified and reproduced with permission from [84].

DRG neurons. Adult rats injected with NGF develop thermal hypoalgesia and high antibody titers to NGF [82], which results in reduced systemic levels of NGF. Under these conditions, there is an attenuation of TTX-resistant sodium currents and SNS mRNA in small DRG neurons [83], suggesting that NGF contributes to the maintenance of steady-state SNS levels in DRG neurons in the adult nervous system *in vivo*. Changes in the expression of type III and SNS sodium channels in DRG neurons following nerve injury thus may reflect, at least in part, the loss of access to peripheral pools of NGF.

In contrast to SNS, NaN expression in DRG neurons appears not to be modulated by NGF, but instead by glial-derived growth factor (GDNF) [84]. NaN is preferentially expressed in DRG neurons that bind the isolectin IB4, a marker for cells that express ret and GFR receptors for GDNF. In an *in vitro* model of axotomy where NaN mRNA levels are reduced with increasing time in culture, Fjell et al. [84] observed that exogenous administration of GDNF to these cultures restored expression of NaN to levels at or above normal in IB4-positive DRG neurons (Fig. 14). GDNF, in addition, also significantly upregulated SNS in IB4-positive, but not IB4-negative, neurons. Consistent with the upregulation of NaN and SNS mRNAs in GDNF-treated neurons, there was also a significant increase in TTX-resistant sodium current in these neurons (Fig. 15). Interestingly, the reduction in conduction velocity of C-fibers that follows sciatic nerve axotomy is ameliorated by intrathecal administration of GDNF [85]. Although the mechanism responsible for the enhancement of conduction velocity is not known, these results are consistent with a role for GDNF in modulating expression of sodium channels in nociceptive-type DRG neurons.

Perspectives

The involvement of sodium channels in the pathogenesis of pain syndromes is clearly complex, encompassing the expression and modulation of multiple sodium channel isoforms, each with differing functional properties, which are influenced by a dynamic interplay of neuroactive molecules, including neurotrophic factors, inflammatory mediators, cytokines and neurotransmitters. Progress has been made to elucidate the exact role(s) of each type of sodium channel in the physiology of intact DRG neurons, and how ensembles of sodium channel isoforms expressed in an individual cell affect electrogenesis. Moreover, substantial advancements have been made towards delineating the injury-induced alterations in the expression and spatial organization of sodium channels in DRG neurons in neuropathic and inflammatory pain syndromes. Primary sensory neurons express several sodium channel genes (SNS, NaN and PN1) that are not found in other neuronal types, and of these genes, SNS and NaN are preferentially expressed in small diameter DRG neurons, which are predominantly nociceptive. The selective expression of specific sodium

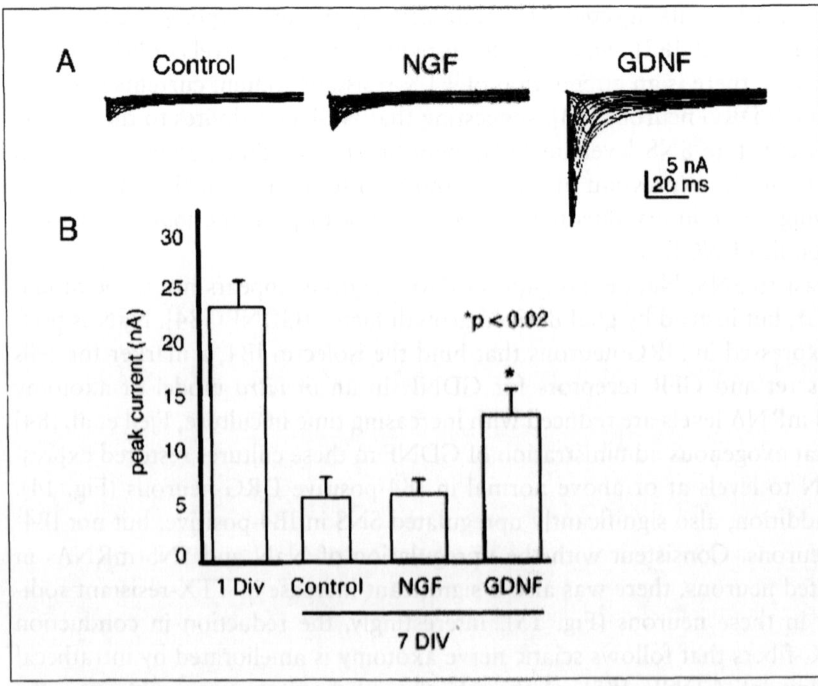

Figure 15
Effects of NGF and GDNF on TTX-resistant currents. (A) Families of TTX-resistant sodium current traces from representative DRG neurons from control, NGF-treated and GDNF-treated conditions. (B) TTX-resistant peak current amplitudes from control, NGF-treated and GDNF-treated DRG neurons at 7 days in culture. GDNF treatment significantly increased TTX-resistant current compared to control neurons at 7 days in culture. Modified and reproduced with permission from [84].

channel isoforms within nociceptive neurons, together with an understanding of the unique molecular architecture of these channels, may present an opportunity for pharmacological manipulation of these channels. Sodium channels within the nociceptive pathway may thus emerge as important therapeutic targets in the management of pain syndromes.

Acknowledgement
Research in the authors' laboratory has been supported in part by grants from the National Multiple Sclerosis Society and the Paralyzed Veterans of America/Eastern Paralyzed Veterans Association, and by the Medical Research Service and Rehabilitation Research Service, Department of Veterans Affairs.

References

1 Nordin M, Nyström B, Wallin U, Hagbarth K-E (1984) Ectopic sensory discharges and paresthesiae in patients with disorders of peripheral nerves, dorsal roots and dorsal columns. *Pain* 20: 231–245

2 Ochoa J, Torebjörk HE (1980) Paresthesiae from ectopic impulse generation in human sensory nerves. *Brain* 103: 835–854

3 Wall PD, Devor M (1981) The effect of peripheral nerve injury on dorsal root potentials and on transmission of afferent signals into the spinal cord. *Brain Res* 209: 95–111

4 Zhang J-M, Donnelly DF, Song X-J, LaMotte RH (1997) Axotomy increases the excitability of dorsal root ganglion cells with unmyelinated axons. *J Neurophysiol* 78: 2790–2794

5 Malhotra JD, Kazen Gillespie, K, Hortsch M, Isom LL (2000) Sodium channel β subunits mediate homophilic cell adhesion and recruit ankyrin to points of cell-cell contact. *J Biol Chem* 275: 11383–11388

6 Morgan K, Stevens EB, Shah B, Cox PJ, Dison AK, Lee K, Pinnock RD, Hughes J, Richardson PJ, Mizuguchi K et al (2000) *Proc Natl Acad Sci USA* 97: 2308– 2313

7 Felts PA, Yokoyama S, Dib-Hajj S, Black J, Waxman SG (1997) Sodium channel α-subunit mRNAs I, II, III, NaG, Na6 and hNE (PN1): different expression patterns in developing rat nervous system. *Mol Brain Res* 45: 71–82

8 Kostyuk PG, Veselovsky NS, Tsyandryenko AY (1981) Ionic currents in the somatic membrane of rat dorsal root ganglion neurons. I. Sodium currents. *Neuroscience* 6: 2423–2430

9 Roy ML, Narahashi T (1992) Differential properties of tetrodotoxin-sensitive and tetrodotoxin-resistant sodium channels in rat dorsal root ganglion neurons. *J Neurosci* 12: 2104–2111

10 Caffrey JM, Eng DL, Black JA, Waxman SG, Kocsis JD (1992) Three types of sodium channels in adult rat dorsal root ganglion neurons. *Brain Res* 592: 283–297

11 Elliott AA, Elliott JR (1993) Characterization of TTX-sensitive and TTX-resistant sodium currents in small cells from adult rat dorsal root ganglia. *J Physiol Lond* 463: 39–56

12 Cummins TR, Waxman SG (1997) Downregulation of tetrodotoxin-resistant sodium currents and up-regulation of a rapidly repriming tetrodotoxin-sensitive sodium current in small spinal sensory neurons after nerve injury. *J Neurosci* 17: 3503–3514

13 Cummins TR, Dib-Hajj SD, Black JA, Akopian AN, Wood JN, Waxman SG (1999) A novel persistent tetrodotoxin-resistant sodium current in small primary sensory neurons. *J Neurosci* 19: RC43

14 Dib-Hajj SD, Tyrell L, Cummins TR, Black JA, Wood PM, Waxman SG (1999) Two tetrodotoxin-resistant sodium channels in human dorsal root ganglion neurons. *FEBS Lett* 462: 117–120

15 Black JA, Dib-Hajj S, McNabola K, Jeste S, Rizzo MA, Kocsis JD, Waxman SG (1996) Spinal sensory neurons express multiple sodium channel α-subunit mRNAs. *Mol Brain Res* 43: 117–132

16 Dib-Hajj SD, Tyrrell L, Black JA, Waxman SG (1998) NaN, a novel voltage-gated Na channel, is expressed preferentially in peripheral sensory neurons and down-regulated after axotomy. *Proc Natl Acad Sci USA* 95: 8963–8969

17 Noda M, Ikeda T, Kayano T, Suzuki H, Takeshima H, Kurasaki M, Takahashi, H, Numa S (1986) Existence of distinct sodium channel messenger RNAs in rat brain. *Nature* 320: 188–192

18 Schaller KL, Krzemien DM, Yarowsky PJ, Krueger BK, Caldwell JH (1995) A novel, abundant sodium channel expressed in neurons and glia. *J Neurosci* 15: 3231–3242

19 Dietrich PS, McGivern JG, Delgado SG, Koch BD, Eglen RM, Hunter JC, Sangeswaran L (1998) Functional analysis of a voltage-gated sodium channel and its splice variant from rat dorsal root ganglia. *J Neurochem* 70: 2262–2272

20 Toledo-Aral JJ, Moss BL, He Z-J, Koszowski AG, Whisenand T, Levinson SR, Wolff JJ, Silos-Santiago I, Halegoua S, Mandel G (1997) Identification of PN1, a predominant voltage-dependent sodium channel expressed principally in peripheral neurons. *Proc Natl Acad Sci USA* 94: 1527–1532

21 Cummins TR, Howe JR, Waxman SG (1998) Slow closed-state inactivation: A novel mechanism underlying ramp currents in cells expressing the hNE/PN1 sodium channel. *J Neurosci* 18: 9607–9619

22 Akopian AN, Sivilotti L, Wood JN (1996) A tetrodotoxin-resistant voltage-gated sodium channel expressed by sensory neurons. *Nature* 379: 257–262

23 Sangameswaran L, Delgado SG, Fish LM, Koch BD, Jakeman LB, Stewart GR, Sze P, Hunter JC, Eglen RM, Herman RC (1996) Structure and function of a novel voltage-gated Tetrotodoxin-resistant sodium channel specific to sensory neurons. *J Biol Chem* 271: 5953–5956

24 Tate S, Benn S, Hick C, Trezise D, John V, Mannion RJ, Costigan M, Plumpton C, Grose D, Gladwell Z et al (1998) Two sodium channels contribute to the TTX-R sodium current in primary sensory neurons. *Nat Neurosci* 1: 653–655

25 Sleeper AA, Cummins TR, Hormuzdiar W, Tyrrell L, Dib-Hajj SD, Waxman SG, Black JA (2000) Changes in expression of two tetrodotoxin-resistant sodium channels and their currents in dorsal root ganglion neurons following sciatic nerve injury, but not rhizotomy. *J Neurosci* 20: 7279–7289

26 Dib-Hajj SD, Tyrrell L, Escayg A, Wood PM, Meisler MH, Waxman SG (1999) Coding sequence, genomic organization, and conserved chromosomal localization of the mouse gene Scn11a encoding the sodium channel NaN. *Genomics* 59: 309–318

27 Amaya F, Decoster I, Samad TA, Plumpton C, Tate S, Mannion RJ, Costigan M, Woolf CJ (2000) Diversity of expression of the sensory neurons-specific TTX-resistant voltage-gated sodium ion channels SNS and SNS2. *Mol Cell Neurosci* 15: 331–342

28 Akopian AN, Souslova V, England S, Okuse K, Ogata N, Ure J, Smith A, Kerr BJ, McMahon SB, Boyce S et al (1999) The tetrodotoxin-resistant sodium channel SNS has a specialized function in pain pathways. *Nat Neurosci* 2: 541–548

29 Stys PK, Ransom BR, Waxman SG (1993) Non-inactivating, TTX-sensitive Na^+ conductance in rat optic nerve axons. *Proc Natl Acad Sci USA* 90: 6976–6980

30 Kapoor R, Li Y-G , Smith J (1997) Slow sodium-dependent potential oscillations contribute to ectopic firing in mammalian demyelinated axons. *Brain Res* 120: 647–652
31 Schwindt PC, Crill WE (1995) Amplification of synaptic current by persistent sodium conductance in apical dendrite of neocortical neurons. *J Neurophysiol* 74: 2220–2224
32 Stafstrom CE, Schwindt PC, Crill WE (1982) Negative slope conductance due to a persistent subthreshold sodium current in cat neocortical neurons *in vitro*. *Brain Res* 236: 221–226
33 Rush AM, Brau ME, Elliott AA, Elliott JR (1998) Electrophysiological properties of sodium current subtypes in small cells from adult rat dorsal root gangia. *J Physiol* 511: 771–789
34 Scholz A, Appel N, Vogel W (1998) Two types of TTX-resistant and one TTX-sensitive Na^+ channel in rat dorsal root ganglion neurons and their blockade by halothane. *Eur J Neurosci* 10: 2547–2556
35 Brock JA, McLachlan EM, Belmonte C (1998) Tetrodotoxin-resistant impulses in single nociceptor nerve terminals in guinea-pig cornea. *J Physiol* 512: 211–217
36 Jeftinija S (1994) The role of tetrodotoxin-resistant sodium channels of small primary afferent fibers. *Brain Res* 639: 125–134
37 Quasthoff S, Grosskreutz J, Schroder JM, Schneider U, Grafe P (1995) Calcium potentials and tetrodotoxin-resistant sodium potentials in unmyelinated C fibres of biopsied human sural nerve. *Neuroscience* 69: 955–965
38 Coward K, Plumpton C, Facer P, Birch R, Carlstedt T, Tate S, Bountra C, Anand P (2000) Immunolocalization of SNS/PN3 and NaN/SNS2 sodium channels in human pain states. *Pain* 85: 41–50
39 Fjell J, Hjelstrom P, Hormuzdiar W, Milenkovic M, Aglieco F, Tyrrell L, Dib-Hajj S, Waxman SG, Black JA (2000) Localization of the tetrodotoxin-resistant sodium channel NaN in nociceptors. *NeuroReport* 11: 199–202
39 Eccles JC, Libet B, Young RR (1958) The behavior of chromatolysed motoneurons studied by intracellular recording. *J Physiol Lond* 143: 11–40
41 Kuno M, Llinas R (1970) Enhancement of synaptic transmission by dendritic potentials in chromatolysed motoneurons of the cat. *J Physiol Lond* 210: 807–821
42 Gurtu S, Smith PA (1988) Electrophysiological characteristics of hamster dorsal root ganglion cells and their response to axotomy. *J Neurophysiol* 59: 408–423
43 Devor M, Wall PD. (1990) Cross-excitation in dorsal root ganglia of nerve-injured and intact rats. *J Neurophysiol* 64: 1733–1746
44 Sernagor E, Yarom Y, Werman R (1986) Sodium dependent regenerative responses in dendrites of axotomized motoneurons in the cat. *Proc Natl Acad Sci USA* 83: 7966–7970
45 Titmus MJ, Faber DS (1986) Altered excitability of goldfish mauthner cell following axotomy. II. localization and ionic basis. *J Neurophysiol* 55: 1440–1454
46 Devor M, Keller CH, Deerinck TJ, Ellisman MH (1989) Na^+ channel accumulation on axolemma of afferent endings in nerve end neuromas in *Apteronotus*. *Neurosci Lett* 102: 149–154

47 England JD, Gamboni F, Ferguson MA, Levinson SR (1994) Sodium channels accumulate at the tips of injured axons. *Muscle & Nerve* 17: 593–598
48 England JD, Happel LT, Kline DG, Gamboni F, Thouron CL, Lui ZP, Levinson SR (1996) Sodium channel accumulation in humans with painful neuromas. *Neurology* 47: 272–276
49 Black JA, Cummins TR, Plumpton C, Chen Y, Clare J, Waxman SG (1999) Upregulation of a previously silent sodium channel after peripheral, but not central, nerve injury in DRG neurons. *J Neurophysiol* 82: 2776–2785
50 Suzuki H, Beckh S, Kubo H, Yahagi N, Ishida H, Kayano T, Noda M, Numa S (1988) Functional expression of cloned cDNA encoding sodium channel III. *FEBS Lett* 228: 195–200
51 Novakovic SD, Tzoumaka E, McGivern JG, Haraguchi M, Sangameswaran L, Gogas KR, Eglen RM, Hunter JC (1998) Distribution of the tetrodotoxin-resistant sodium channel PN3 in rat sensory neurons in normal and neuropathic conditions. *J Neurosci* 18: 2174–2187
52 Chabal C, Russell LC, Burchiel KJ (1989) The effect of intravenous lidocaine, tocainide, and mexiletine on spontaneously active fibers originating in rat sciatic neuromas. *Pain* 38: 333–338
53 Devor M, Wall PD, Catalan N (1992) Systemic lidocaine silences ectopic neuroma and DRG discharge without blocking nerve conduction. *Pain* 48: 261–268
54 Omana-Zapata I, Khabbaz MA, Hunter JC, Bley KR (1997) QX-314 inhibits ectopic nerve activity associated with neuropathic pain. *Brain Res* 771: 228–237
55 Rizzo MA (1997) Successful treatment of painful traumatic mononeuropathy with carbamazepine: insights into a possible molecular pain mechanism. *J Neurol Sci* 152: 103–106
56 Waxman SG, Kocsis JK, Black JA (1994) Type III sodium channel mRNA is expressed in embryonic but not adult spinal sensory neurons, and is reexpressed following axotomy. *J Neurophysiol* 72: 466–471
57 Dib-Hajj S, Black JA, Felts P, Waxman SG (1996) Down-regulation of transcripts for Na channel α-SNS in spinal sensory neurons following axotomy. *Proc Natl Acad Sci USA* 93: 14950–14954
58 Scadding JW (1981) Development of ongoing activity, mechanosensitivity, and adrenalin sensitivity in severed peripheral nerve axons. *Exper Neurol* 73: 345–364
59 Burchiel KJ (1988) Carbamazepine inhibits spontaneous activity in experimental neuromas. *Exp Neurol* 102: 249–253
60 Matzner O, Devor M (1994) Hyperexcitability at sites of nerve injury depends on voltage-sensitive Na^+ channels. *J Neurophysiol* 72: 349–359
61 Okuse K, Chaplan SR, McMahon SB, Luo ZD, Calcutt NA, Scott BP, Akopian AN, Wood JN (1997) Regulation of expression of the sensory neurons-specific sodium channel SNS in inflammatory and neuropathic pain. *Mol Cell Neurosci* 10: 196–207
62 Himes BT, Tessler A (1982) Death of some dorsal root ganglion neurons and plasticity

of other following sciatic nerve section in adult and neonatal rats. *J Comp Neurol* 248: 215–230
63 Rizzo MA, Kocsis JD, Waxman SG (1995) Selective loss of slow and enhancement of fast Na^+ currents in cutaneous afferent dorsal root ganglion neurones following axotomy. *Neurobiol Dis* 2: 87–96
64 Dib-Hajj SD, Fjell J, Cummins TR, Zheng Z, Fried K, LaMotte R, Black JA, Waxman SG (1999) Plasticity of sodium channel expression in DRG neurons in the chronic constriction injury model of neuropathic pain. *Pain* 83: 591–600
65 Kral MG, Xiong Z, Study RE (1999) Alterantion of Na^+ currrents in dorsal root ganglion neurons from rats with a painful neuropathy. *Pain* 81: 15–24
66 Carlton SM, Dougherty PM, Pover CM, Coggeshall RE (1991) Neuroma formation and numbers of axons in a rat model of experimental peripheral neuropathy. *Neurosci Lett* 131: 88–92
67 Waxman SG, Brill MH (1978) Conduction through demyelinated plaques in multiple sclerosis: computer simulations of facilitation by short internodes. *J Neurol Neurosurg Psychiat* 41: 408–417
68 Matzner O, Devor M (1992) Na^+ conductance and the threshold for repetitive neuronal firing. *Brain Res* 597: 92–98
69 Rizzo MA, Kocsis JD, Waxman SG (1996) Mechanisms of paresthesiae, dysesthesiae, and hyperesthesiae: Role of Na^+ channel heterogeneity. *Eur Neurol* 36: 3–12
70 Gold MS, Reichling DB, Shuster MJ, Levine JD (1996) Hyperalgesic agents increase a tetrodotoxin-resistant Na^+ current in nociceptors. *Proc Natl Acad Sci USA* 93: 1108–1112
71 England S, Bevan S, Docherty RJ (1996) PGE_2 modulates the tetrodotoxin-resistant sodium current in neonatal rat dorsal root ganglion neurones *via* the cyclic AMP-protein kinase A cascade. *J Physiol Lond* 495: 429–440
72 Gold MS, Levine JD, Correa AM (1998) Modulation of TTX-R I_{Na} by PKC and PKA and their role in PGE_2-induced sensitization of rat sensory neurons *in vitro*. *J Neurosci* 18: 10345–10355
73 Tanaka M, Cummins TR, Ishikawa K, Dib-Hajj SD, Black JA, Waxman SG (1998) SNS Na^+ channel expression increases in dorsal root ganglion neurons in the carrageenan inflammatory pain model. *NeuroReport* 9: 967–972
74 Gould HJ III, England JD, Liu ZP, Levinson SR (1998) Rapid sodium channel augmentation in response to inflammation induced by complete Freund's adjuvant. *Brain Res* 802: 69–74
75 Schwab ME, Heurman R, Thoenen H (1982) Communication between target organs and nerve cells: retrograde axonal transport and site of action of nerve growth factors. *Cold Spring Harb Symp Quant Biol* 46: 125–134
76 Weskamp G, Otten U (1978) An enzyme-linked immunoassay for nerve growth factor (NGF): A tool for studying regulatory mechanisms involved in NGF production in brain and in peripheral tissues. *J Neurochem* 48: 1779–1786
77 Woolf CJ, Safieh-Garabedian B, Ma Q-P, Crilly P, Winters J (1994) Nerve growth fac-

tor contributes to the generation of inflammatory sensory hypersensitivity. *Neuroscience* 62: 327–331
78 Aguayo LG, White G (1992) Effects of nerve growth factor on TTX- and capsaicin-sensitivity in adult rat sensory neurons. *Brain Res* 570: 61–67
79 Zur KB, Oh Y, Waxman SG, Black JA (1995) Different up-regulation of sodium channel α- and β-subunit mRNAs in cultured embryonic DRG neurons following exposure to NGF. Molec *Brain Res* 30: 97–103
80 Black JA, Langworthy K, Hinson AW, Dib-Hajj SD, Waxman SG (1997) NGF has opposing effects on Na^+ channel III and SNS gene expression in spinal sensory neurons. *NeuroReport* 8: 2331–2335
81 Dib-Hajj SD, Black JA, Cummins TR, Kenney AM, Kocsis JD, Waxman SG (1998) Rescue of α-SNS sodium channel expression in small dorsal root ganglion neurons after axotomy by nerve growth factor *in vivo*. *J Neurophysiol* 79: 2668–2678
82 Chundler EH, Anderson LC, Byers MR (1997) Nerve growth factor depletion by autoimmunization produces thermal hypoalgesia in adult rats. *Brain Res* 765: 327–330
83 Fjell J, Cummins TR, Fried K, Black JA, Waxman SG (1999) *In vivo* NGF deprivation reduces SNS/PN3 expression and TTX-R sodium currents in IB4-negative DRG neurons. *J Neurophysiol* 81: 803–810
84 Fjell J, Cummins TR, Dib-Hajj SD, Fried K, Black JA, Waxman SG (1999) Differential role of GDNF and NGF in the maintenance of two TTX-resistant sodium channels in adult DRG neurons. Mol *Brain Res* 67: 267–282
85 Bennett DL, Michael GJ, Ramachandran N, Munson JB, Averill S, Yan Q, McMahon SB, Priestly JV (1998) A distinct subgroup of small DRG cells express GDNF receptor components and GDNF is protective for these neurons after nerve injury. *J Neurosci* 18: 3059–3072
86 Waxman SG, Cummins TR, Dib-Hajj S, Fjell J, Black JA (1999) Sodium channels, excitability of primary sensory neurons, and the molecular basis of pain. *Muscle & Nerve* 22: 1177–1187

Changes in DRG neurons after injury: possible involvement in the development and maintenance of neuropathic pain

Valerie M.K. Verge[1], Tracy D. Wilson-Gerwing[1], Laurie A. Karchewski[2] and Kelly A. Gratto[1]

[1]Department of Anatomy & Cell Biology and Cameco MS Neuroscience Research Center, University of Saskatchewan, Saskatoon, SK, Canada S7N 5E5; [2]Neural Plasticity Research Group, Department of Anesthesia and Critical Care, Massachusetts General Hospital and Harvard Medical School, 149 13th St., Room 4309, Charlestown, MA 02129, USA

Introduction

Peripheral nerve injury evokes a series of molecular, biochemical, and cyto-architectural changes in primary sensory neurons. These events, together with changes in the injured nerve and corresponding regions of the spinal cord, contribute to the repair processes and the neuropathic pain states that develop. Neuropathic pain is a tremendous clinical problem, whereby normally innocuous or low-intensity stimuli acquire the ability to activate pain pathways. The goal of basic research in this area is to elucidate the diverse events and mechanisms responsible for this state so that more effective therapeutic interventions and treatments may be developed. The purpose of this chapter is to highlight aspects of the sensory neuron response to injury believed to be involved in the transition from normal nociception to the neuropathic pain state. The findings to be discussed stem largely from two models of peripheral nerve injury – either complete sciatic nerve transection or the most commonly used model of neuropathic pain, chronic constriction injury (CCI), which involves a partial lesion of the sciatic nerve induced by loosely tying the sciatic nerve with chromic gut sutures [1]. The latter results in the slow, edematous axotomy of predominantly large diameter myelinated axons (90% of Aβ and Aδ fibers and 30% of C fibers by three days post-ligation [2]), a reproducible thermal and mechanical allodynia, and the exposure of the uninjured axons to the Wallerian degenerating nerve in which many proinflammatory molecules are produced.

Neuropathic pain sensations arise from alterations in the activity and phenotype of predominantly three classes of sensory afferents – the myelinated Aδ and Aβ fibers which appear largely responsible for the mechanical allodynia observed, and the unmyelinated C fibers which contribute to the hyperalgesic state. These fibers have been shown to innervate both cutaneous and non-cutaneous tissue (reviewed in [3]). Biochemically and anatomically it is now recognized that two major classes of C fibers exist – those which are responsive to nerve growth factor (NGF) and

Mechanisms and Mediators of Neuropathic Pain, edited by Annika B. Malmberg and Sandra R. Chaplan

express calcitonin gene related peptide (CGRP) and substance P (SP), and those which are responsive to glial-derived neurotrophic factor (GDNF) and express a cell surface α-galactosyl epitope that can be stained with the lectin IB-4 [4]. The myelinated Aδ and Aβ fibers are most likely to respond to either neurotrophin-3 (NT-3) or brain-derived neurotrophic factor (BDNF) that bind preferentially to trkC and trkB receptors respectively (see refs in [5]), but there is potential for modulation by multiple neurotrophins in sensory neurons due to an extensive overlap in neuronal trk expression [5].

Axotomy induces alterations in electrical activity, ion channel expression and chemical sensitivity

One of the early measurable events associated with nerve injury is the generation of ectopic spontaneous activity or bursting in injured neurons which can induce excitatory changes in the spinal cord and enhance painful dysesthesias [6–8]. Early work revealed that the discharges arise from many of the injured nerves' Aβ and Aδ afferents, while recent studies have furthered these findings and show that discharges which originate in the DRG following injury are primarily from muscle (which also includes some nociceptors) and not skin afferents [9]. These alterations in firing patterns of neurons can lead to an intense rapid central discharge lasting for prolonged periods that may underlie the process of central sensitization. The molecular mechanisms underlying these changes in electrical activity have only recently begun to be elucidated [10]. The abnormal accumulation of sodium channels at the tips of injured axons makes them likely candidates for the generation of this ectopic activity [11, 12]. DRG neurons express at least six known sodium channel transcripts [13], characterized as producing tetrodotoxin (TTX)-sensitive (αI, Na6, PN1/hNE), TTX-resistant (αSNS-PN3, NaG) or TTX-insensitive (NaN) sodium currents [10]. Expression of αSNS-PN3 and NaN sodium channels (which is restricted to small sensory neurons likely to be involved in nociception) is down-regulated with spinal nerve ligation. Thus, it is not surprising that there is a corresponding long-lasting decrease in TTX-resistant Na currents in these injured cells [10, 14]. The observed tactile allodynia may stem from the significant up-regulation of expression for a previously silent α-III sodium channel in injured [10], or be influenced by the elevated expression of αSNS-PN3 receptors in large neurons of adjacent uninjured DRG [15]. Axotomy-associated reductions in the retrograde supply of the trophic molecules that maintain the functional properties of this subset of neurons, namely NGF and GDNF, are implicated in the injury-associated reductions of TTX-resistant channels. Exogenous application of either factor can increase expression of these channels in the appropriate responsive sub-populations of neurons [10, 16, 17], while NGF can effect a down-regulation in α-III channel expression *in vitro* [13]. In contrast to sciatic nerve transection, CCI-induced neuropathic pain is associated with a shift in the subcellular localization of SNS-PN3 Na channels from the cell body to axon tips proximal to the injury site, with no discernible changes in expression levels. This change in distribution may contribute to the generation of ectopic activity at the lesion site [12, 15]. Compelling evidence for a role of SNS-PN3 Na in the mediation of neuropathic and chronic inflammatory pain comes from experiments in which inhibition of SNS-PN3 protein expression by administration of SNS-PN3 antisense oligonucleotides was effective in reversing the hyperalgesia and allodynia associated with traumatic nerve injury and chronic inflammation, but not normal nociception on the contralateral side [15].

There are two other classes of ion channels that have been implicated in nociceptive function, namely P2X3, the ligand gated ion channel that is activated by ATP [18] and VR1, the capsaicin (vanilloid), proton or heat-activated cation channel which transduces noxious thermal stimuli [19]. The recent generation of mice lacking the VR1 receptor has confirmed that this protein plays a key role in the detection of painful heat and the thermal hypersensitivity associated with inflammation; but interestingly, injury-associated changes in thermal or mechanical sensitivity were indistinguishable from control mice [20]. Whether these aspects of thermosensation might involve other members of this receptor family, for example, the recently identified VR1 homologue VRL-1, remains to be shown [21]. ATP, on the other hand, appears to have a role in neuropathic pain associated with several pathologies. The restriction of ATP P2X3 receptor expression to the IB-4, GDNF-responsive subpopulation of DRG neurons that terminate in inner lamina II, as opposed to the trkA/CGRP expressing neurons which project heavily to lamina I and outer lamina II [22], suggests modality specialization. The inner part of lamina II has been recently shown to express the gamma isoform of protein kinase C. It is speculated that there is an important role for activation of this pathway by injured C fibers in the development of hyperalgesic responses to nerve injury as evidenced by the significant decrease in the magnitude of the mechanical and thermal allodynia that developed following partial nerve lesion in PKCγ KO mice [23]. P2X3 expression in small to medium size neurons is down-regulated with sciatic nerve transection [24], yet up-regulated with CCI injury [25]. This implies select injury-dependant roles for ATP, which has been shown to have pronociceptive actions [26] *via* P2X3, effecting glutamate release from sensory afferents [27]. Perhaps the reduced expression of this ATP receptor following injury allows for enhanced metabolic breakdown of ATP to adenosine, which can play a role in alleviating injury-induced mechanical allodynia [28]. Exogenous GDNF is capable of reversing injury-induced decreases in P2X3 expression [24], but whether it is involved in the CCI-associated increases in P2X3 expression is not known.

Changes in nociceptive input following injury may also be derived from alterations in metabotropic receptor expression, such as the B1 and B2 bradykinin receptors, which are found on close to 90% of injured neurons [29, 30]. Bradykinin, a potent algogenic mediator, is produced as a result of inflammation such as that found at the site of nerve injury and in the Wallerian degenerating nerve (see references in [30]). Absence or delay in the onset of Wallerian degeneration has been linked to reduced neuropathic pain syndromes [31], while activation of bradykinin receptors is believed to promote hyperalgesia *via* the protein kinase C pathway [32]. Both sciatic nerve transection and CCI result in elevated expression of the B2 receptor and a *de novo* expression of the B1 receptor. Specific pharmacological antagonists of bradykinin are effective at alleviating the CCI-associated hyperalgesia [30].

Axotomy results in increased γ-amino butyric acid-A ($GABA_A$)-receptor mediated conductance in cutaneous, but not muscle, afferents [33]. Exogenous BDNF, but not NGF, is effective in attenuating the injury-induced increase in $GABA_A$ response. This supports a role for these factors in maintaining the normal electrical properties of these fibers, but the precise mechanism by which it does this is unknown [16]. NGF did however, limit the injury-induced loss of slow sodium current and inflected action potentials in the cutaneous afferent neurons.

NO – an unconventional mediator of neuropathic pain

The gas, nitric oxide (NO), has also been implicated in nociception (reviewed in [3]). Peripheral nerve injury and inflammation result in an up-regulation in DRG neuronal expression of nitric oxide synthase, the enzyme responsible for the catabolism of L-arginine to L-citrulline to generate NO [34–36]. In addition, there is a corresponding reduction in NOS expression in the interneurons of the dorsal horn where nociceptive afferents terminate, suggesting an injury-associated change in spinal processing of nociceptive input in response to NO [35]. Reduced retrograde supply of NGF to sensory neurons is implicated in the elevation of neuronal NOS expression following peripheral nerve injury [37]. NO release from sensory neurons has been linked to the increased spontaneous activity that occurs with nerve injury [38], but there are conflicting reports on the role of NO in neuropathic pain [39, 40].

Anatomical rearrangements of afferents

Anatomical rearrangements of sensory fibers in the spinal cord may also contribute to the development of neuropathic pain. Woolf and colleagues [41] noted that peripheral nerve injury resulted in the sprouting of normally innocuous mechanoreceptors from deeper laminae into lamina II (the region of the spinal cord that normally receives nociceptive input). This rearrangement is also seen following CCI [42]. This may allow inappropriate activation of C fibers and contribute to the development of mechano-allodynic pain states mediated by large myelinated sensory afferents following nerve lesion. The interaction between mechano-receptor and nociceptor terminals is further facilitated by an expansion of the territory that C fibers occupy following peripheral axotomy [43]. One of the proposed mechanisms by which these sprouting events is speculated to occur is in response to the vacated synaptic sites caused by the death of some DRG neurons following injury or trans-ganglionic degenerative atrophy of the central terminals of damaged C fibers [44]. In support of this is the ability of exogenous NGF to prevent axotomy-associated C fiber degeneration and cell death [45, 46] and for both NGF and GDNF to prevent A-fiber sprouting into lamina II [47]. It would appear, however, that it is probably the ability of these factors to counteract injury-induced neurochemical changes in these fibers that is critical (see below). Recent work by Mannion et al. [48] indicates that peripheral (not central) deafferentation of C fibers and the injury-induced changes in the superficial dorsal horn are prerequisites for this sprouting, but deafferentation of A fibers is not. As a cautionary note, when evaluating studies that examine injury-induced A-fiber sprouting using the tracer cholera toxin-B, it should be noted that the percentage of injured sensory neurons which internalize and transport the tracer is increased as compared to uninjured controls [49] and thus the degree of sprouting may be overestimated.

Changes in peptidergic phenotype following injury: functional implications

Axotomy induces dramatic changes in the levels and cellular distribution of neuronal peptide and receptor expression, many of which have an impact on the development and maintenance of neuropathic pain states (see [3] for review). Sciatic nerve transection results in a marked decrease in SP [50], somatostatin (SOM) [51], CGRP [52], and pituitary adenylate cyclase activating peptide (PACAP) expression [53, 54] in the C fiber and Aδ fiber subpopulations of DRG neurons. In contrast, a dramatic upregulation of vasoactive intestinal peptide (VIP) [55], galanin (GAL) [55], neuropeptide tyrosine (NPY) [56, 57], cholecystokinin (CCK) [58], PACAP [54, 59], SP [60], and CGRP [61] are observed in the same ganglia. What is known about the potential functional and physiological implications for the injury-induced peptides with regards to nociception is briefly outlined below.

VIP is one of the more rapidly induced injury-associated peptides, with expression restricted to primarily small DRG neurons projecting to laminas I and II of the dorsal horn. VIP is proposed to take over the major excitatory role in pain transmission that SP and CGRP hold in intact C fibers, as the injured neurons now have reduced SP and CGRP expression [62]. VIP antagonists have been shown to attenuate the induction of central hyperexcitation of injured, but not intact, afferents (see references in [62]). Instead of a C fiber function, SP expression is induced in injured myelinated Aβ fibers and released upon their stimulation [63] where it is believed to contribute to the mechanical allodynic response.

NPY, which is virtually not detectable before injury, is upregulated in up to 50% of DRG neurons (primarily medium to large in size), with approximately 40% of these neurons expressing NPY Y1 and Y2 receptors [64]. There is also a modest increase of NPY observed in C fibers, which is proposed to result from the ability of VIP to up-regulate NPY expression [3]. In addition, the localization of NPY Y1 and post-injury up-regulation of NPY Y2 receptors on the subpopulation of CGRP/SP-positive nociceptors provides an anatomical circuit in support of potential paracrine/autocrine influences of NPY on nociception [65, 66]. Physiological evidence of such a role is supported by the ability of NPY to inhibit calcium current and potassium-evoked SP release from DRG neurons *in vitro* [67] and inhibit *in vivo* release of SP following C fiber stimulation [68]. Work employing either the use of selective Y1 and Y2 agonists or NPY demonstrates an ability for these compounds to suppress the nociceptive flexor reflex after injury [69]. Thus, the evidence suggests that NPY might reduce neuropathic pain associated with complete nerve transection. There are instances, however, when both NPY [70], and GAL [71] have been reported to promote algesia. This is observed with only a partial nerve injury (i.e. CCI) and may reflect the exposure of the spared axons to the Wallerian degenerating distal nerve, which is tightly linked to the development of pain following CCI [31]. Interestingly, it would appear that the severity of allodynic state observed with partial nerve injury is inversely correlated with the level of GAL expression observed in the injured DRG [72]. A similar relationship cannot be attributed to NPY since the level of up-regulation did not differ significantly between allodynic and non-allodynic rats [72]. Thus, it would appear important to correlate behavioral responses with the level of GAL expression to accurately predict GAL's role following partial nerve lesions.

With complete sciatic nerve transection, GAL expression is up-regulated in small to medium sized neurons projecting to the superficial dorsal horn and gracile nuclei. In this form of injury, there is enhanced release of GAL in the dorsal horn [73] and GAL has been proposed to have a predominantly antinociceptive function. Indeed, GAL appears to be tonically active in inhibiting nociceptive input from injured peripheral nerves [74–76]. It is possible that the injury-associated down-regulation in GAL1 and GAL2 receptors in nociceptive neurons may factor into GAL's analgesic role following injury [77], as this would reduce the pronociceptive functions that it is reported to have on dorsal horn afferents associated with intact nerves [3].

PACAP expression is down-regulated in small, and up-regulated in medium to large neurons following complete sciatic nerve transection [53, 54]. A slightly different pattern occurs with CCI injury. Here it was noted that more small neurons had an elevated level of PACAP expression as compared to those with complete nerve transection [78]. This may reflect the exposure of the spared C fibers to a pool of NGF in the Wallerian degenerating nerve (see [31]), which has recently been shown to modulate neuronal PACAP expression [53]. The exact role of PACAP in modulating neuropathic pain states remains unresolved. In intact and inflamed sensory neurons it has been endowed with both nociceptive [79, 80] and anti-nociceptive [59, 81] functions.

The elevated expression of CCK in sensory neurons following injury [58, 82], while readily detectable at the message level in approximately 30% of injured neurons (primarily small to medium in size), is not easily detected at the protein level. Injury also elevates expression of CCK2 receptor mRNA and protein in sensory neurons, with a marked increase in the binding of CCK2 receptor agonist to the superficial layers of the dorsal horn. One of the functional consequences of this plasticity appears to be a CCK-driven and CCK R dependant antagonization of opioid interactions at the level of the spinal cord. This would explain the relative inefficacy of opioids in managing neuropathic pain (reviewed in [82]).

In summary, nerve injury results in a general down-regulation of peptides normally expressed in the nociceptive population and an up-regulation of peptides that exert both facilitatory and inhibitory affects on neuropathic pain states. There is also a novel appearance of many peptides in the largely non-peptidergic A fiber afferent population that serve to potentially dampen and promote neuropathic pathways in the regions where these fibers normally terminate (i.e. gracile nucleus) and aberrantly sprout (i.e. from laminas III and IV into lamina II). This subpopulation of neurons goes on to acquire a progressive tactile hypersensitivity.

Molecular mechanisms underlying alterations in sensory neuron phenotype following injury – roles of cytokines and neurotrophins

Multiple factors underlie the changes in neuronal function and phenotype observed after axotomy. Reduction in the retrograde supply of peripherally derived neurotrophins and growth factors is not the only signal responsible for these changes [83, 84]. Expression and availability of intraganglionic-, endoneurial- and target-derived neurotrophins and cytokines change. An adequate supply of the right trophic molecules to the injured neurons is necessary to prevent cell death, regulate regeneration programs and restore intact phenotype. However, the frequent development of neuropathic pain syndromes is an unwelcome consequence. Chronic sciatic nerve transection results in a heterogeneous and rapid increase in the concentration of BDNF, bFGF, and interleukin-6 (IL-6) mRNA in neurons of the corresponding DRG [85–87]. The neuronal profiles which express these factors include the Aβ group of myelinated afferents. Indeed, BDNF mRNA colocalizes with NPY and GAL expression and may influence this expression. At low, but not at high doses, BDNF attenuates high-threshold mechanosensitivity resulting from injury, but no affect on mechanical allodynia or thermal sensitivity is observed [88]. Interestingly, the ecotopic electrical activity in injured neurons may contribute to the ongoing elevated expression of BDNF observed [89]. IL-6 is also believed to affect nociception. IL-6 infusion can upregulate GAL expression [90], while IL-6 KO mice with CCI do not develop thermal or mechanical allodynia and have significantly less robust SP and GAL responses as compared to wild type controls [90]. There is also a post-axotomy increase in retrograde transport of ciliary neurotrotrophic factor (CNTF) and leukemia inhibitory factor (LIF) to the neurons. This correlates with enhanced expression of LIF in proximal and distal segments of the crushed nerve and increased extracellular CNTF at the lesion site, despite a decrease in synthesis in the degenerating nerve (see [85] for references). CNTF and LIF antagonize the effects of NGF on neuropeptide expression in sensory neurons *in vitro* [91]. LIF is most closely linked with regulation of GAL expression following injury [92], but appears to require reduced access to NGF as well [93]. While the cytokines appear to drive injury-associated peptide expression, exogenous neurotrophins, NGF and NT-3, play significant roles in mitigating this expression in injured neurons. Specifically, delayed infusion of NGF can down-regulate expression of NPY, CCK, VIP, PACAP, and GAL in injured neurons [53, 94]. NT-3 appears to have a greater affect on the medium to large size sensory neurons and mitigates increases in PACAP, GAL, and NPY expression in injured neurons [53, 95, 96]. More recent work reveals a similar influence of NT-3 on injury phenotype associated with CCI [78]. Thus, NGF and NT-3 through their modulation of peptide expression and afferent sprouting events (see above), may ameliorate neuropathic pain states. Exactly how NGF and NT-3 effect down-regulation of injury-associated gene expression is unclear. However, as many of the injury-associated cytokines and growth factors positively regulate the injury response it is possible that NGF and NT-3 might reverse this regulation by effecting down-regulation of these injury-associated factors [97, 98].

Conclusions and perspectives

Identifying and understanding the exact mechanisms involved in the pathogenesis of neuropathic pain is an enormous challenge that will involve an integration of the multiple changes that occur in the different regions of the somatosensory pathway following nerve injury. This chapter has touched on just some of the DRG neuronal responses to injury that may factor into this pathology. While it is evident that tremendous advances have been made in elucidating the events, the entire answer has not yet been found.

Acknowledgments

We thank Marlin Korczak for technical assistance. Research in this lab is supported by the Canadian Institutes of Health Research, Canadian Neurotrauma Research Program, Saskatchewan Neurotrauma Initiative, University of Saskatchewan, and Regeneron Pharmaceuticals (Tarrytown, New York).

References

1 Bennett GJ, Xie YK (1988) A peripheral mononeuropathy in rat that produces disorders of pain sensations like those seen in man. *Pain* 33: 87–107

2 Kajander KC, Bennett GJ (1992) Onset of a painful peripheral neuropathy in rat: a partial and differential deafferentation and spontaneous discharge in Aβ and Aδ primary afferent neurons. *J Neurophysiol* 68: 734–744

3 Millan Mark J (1999) The induction of pain: An integrative review. *Prog Neurobiol* 57: 1–164

4 Molliver DC, Wright DE, Leitner ML, Parsadanian A Sh, Doster K, Wen D, Yan Q, Snider WD (1997) IB4-binding DRG neurons switch from NGF to GDNF dependence in early postnatal life. *Neuron* 19: 849–861

5 Karchewski LA, Kim FA, Johnston J, McKnight RM, Verge VMK (1999) Anatomical evidence supporting the potential for modulation by multiple neurotrophins in the majority of adult lumbar sensory neurons. *J Comp Neurol* 413: 327–341

6 Wall PD, Waxman S, Basbaum AI (1974) Ongoing activity in peripheral nerve: injury discharge. *Exp Neurol* 45: 576–589

7 Wall PD, Devor M (1981) The effect of peripheral nerve injury on dorsal root potentials and on transmission of afferent signals into the spinal cord. *Brain Res* 209(1): 95–111

8 Ochoa JL, Torebjork HE (1980) Paraesthesiae from ectopic impulse generation in human sensory nerves. *Brain* 103: 835–853

9 Michaelis M, Liu X, Janig W (2000) Axotomized and intact muscle afferents but no skin afferents develop ongoing discharges of dorsal root ganglion origin after peripheral nerve lesion. *J Neurosci* 20: 2742–2748

10 Waxman SG (1999) The molecular pathophysiology of pain: abnormal expression of sodium channel genes and its contributions to hyperexcitability of primary sensory neurons. *Pain* Suppl 6: S133–S140

11 Devor M, Keller CH, Deerinck TJ, Levinson SR, Ellisman MH (1989) Na^+ channel accumulation on axolemma of afferent endings in nerve end neuromas in Apteronotus. *Neurosci Lett* 102: 149–154

12 Novakovic SD, Tzoumaka E, McGivern JG, Haraguchi M, Sangameswaran L, Gogas KR, Eglen RM, Hunter JC (1998) Distribution of the tetrodotoxin-resistant sodium channel PN3 in rat sensory neurons in normal and neuropathic conditions. *J Neurosci* 18: 2174–2187

13 Black JA, Dib-Hajj S, McNabola K, Jeste S, Rizzo MA, Kocsis JD, Waxman SG (1996) Spinal sensory neurons express multiple sodium channel alpha-subunit mRNAs. *Brain Res Mol Brain Res* 43: 117–131

14 Rizzo MA, Kocsis JD, Waxman SG (1995) Selective loss of slow and enhancement of fast Na^+ currents in cutaneous afferent DRG neurons following axotomy. *Neurobiol Dis* 2: 87–96

15 Porreca F, Lai J, Bian D, Wegert S, Ossipov MH, Eglen RM, Kassotakis L, Novakovic S, Rabert DK, Sangameswaran L et al (1999) A comparison of the potential role of the

tetrodotoxin-insensitive sodium channels, PN3/SNS and NaN/SNS2, in rat models of chronic pain. *Proc Natl Acad Sci USA* 96: 7640–7644
16 Oyelese AA, Rizzo MA, Waxmann SG, Kocsis JD (1997) Differential effects of NGF and BDNF on axotomy-induced changes in $GABA_A$-receptor-mediated conductance and sodium currents in cutaneous afferent neurons. *J Physiol* 78: 31–42
17 Fjell J, Cummins TR, Dib-Hajj SD, Fried K, Black JA, Waxman SG (1999) Differential role of GDNF and NGF in the maintenance of two TTX-resistant sodium channels in adult DRG neurons. *Brain Res Mol Brain Res* 67: 267–282
18 Chen CC, Akopian AN, Sivilotti L, Colquhoun D, Burnstock G, Wood JN (1995) A P2X purinoceptor expressed by a subset of sensory neurons. *Nature* 377: 428–431
19 Caterina MJ, Schumacher MA, Tominaga M, Rosen TA, Levine JD, Julius D (1997) The capsaicin receptor: a heat-activated ion channel in the pain pathway. *Nature* 389: 816–824
20 Caterina MJ, Leffler A, Malmberg AB, Martin WJ, Trafton J, Petersen-Zeitz KR, Koltzenburg M, Basbaum AI, Julius D (2000) Impaired nociception and pain sensation in mice lacking the capsaicin receptor. *Science* 288: 306–313
21 Caterina MJ, Rosen TA, Tominaga M, Brake AJ, Julius D (1999) A capsaicin-receptor homologue with a high threshold for noxious heat. *Nature* 398: 436–441
22 Vulchanova L, Riedl MS, Shuster SJ, Stone LS, Hargreaves KM, Buell G, Surprenant A, North RA, Elde R (1998) P2X3 is expressed by DRG neurons that terminate in inner lamina II. *Eur J Neurosci* 10: 3470–3478
23 Malmberg AB, Chen C, Tonegawa S, Basbaum AI (1997) Preserved acute pain and reduced neuropathic pain in mice lacking PKC gamma. *Science* 278: 279–283
24 Bradbury EJ, Burnstock G, McMahon SB (1998) The expression of P2X3 purinoreceptors in sensory neurons: effects of axotomy and glial-derived neurotrophic factor. *Mol Cell Neurosci* 12: 256–268
25 Novakovic SD, Kassotakis LC, Oglesby IB, Smith JA, Eglen RM, Ford AP, Hunter JC (1999) Immunocytochemical localization of P2X3 purinoceptors in sensory neurons in naive rats and following neuropathic injury. *Pain* 80: 273–282
26 Cook SP, Vulchanova L, Hargreaves KM, Elde R, McCleskey EW (1997) Distinct ATP receptors on pain-sensing and stretch-sensing neurons. *Nature* 387: 505–508
27 Gu JG, MacDermott AB (1997) Activation of ATP P_{2X} receptors elicits glutamate release from sensory neuron synapses. *Nature* 389: 749–753
28 Cui JG, Sollevi A, Linderoth B, Meyerson BA (1997) Adenosine receptor activation suppresses tactile hypersensitivity and potentiates spinal cord stimulation in mononeuropathic rats. *Neurosci Lett* 223: 173–176
29 Petersen M, Eckert AS, Segond von Banchet G, Heppelmann B, Klusch A, Kniffki KD (1998) Plasticity in the expression of bradykinin binding sites in sensory neurons after mechanical nerve injury. *Neurosci* 83: 949–959
30 Levy D, Zochodne DW (2000) Increased mRNA expression of the β1 and β2 bradykinin receptors and antinociceptive effects of their antagonists in animal model of neuropathic pain. *Pain* 86: 265–271

31 Ramer MS, French GD, Bisby MA (1997) Wallerian degeneration is required for both neuropathic pain and sympathetic sprouting into the DRG. *Pain* 72: 71–78
32 Burgess GM, Mullaney I, McNeill M, Dunn PM, Rang HP (1989) Second messengers involved in the mechanisms of action of bradykinin in sensory neurons in culture. *J Neurosci* 9: 3314–3325
33 Oyelese AA, Kocsis JD (1996) $GABA_A$ receptor-mediated conductance and action potential waveform in cutaneous and muscle afferent neurons of the adult rat: differential expression and response to nerve injury. *J Neurophysiol* 76: 2383–2392
34 Verge VMK, Zhang X, Xu X-J, Wiesenfeld-Hallin Z, Hökfelt T (1992) Marked increase in nitric oxide synthase mRNA levels in rat dorsal root ganglia after peripheral axotomy – *in situ* hybridization and functional studies. *Proc Natl Acad Sci USA* 89: 11617–11621
35 Zhang X, Verge VMK, Wiesenfeld-Hallin Z, Ju G, Bredt D, Snyder SH, Hökfelt T (1993) Nitric oxide synthase-like immunoreactivity in lumbar dorsal root ganglia and spinal cord of rat and monkey and effect of peripheral axotomy. *J Comp Neurol* 335: 563–575
36 Choi Y, Raja SN, Moore LC, Tobin JR (1996) Neuropathic pain in rats is associated with altered nitric oxide synthase activity in neural tissue. *J Neurol Sci* 138: 14–20
37 Verge VMK, Wilson TD, Johnston J (1999) NGF modulates expression of nitric oxide synthase in injured DRG neurons. *J Periph Nervous Sys* 4: 190
38 Wiesenfeld-Hallin Z, Hao JX, Xu XJ, Hökfelt T (1993) Nitric oxide mediates ongoing discharges in dorsal root ganglion cells after peripheral nerve injury. *J Neurophysiol* 70: 2350–2353
39 Thomas DA, Ren K, Besse D, Ruda MA, Dubner R (1996) Application of nitric oxide synthase inhibitor, N omega-nitro-L-arginine methyl ester, on injured nerve attenuates neuropathy-induced thermal hyperalgesia in rats. *Neurosci Lett* 210: 124–126
40 Luo ZD, Chaplan SR, Scott BP, Cizkova D, Calcutt NA, Yaksh TL (1999) Neuronal nitric oxide synthase mRNA upregulation in rat sensory neurons after spinal nerve ligation: lack of a role in allodynia development. *J Neurosci* 19: 9201–9208
41 Woolf CJ, Shortland P, Coggeshall RE (1992) Peripheral nerve injury triggers central sprouting of myelinated afferents. *Nature* 355: 75–78
42 Nakamura S, Myers RR (1999) Myelinated afferents sprout into lamina II of L3-5 dorsal horn following chronic constriction nerve injury in rats. *Brain Res* 818: 285–290
43 Cameron AA, Pover CM, Willis WD, Coggeshall RE (1992) Evidence that fine primary afferent axons innervate a wider territory in the superficial dorsal horn following peripheral axotomy. *Brain Res* 575: 151–154
44 Arvidsson J, Ygge J, Grant G (1986) Cell loss in lumbar dorsal root ganglia and transganglionic degeneration after sciatic nerve resection in the rat. *Brain Res* 373: 15–21
45 Csillik B, Schwab ME, Thoenen H (1985) Transganglionic regulation of central terminals of dorsal root ganglion cells by nerve growth factor (NGF). *Brain Res* 331: 11–15
46 Rich KM, Luszczynski JR, Osborne PA, Johnson EM Jr (1987) Nerve growth factor pro-

tects adult sensory neurons from cell death and atrophy caused by nerve injury. *J Neurocytol* 16: 261–268
47 Bennett DL, French J, Priestley JV, McMahon SB (1996) NGF but not NT-3 or BDNF prevents the A fiber sprouting into lamina II of the spinal cord that occurs following axotomy. *Mol Cell Neurosci* 8: 211–220
48 Mannion RJ, Doubell TP, Gill H, Woolf CJ (1998) Deafferentation is insufficient to induce sprouting of A-fibre central terminals in the rat dorsal horn. *J Comp Neurol* 393: 135–144
49 Tong YG, Wang F, Ju G, Grant G, Hökfelt T, Zhang X (1999) Increased uptake and transport of cholera toxin β-subunit in dorsal root ganglion neurons after peripheral axotomy: possible implications for sensory sprouting. *J Comp Neurol* 404: 143–158
50 Jessell T, Tsunoo A, Kanasawa I, Otsuka M (1979) Substance P: Depletion in the dorsal horn of rat spinal cord after section of the peripheral processes of primary sensory neurons. *Brain Res* 168: 247–259
51 Noguchi K, De Léon M, Nahin RL, Ruda MA (1993) Quantification of axotomy-induced alteration of neuropeptide mRNAs with special reference to neuropeptide Y mRNA and the effects of neonatal capsaicin treatment. *J Neurosci Res* 35: 54–66
52 Noguchi K, Senba E, Morita Y, Sato M, Tohyama M (1990) a-CGRP and b-CGRP mRNAs are differentially regulated in the rat spinal cord and dorsal root ganglion. *Mol Br Res* 7: 299–304
53 Jongsma H, Danielsen N, Gratto KA, Karchewski LA, Sundler F, Kanje M, Verge VMK (1999) Exogenous NT-3 & NGF differentially modulate PACAP expression in adult sensory neurons, suggesting distinct roles in injury and inflammation. *Soc Neurosci Abstr* 25: 512
54 Jongsma H, Danielsen N, Sundler F, Kanje M (2000a) Alteration of PACAP distribution and PACAP receptor binding in the rat sensory nervous system following sciatic nerve transection. *Brain Res* 853: 186–196
55 Villar MJ, Cortes R, Theodorsson E, Wiesenfeld-Hallin Z, Schalling M, Fahrenkrug J, Emson PC, Hökfelt T (1989) Neuropeptide expression in rat dorsal root ganglion cells and spinal cord after peripheral nerve injury with special reference to galanin. *Neurosci* 33: 587–604
56 Wakisaka S, Kajander KC, Bennett GJ (1991) Increased neuropeptide (NPY)-like immunoreactivity in rat sensory neurons following peripheral axotomy. *Neurosci Lett* 124: 200–203
57 Ma W, Bisby MA (1998) Partial and complete sciatic nerve injuries induce similar increases of neuropeptide Y and vasoactive intestinal peptide immunoreactivities in primary sensory neurons and their central projections. *Neurosci* 86: 1217–1234
58 Verge VMK, Merlio JP, Grondin J, Ernfors P, Persson H, Riopelle RJ, Hökfelt T, Richardson PM (1992) Colocalization of NGF binding sites, trk mRNA, and low-affinity NGF receptor mRNA in primary sensory neurons: responses to injury and infusion of NGF. *J Neurosci* 12: 4011–4022
59 Zhang X, Hannibal J, Zhao Q, Moller K, Danielsen N, Fahrenkrug J, Sundler F (1996)

Pituitary adenylate cyclase activating peptide expression in the ray dorsal root ganglia: up-regulation after peripheral nerve injury. *Neurosci* 74: 1099–1110

60 Noguchi K, Kawai Y, Fukuoka T, Senba E, Miki K (1995) Substance P induced by peripheral nerve injury in primary afferent sensory neurons and its effect on dorsal column nucleus neurons. *J Neurosci* 15(11): 7633–7643

61 Ma W, Bisby MA (1998) Increase of calcitonin gene-related peptide immunoreactivity in the axonal fibers of the gracile nuclei of adult and aged rats after complete and partial sciatic nerve injuries. *Exp Neurol* 152: 137–149

62 Hökfelt T, Zhang X, Wiesenfeld-Hallin Z (1994) Messenger plasticity in primary sensory neurons following axotomy and its functional implications. *Trends Neurosci* 17: 22–30

63 Malcangio M, Ramer MS, Jones MG, McMahon SB (2000) Abnormal substance P release from the spinal cord following injury to primary sensory neurons. *Eur J Neurosci* 12: 397–9

64 Landry M, Holmberg K, Zhang X, Hökfelt T (2000) Effect of axotomy on expression of NPY, galanin, and NPY Y_1 and Y_2 receptors in dorsal root ganglia and the superior cervical ganglion studied with double-labeling *in situ* hybridization and immunohistochemistry. *Exp Neurobiol* 162: 361–384

65 Mantyh PW, Allen CJ, Rogers S, DeMaster E, Ghilardi JR, Mosconi T, Kruger L, Manno PJ, Taylor IL, Vigna SR (1994) Some sensory neurons express neuropeptide Y receptors: potential paracrine inhibition of primary afferent nociceptors following peripheral nerve injury. *J Neurosci* 14: 3958–3968

66 Zhang X, Shi TJ, Holmberg K, Landry M, Huang W, Xiao H, Ju G, Hokfelt T (1997) Expression and regulation of the neuropeptide Y (Y_2) receptors in sensory ganglia and autonomic regions. *Proc Natl Acad Sci USA* 94: 729–734

67 Bleakman D, Colmers WF, Fournier A, Millar RJ (1991) Neuropeptide Y inhibits Ca^{2+} influx into cultured dorsal root ganglion neurones of the rat *via* a Y_2 receptor. *Br J Pharmacol* 103: 1781–1789

68 Duggan AW, Hope PJ, Lang CW (1991) Microinjection of neuropeptide Y into the superficial dorsal horn reduces stimulus-evoked release of immunoreactive substance P in the anaesthetized cat. *Neurosci* 44: 733–740

69 Xu IS, Hao JX, Xu XJ, Hokfelt T, Wiesenfeld-Hallin Z (1999) The effect of intrathecal selective agonists of Y_1 and Y_2 neuropeptide Y receptors on the flexor reflex in normal and axotomized rats. *Brain Res* 833: 251–257

70 White DM (1997) Intrathecal neuropeptide Y exacerbates nerve injury-induced mechanical hyperalgesia. *Brain Res* 750: 141–146

71 Ma W, Bisby MA (1997) Differential expression of galanin immunoreactivities in the primary sensory neurons following partial and complete sciatic nerve injuries. *Neurosci* 79: 1183–1195

72 Sten Shi TJ, Cui JG, Meyerson BA, Linderoth B, Hökfelt T (1999) Regulation of galanin and neuropeptide Y in dorsal root ganglia and dorsal horn in rat mononeuropathic models: possible relation to tactile hypersensitivity. *Neurosci* 93: 741–757

73 Colvin LA, Mark MA, Duggan AW (1997) The effect of a peripheral mononeuropathy on immunoreactive (ir)-galanin release in the spinal cord of the rat. *Brain Res* 766: 259–261
74 Wiesenfeld-Hallin Z, Xu XJ, Langel U, Bedecs K, Hökfelt T, Bartfai T (1992) Galanin-mediated control of pain: enhanced role after nerve injury. *Proc Natl Acad Sci USA* 89: 3334–3337
75 Verge VMK, Xu XJ, Langel U, Hökfelt T, Wiesenfeld-Hallin Z, Bartfai T (1993) Evidence for endogenous inhibition of autotomy by galanin in the rat after sciatic nerve section: demonstrated by chronic intrathecal infusion of a high affinity galanin receptor antagonist. *Neurosci Lett* 149: 193–197
76 Hao JX, Shi TJ, Xu IS, Kaupilla T, Xu XJ, Hökfelt T, Bartfai T, Wiesenfeld-Hallin Z (1999) Intrathecal galanin alleviates allodynia-like behaviour in rats after partial peripheral nerve injury. *Eur J Neurosci* 11: 427–432
77 Xu ZQ, Shi TJ, Landry M, Hökfelt T (1996) Evidence for galanin receptors in primary sensory neurones and effect of axotomy and inflammation. *NeuroReport* 8: 237–242
78 Wilson TD, Verge VMK (2000) NT-3 alters the phenotypic changes associated with chronic constriction injury (CCI) in adult DRG. *Soc Neurosci Abstr* 698.8
79 Narita M, Dun SL, Dun NJ, Tseng LF (1996) Hyperalgesia induced by pituitary adenylate cyclase-activating polypeptide in the mouse spinal cord. *Eur J Pharmacol* 311: 121–126
80 Xu XJ, Wiesenfeld-Hallin Z (1996) Intrathecal pituitary adenylate cyclase activating polypeptide facilitates the spinal nociceptive flexor reflex in the rat. *Neurosci* 72: 801–804
81 Zhang Y-Z, Sjölund B, Moller K, Håkanson R, Sundler F (1993) Pituitary adenylate cyclase activating peptide produces a marked and long-lasting depression of a C-fibre-evoked flexion reflex. *Neuroscience* 57:733–737
82 Wiesenfeld-Hallin Z, Lucas GA, Alster P, Xu XJ, Hökfelt T (1999) Cholecystokinin/opioid interactions. *Brain Res* 848: 78–89
83 Richardson PM, Lu X (1994) Inflammation and axonal regeneration. *J Neurol* 242 (Suppl 1): S57–S60
84 Murphy PG, Borthwick LS, Johnston RS, Kuchel G, Richardson PM (1999) Nature of the retrograde signal from injured nerves that induces interleukin-6 mRNA in neurons. *J Neurosci* 19: 3791–800
85 Verge VMK, Gratto KA, Karchewski LA, Richardson PM (1996) Neurotrophins and nerve injury in the adult. *Phil Trans R Soc Lond B* 351: 423–430
86 Tonra JR, Curtis R, Wong V et al (1998) Axotomy upregulates the anterograde transport and expression of brain-derived neurotrophic factor by sensory neurons. *J Neurosci* 18: 4373–4383
87 Murphy PG, Grondin J, Altares M, Richardson PM (1995) Induction of interleukin-6 in axotomized sensory neurons. *J Neurosci* 15: 5130–5138
88 Miki K, Fukuoka T, Tokunaga A, Kondo E, Dai Y, Noguchi K (2000) Differential effect

of brain-derived neurotrophic factor on high-threshold mechanosensitivity in a rat neuropathic pain model. *Neurosci Lett* 278: 85–88

89 Mannion RJ, Costigan M, Decosterd I, Amaya F, Ma QP, Holstege JC, Ji RR, Acheson A, Lindsay RM, Wilkinson GA, Woolf CJ (1999) Neurotrophins: peripherally and centrally acting modulators of tactile stimulus-induced inflammatory pain hypersensitivity. *Proc Natl Acad Sci USA* 96: 9385–9390

90 Murphy PG, Ramer MS, Borthwick L, Gauldie J, Richardson PM, Bisby MA (1999) Endogenous interleukin-6 contributes to hypersensitivity to cutaneous stimuli and changes in neuropeptides associated with chronic nerve constriction in mice. *Eur J Neurosci* 11: 2243–2253

91 Mulderry PK (1994) Neuropeptide expression by newborn and adult rat sensory neurons in culture: effects of nerve growth factor and other neurotrophic factors. *Neuroscience* 59: 673–688

92 Sun Y, Zigmond RE (1996) Leukaemia inhibitory factor induced in the sciatic nerve after axotomy is involved in the induction of galanin in sensory neurons. *Eur J Neurosci* 8(10): 2213–2220

93 Shadiack AM, Vaccariello SA, Sun Y, Zigmond RE (1998) Nerve growth factor inhibits sympathetic neurons' response to an injury cytokine. *Proc Natl Acad Sci USA* 95: 7727–7730

94 Verge VMK, Richardson PM, Wiesenfeld-Hallin Z, Hökfelt T (1995) Differential influence of nerve growth factor on neuropeptide expression *in vivo*: a novel role in peptide suppression in adult sensory neurons. *J Neurosci* 15: 2081–2096

95 Gratto KA, Friedman B, Liu L, Lindsay RM, Verge VMK (1995) Further studies on the role of NT-3 in intact and injured primary sensory neurons. *Soc Neuroscience Abstr* 21: 1056

96 Sterne GD, Brown RA, Green CJ, Terenghi G (1998) NT-3 modulates NPY expression in primary sensory neurons following peripheral nerve injury. *J Anat* 193: 273–281

97 Karchewski LA, Gratto KA, Verge VMK (2000) BDNF expression is differentially regulated by NGF and NT-3 in adult primary sensory neurons. *Soc Neurosci Abstr* 221.6

98 Verge VMK, Johnston JM (2000) Infusion of NGF or NT-3 results in reduced levels of IL-6 and bFGF mRNAs in injured adult primary sensory neurons. *Soc Neurosci Abstr* 698.6

Neuroinflammation, cytokines and neuropathic pain

Linda S. Sorkin

Department of Anesthesiology, University of California, San Diego, La Jolla, CA 92103-0818, USA

Introduction

Many forms of neuropathic-like pain are thought to have inflammatory components. Neuro-inflammation can result secondary to nerve damage, from an immune response to an antigen located on or near the peripheral nerve or in response to a frank neuritis. These conditions have several elements in common including macrophage infiltration, increased numbers of activated T cells and up-regulation of several of the proinflammatory cytokines. The most studied neuropathic model with respect to immune activation is the chronic constriction injury (CCI) or Bennett model [1]. In the original model, four loose ligatures of chromic gut are placed around the sciatic nerve just above the knee. The ligatures are tied just tightly enough to barely constrict the nerve and importantly they reduce, but do not halt, blood flow through the epineurial vasculature. This model produces thermal hyperalgesia, mechanical allodynia and/or spontaneous pain of duration varying from one week to one month depending on the lab.

Local neuritis and pain

Several years later, it was proposed that a local inflammatory response arising from the use of chromic gut suture material resulted in a change in the chemical mileau around the nerve trunk; the inflammation was hypothesized to be the causal factor for the generation of neuropathic pain [2]. A dose response relationship was shown between the quantity of chromic gut suture used and the degree of nociceptive behavior. These authors found a synergy between the gut suture material and the ligation, but found no hyperalgesia when a non inflammation-producing substance such as surgical silk was used to produce the ligature. This last observation has been

Mechanisms and Mediators of Neuropathic Pain, edited by Annika B. Malmberg and Sandra R. Chaplan

disputed due to the fact that the ligatures were tied less tightly than is common, resulting in less or no restriction in blood flow. This is important as peripheral nerve ischemia is also thought to contribute to the production of hyperesthesia in the CCI model [3]. However, it is now commonly accepted that immune responses at the peripheral nerve due to the chromic gut ligatures play a major role in the etiology of the pain state. Confirmation of this hypothesis has been achieved by performing CCI in animals with suppressed immune systems. Daily systemic treatment either with the corticosteroid, dexamethosone, [4] or the immunosuppressant, cyclosporin A [5], dose dependently reduces the hyperesthesia. Alternatively, thermal hyperalgesia can be exacerbated by pre-soaking the ligatures in complete Freunds adjuvant (CFA) prior to their placement [4]. Immune activation around the nerve trunk in the absence of the ligatures or other overt damage-initiating procedures, such as was suggested by Maves et al. [2], is now the basis for at least two new neuropathic pain models. In the first, a surgical sponge treated with either CFA or carrageenan is wrapped around the sciatic nerve at mid thigh level. This induces behavioral sensitivity to a wide range of sensory stimuli applied to the hindpaw including heat, cold, light touch and pinprick that is not present in saline soaked sponge treated animals [6]. Note, in this model the heat hyperalgesia is the most transient, disappearing on the 4th post-surgical day while sensitivity to the other three modalities is present through day 5. Chemical neuritis induced pain behavior is dose dependently decreased by ongoing treatment with cyclosporin A [5]. A similarly designed model was developed more recently by Watkins and Maier's group. Sciatic Inflammatory Neuritis (SIN) involves placing a sponge attached to an exteriorized catheter around the nerve. Zymogen, rather than CFA or carrageenan, is then injected several days after surgery. Separation of surgery and neuroinflammation eliminates potentially confounding immunosuppresent effects of anesthesia [7, 8]. The procedure results in a rapid onset mechanical allodynia reflected to the ipsilateral hindpaw that lasts for over a week [9]. Unlike the acutely placed CFA or carrageenan or the CCI injury for that matter, there is no thermal hyperalgesia. Increasing the amount of injected zymogen results in a contralateral allodynia, but still no change in response to thermal stimulation. Interestingly, this is a circumscribed mirror image spread and does not appear to be due to zymogen leakage into the systemic circulation. Controls have been performed for both the acutely initiated inflammatory neuritis (CFA and carrageenan) and the SIN models illustrating that delivery of the immune activators to the nearby muscle to produce a local myositis does not result in any change in nociceptive behavior. Thus, the allodynia or hyperalgesia is due to immune/inflammatory processes occurring at the nerve. There are two additional animal pain models that manifest as mechanical allodynia with no discernable thermal hyperalgesia. These are epineurial administration of tumor necrosis factor [10] and systemic or intrathecal administration of an antibody to the GD2 ganglioside [11]. The GD2 ganglioside is a minor component of peripheral nerve and its antibody is thought to trigger a complement dependent cytotoxicity upon binding. Other than proximity to anesthetic exposure, differences between the models that might explain the discrepancy in the generation of thermal hyperalgesia have not been identified.

Light microscopic examination of CFA-treated nerves reveals endoneurial edema associated with a small amount of axonal degeneration and demyelination [6]. Infiltration of immune cells into the nerve at the level of the CFA-soaked sponge is much more remarkable. Invading cells are primarily macrophages, granulocytes and T cells that stain for both CD4 and CD8. The pattern of staining seen is consistent with transport *via* the endoneurial vasculature rather than by simple diffusion across the epineurium. Two related questions arise: (1) Is the axonal degeneration and/or demyelination a necessary trigger, and (2), which action or product of these infiltrating cell types, if any, is necessary for or contributes to the sensory abnormalities?

Wallerian degeneration or macrophage infiltration as a cause of neuropathic pain

One theory is that macrophage-mediated Wallerian degeneration is the necessary triggering event. Certainly, the thermal hyperalgesia following CCI in normal animals peaks at the time of maximum activated macrophage infiltration of the nerve [12]. CCI injury in mice with delayed macrophage recruitment and Wallerian degeneration (C57BL/WLD) results in a correspondingly delayed as well as reduced thermal hyperalgesia [13, 14]. The time profile of mechanical allodynia reported by Sommer and Schäfers is, however, quite different. Although the onset latency and initial magnitude of the allodynia are identical in control and C57BL/WLD mice, the duration of the allodynia lasts far longer in the animals with the defect. Allodynia appeared to be temporally related to nerve fiber regeneration in this study. In contrast, Ramer et al. found neither thermal hyperalgesia nor mechanical allodynia in C57BL/WLD mice with a CCI injury [15]. Thus, while this confirms the macrophages role in thermal hyperalgesia, it makes it hard to come to any conclusion about the conditions necessary for development and maintenance of the allodynia.

Depletion of circulating phagocytosing monocytes by systemic treatment with liposomes containing CL2MDP decreases thermal hyperalgesia following partial tight sciatic nerve ligation [16]. This further links infiltration of activated macrophages into the nerve with the development of neuropathic pain. The strongest evidence for this association takes advantage of the fact that in both the CCI and partial sciatic ligation [17] models, not all animals subjected to the procedure develop hyperesthesia. Animals that display post-surgical allodynia, have significantly more macrophages in the injured sciatic nerve than non-allodynic animals subjected to the same procedure. There is a clear correlation between the number of macrophages in the nerve and the withdrawal threshold to mechanical stimulation [18]. Interestingly, the increase in number of macrophages in the non-allodynic animals is the same as that in animals with a sciatic nerve axotomy, animals that presumably have the maximum amount of Wallerian degeneration. Thus, the essential relationship appears to be between neuropathic pain and increased infiltration by macrophages.

Cytokines

Activated macrophages are sources of many nociceptive substances, including the proinflammatory cytokines, tumor necrosis factor-α (TNF), interleukin-1β (IL-1) and interleukin-6 (IL-6). In addition to their many effects on the immune system, these interactive cytokines all influence nerve regeneration and TNF induces up-regulation of nerve growth factor [19]. Results concerning cytokine effects on hyperalgesia need to be interpreted with this in mind. TNF is released from macrophages activated by endotoxin or zymosan and was originally identified for its toxic effect on tumor cells [20]. Since then it has become apparent that it is central to the development of inflammation and the resultant inflammatory pain [19, 21, 22]. TNF up-regulation and release also appears to play equally important roles in demyelination, Wallerian degeration [23, 24] and neuropathic pain.

In normal animals, a modest amount of TNF-immunoreactivity is found in DRG neurons and to a lesser degree along the axons [25]. The TNF is located predominantly in large DRG neurons; small cells display patchy non-nuclear staining [25]. TNF receptor I protein is present at low levels in the naïve animal, and like the protein, there is more in the larger cells [25, 26]. After injection of biotinylated TNF around the sciatic nerve trunk, fast retrograde transport moves it to the DRG within 6 h. The increased TNF elicits increased TNF receptor I and II in DRG neurons of all sizes.

Following CCI nerve injury, TNF content in the nerve increases dramatically. Schwann cells become an additional, but probably still minor, source of endogenous TNF [25, 27, 28]. Around the injury site, transmembrane TNF precursor and the active soluble TNF monomer are up-regulated as soon as 3 h after the placement of the CCI ligatures; the initial peak for both proteins occurs 6 h post-injury [26]. A second peak, usually larger than the first, occurs 5 days following the CCI procedure. These maximum levels of TNF correspond to periods of maximal hyperalgesia. The first peak is thought to be due to increased TNF expression in resident mast cells as well as in the Schwann cells [26, 27]. The second peak is more likely due to production and release of TNF from infiltrating macrophages. After CCI injury, uptake and transport of TNF to the DRG is inhibited and TNF becomes concentrated distal to the cell body [25]. Following CCI injury, there is a huge increase in TNF receptor 1 protein that peaks 6 h post-surgery, coincident with the first peak of the active TNF monomer [26]. A second peak in the receptor protein develops three days after the injury, preceding the second TNF peak [26].

As with the macrophages, there is a positive correlation between number of TNF staining cells and withdrawal threshold [18]. Significantly more cells stain for TNF in the CCI and sciatic nerve tight ligation models in animals with allodynia compared to those that do not display changes in mechanical sensitivity [18], thus indicating a clear linkage between this macrophage product and hyperesthesia. However, axotomy increases TNF as much as any of the allodynia-producing lesions, as does ischemic insult to the nerve.

As would be expected if TNF were playing a causal role in neuropathic pain, reduction or sequestering of TNF or reduction of available TNF receptor reduces mechanical allodynia and thermal hyperalgesia in the majority of, but not all, neuropathic pain models. Daily systemic treatment with thalidomide during the first week following CCI injury reduces TNF-like immunoreactivity of the sciatic nerve without affecting macrophage infiltration, IL-1 or IL-6 [29]. Thalidomide treatment results in reduced sensitization of all modalities normally sensitized following CCI injury, but has no effect on the CFA neuritis-induced sensitization [5, 30]. It may be that the role of TNF is less important when the immune system's function is to combat bacterial infection than when it is initiating phagocytosis. Alternative methods of reducing free TNF by epineurial injection of TAPI, a metalloprotease inhibitor, or a TNF neutralizing antibody are both effective in reducing the hyperesthesia induced by CCI [28, 31]. Importantly, TAPI injection is also anti-hyperalgesic without affecting either the number of degenerating axons or the number of infiltrating macrophages [28]. These data in conjunction with the thalidomide results imply that the role of the macrophages in producing hyperesthesia is cytokine production and that TNF is necessary for some portion of the sensitization [32]. Epineurial injection of neutralizing antibodies to TNF receptor I, but not TNF receptor II, also reduce the enhanced responses to both thermal and mechanical stimulation [33].

Direct endoneurial injection of TNF results in thermal hyperalgesia and mechanical allodynia that lasts for days [34]. This may involve a positive feedback component as macrophage infiltration and activation also occur within this time frame as do endoneurial inflammation and primary demyelination. Acute epineurial application of TNF to the sciatic nerve in awake animals results in transient mechanical allodynia, without affecting thermal sensitivity [10]. Direct application of TNF within a restricted dose range to the sciatic nerve results in the development of ectopic activity in identified Aδ and C nociceptors [35]. Administration of TNF to the dorsal root ganglia (DRG) *in vitro*, within the same dose range, also initiates ectopic bursting in C fibers [36]; when higher concentrations of TNF are used, this response is lost. However, at these higher doses TNF increases the response of small DRG neurons to capsaicin [37]. Animals with chronic compression injuries to their L5 DRG display a marked allodynia and hyperalgesia [38]. *In vitro* teased fiber recordings obtained from these hyperalgesic animals are more likely to show increased ongoing discharge rates following TNF exposure than recordings obtained from control animals [39]. This indicates a synergy between the exogenous TNF and unknown products generated in the neurons in response to injury. Whether or not any of these same agents are produced in response to neuritis or by immune activation is unknown, but likely. Interestingly, in these injured animals background discharge in a significant number of Aβ fibers is suppressed [39]. The impact that these bi-directional TNF-induced changes have on dorsal horn activity has not been explored.

Much less work has explored the roles of IL-1 and IL-6 within the peripheral nerve in the generation of neuropathic pain. Cue et al [18] found a stronger correlation of IL-1β positive cells with Wallerian degeneration than with hyperalgesia. TAPI, a metalloprotease inhibitor [28], reduces hyperalgesia and blocks the CCI-induced increase in TNF without altering the CCI-induced increase in IL-1β. Thus, unlike what has been reported for peripheral inflammation, there is not a clear pro-inflammatory cascade leading from TNF to IL-1 [21]. However, treatment of the nerve with neutralizing antibodies to IL-1 receptor not only reduces hyperalgesia, but decreases TNF immunoreactivity in the nerve.

IL-6 is found at low levels in uninjured sciatic nerve [40] and increases following nerve injury, but it increases more in animals with allodynia than in either non-allodynic or axotomized animals [18]. Rats with sciatic nerve ischemia and allodynia also have a high level of IL-1 positive cells. Thus, IL-6 seems to be the most specific of the above cytokines for neuropathic pain. IL-6 knockout mice display neither thermal hyperalgesia nor mechanical allodynia following CCI injury [41]. This would seem to be strong evidence for the necessity of IL-6 for neuropathic pain. However, there is a report that IL-6 deficient mice have lower nociceptive thresholds than wild type mice [42] which complicates interpretation of the data. Intraneural injection of IL-6 into the sciatic nerve causes severe inflammation and demyelination as well as T cell and macrophage infiltration [43]. Although nociceptive measures were not made, it is likely that these animals would have been hyperalgesic. The histopathologic and immunological sequelae to IL-6 injection would make it hard to assess the effect of increases in IL-6 alone.

Cytokine release secondary to macrophage infiltration appears to be critical for the development of neuropathic pain in several of the neuropathic pain models. The sequence of the proinflammatory cascade, the necessity for any one agent and whether the various cytokines are working in series or in parallel to produce hyperalgesia/allodynia is not clear. Other remaining questions include why reduction of available TNF has no effect on the neuritis model or why some immune challenge based neuropathic pains result in alteration of only mechanical allodynia and not thermal hyperalgesia.

Acknowledgements
Supported by National Institutes of Health grant NS35630.

References

1 Bennett GJ, Xie YK (1988) A peripheral mononeuropathy in rat that produces disorders of pain sensation like those seen in man [see comments]. *Pain* 33: 87–107

2 Maves TJ, Pechman PS, Gebhart GF, Meller ST (1993) Possible chemical contribution from chronic-gut sutures produces disorders of pain sensation like those seen in man. *Pain* 54: 57–69

3 Myers RR, Yamamoto T, Yaksh TL, Powell HC (1993) The role of focal nerve ischemia and Wallerian degeneration in peripheral nerve injury producing hyperesthesia. *Anesthesiology* 78: 308–316

4 Clatworthy AL, Illich PA, Castro GA, Walters ET (1995) Role of peri-axonal inflammation in the development of thermal hyperalgesia and guarding behavior in a rat model of neuropathic pain. *Neurosci Lett* 184: 5–8

5 Bennett GJ (1999) Does a neuroimmune interaction contribute to the genesis of painful peripheral neuropathies? *Proc Natl Acad Sci USA* 96: 7737–7738

6 Eliav E, Herzberg U, Ruda MA, Bennett GJ (1999) Neuropathic pain from an experimental neuritis of the rat sciatic nerve. *Pain* 83: 169–182

7 Miller LS, Morita Y, Rangan U, Kondo S, Clemens MG, Bulkley GB (1996) Suppression of cytokine-induced neutrophil accumulation in rat mesenteric venules *in vivo* by general anesthesia. *Intern J Microcirc Clin Exp* 16: 147–154

8 Lockwood LL, Silbert LH, Laudenslager ML, Watkins LR, Maier SF (1993) Anesthesia-induced modulation of *in vivo* antibody levels: a study of pentobarbital, chloral hydrate, methoxyflurane, halothane, and ketamine/xylazine. *Anesth Analg* 77: 769–774

9 Watkins L, Milligan E, Maier S (2002) Glial proinflammatory cytokines mediate exaggerated pain states: Implications for clinical pain. In: M Malchelska, C Stein (eds): *Immune mechanisms of analgesia*. Landes Biosciences, New York, NY, *in press*

10 Sorkin L, Doom C (2000) Epineurial application of TNF elecits an acute mechanical hyperalgesia in the awake rat. *J Peripheral Nervous Sys* 5: 96–1000

11 Slart R, Yu AL, Yaksh TL, Sorkin LS (1997) An animal model of pain produced by systemic administration of an immunotherapeutic anti-ganglioside antibody. *Pain* 69: 119–25

12 Sommer C, Lalonde A, Heckman HM, Rodriguez M, Myers RR (1995) Quantitative neuropathology of a focal nerve injury causing hyperalgesia. *J Neuropathol Exp Neurol* 54: 635–643

13 Myers RR, Heckman HM, Rodriguez M (1996) Reduced hyperalgesia in nerve-injured WLD mice: Relationship to nerve fiber phagocytosis, axonal degeneration and regeneration in normal mice. *Exp Neurol* 141: 94–101

14 Sommer C, Schäfers M (1998) Painful mononeuropathy in C57BL/Wld mice with delayed wallerian degeneration: differential effects of cytokine production and nerve regeneration on thermal and mechanical hypersensitivity. *Brain Res* 784: 154–162

15 Ramer MS, French GD, Bisby MA (1997) Wallerian degeneration is required for both neuropathic pain and sympathetic sprouting into the DRG. *Pain* 72: 71–78

16 Liu T, van Rooijen N, Tracey DJ (2000) Depletion of macrophages reduces axonal degeneration and hyperalgesia following nerve injury. *Pain* 86: 25–32
17 Seltzer Z, Dubner R, Shir Y (1990) A novel behavioral model of neuropathic pain disorders produced in rats by partial sciatic nerve injury. *Pain* 43: 205–218
18 Cui J-G, Holmin S, Mathiesen T, Meyerson B, Linderoth B (2000) Possible role of inflammatory mdiators in tactile hypersensitivity in rat models of mononeuropathy. *Pain* 88: 239–248
19 Woolf CJ, Allchorne A, Safieh-Garabedian B, Poole S (1997) Cytokines, nerve growth factor and inflammatory hyperalgesia: the contribution of tumour necrosis factor alpha. *Br J Pharmacol* 121: 417–424
20 Carswell EA, Old LJ, Kassel RL, Green S, Fiore N, Williamson B (1975) An endotoxin-induced serum factor that causes necrosis of tumors. *Proc Natl Acad Sci USA* 72: 3666–3670
21 Cunha FQ, Poole S, Lorenzetti BB, Ferreira SH (1992) The pivotal role of tumour necrosis factor alpha in the development of inflammatory hyperalgesia. *Br J Pharmacol* 107: 660–664
22 Junger H, Sorkin LS (2000) Nociceptive and inflammatory effects of subcutaneous TNF. *Pain* 85: 145–151
23 Redford EJ, Hall SM, Smith KJ (1995) Vascular changes and demyelination induced by the intraneural injection of tumour necrosis factor. *Brain* 118: 869–878
24 Stoll G, Jung S, Jander S, van der Meide P, Hartung HP (1993) Tumor necrosis factor-alpha in immune-mediated demyelination and Wallerian degeneration of the rat peripheral nervous system. *J Neuroimmunol* 45: 175–182
25 Shubayev V, Myers R (2001) Axonal transport of TNF-α in painful neuropathy: Altered distribution of ligand and receptors. *Neuroimmunol* 114: 48–56
26 Shubayev VI, Myers RR (2000) Upregulation and interaction of TNFalpha and gelatinases A and B in painful peripheral nerve injury. *Brain Res* 855: 83–89
27 Wagner R, Myers RR (1996) Schwann cells produce tumor necrosis factor alpha: Expression in injured and non-injured nerves. *Neurosci* 73: 625–629
28 Sommer C, Schmidt C, George A, Toyka KV (1997) A metalloprotease-inhibitor reduces pain associated behavior in mice with experimental neuropathy. *Neurosci Lett* 237: 45–48
29 George A, Marziniak M, Schafers M, Toyka K, Sommer C (2000) Thalidomide treatment in chronic constrictive neuropathy decreases endoneurial tumor necrosis factor-a, increases interleukin-10 and has long term effects on spinal cord dorsal horn met-enkephalin. *Pain* 88: 267–275
30 Sommer C, Myers RR (1994) Thalidomide inhibition of TNF reduces hyperalgesia in neuropathic rats. *Reg Anesth* 19: 1
31 Lindenlaub T, Teuteberg P, Hartung T, Sommer C (2000) Effects of neutralizing antibodies to TNF-alpha on pain-related behavior and nerve regeneration in mice with chronic constriction injury. *Brain Res* 866: 15–22
32 Wagner R, Janjigian M, Myers RR (1998) Anti-inflammatory interleukin-10 therapy in

CCI neuropathy decreases thermal hyperalgesia, macrophage recruitment, and endoneurial TNF-alpha expression. *Pain* 74: 35–42
33 Sommer C, Schmidt C, George A (1998) Hyperalgesia in experimental neuropathy is dependent on the TNF receptor 1. *Exp Neurol* 151: 138–42
34 Wagner R, Myers RR (1996) Endoneurial injection of TNF-alpha produces neuropathic pain behaviors. *Neuroreport* 7: 2897–901
35 Sorkin LS, Xiao WH, Wagner R, Myers RR (1997) Tumour necrosis factor-alpha induces ectopic activity in nociceptive primary afferent fibres. *Neuroscience* 81: 255–62
36 Liu B-G, Hastings S, Li H, Jia Y-P, Brull S, Zhang J-M (2000) Tumor necrosis factor enhances the excitability of dorsal root ganglion cells with unmyelinated axons. *Soc Neurosci Abs* 26: 1695
37 Nicol GD, Lopshire JC, Pafford CM (1997) Tumor necrosis factor enhances the capsaicin sensitivity of rat sensory neurons. *J Neurosci* 17: 975–982
38 Song XJ, Hu SJ, Greenquist KW, Zhang JM, LaMotte RH (1999) Mechanical and thermal hyperalgesia and ectopic neuronal discharge after chronic compression of dorsal root ganglia. *J Neurophysiol* 82: 3347–3358
39 Zhang J-M, Li H, Brull S (2000) Tumor necrosis factor α mediates spontaneous activity in chronically compressed dorsal root ganglia neurons in the rat. *Anesthesiology* Abstracts 13: A-978
40 Reichert F, Levitzky R, Rotshenker S (1996) Interleukin 6 in intact and injured mouse peripheral nerves. *Eur J Neurosci* 8: 530–535
41 Murphy PG, Ramer MS, Borthwick L, Gauldie J, Richardson PM, Bisby MA (1999) Endogenous interleukin-6 contributes to hypersensitivity to cutaneous stimuli and changes in neuropeptides associated with chronic nerve constriction in mice. *Eur J Neurosci* 11: 2243–2253
42 Xu XJ, Hao JX, Andell-Jonsson S, Poli V, Bartfai T, Wiesenfeld-Hallin Z (1997) Nociceptive responses in interleukin-6-deficient mice to peripheral inflammation and peripheral nerve section. *Cytokine* 9: 1028–1033
43 Deretzi G, Pelidou SH, Zou LP, Quiding C, Zhu J (1999) Local effects of recombinant rat interleukin-6 on the peripheral nervous system. *Immunology* 97: 582–587

Spinal cord plasticity following peripheral nerve injury

Pharmacological plasticity associated with neuropathic pain states

Victoria Chapman[1] and Anthony H. Dickenson[2]

[1]School of Biomedical Sciences, University of Nottingham, E-Floor Medical School, Queen's Medical Centre, Nottingham, NG7 2UH, UK; [2]Department of Pharmacology, University College London, Gower Street, London, WC1 E 6BT, UK

Introduction

Chronic pain is a major clinical problem with considerable impact on the quality of life. There is much evidence for plasticity of somatosensory processing in chronic pain states which contributes to variations in the duration, intensity, physiology and pharmacology of these conditions. Brain mapping with positron emission tomography in experimental human studies of pain [1] suggests that different brain areas subserve different components of pain experience [2] and that distinct somatosensory pathways and CNS areas underlie experimental pathophysiological touch evoked pain (allodynia) and physiological pain [3]. The pathology associated with chronic pain states often leads to alterations in both peripheral and central pain systems, with vast implications for the treatment of these conditions. Conventional analgesics, such as the opioids, can have reduced effectiveness in some chronic pain states, in particular some neuropathic pains. The diversity of maladaptive plasticity within peripheral nerves, dorsal root ganglia (DRG), spinal cord and higher brain centres must contribute to the range of neuropathic syndromes. Molecular biology studies of animal models of chronic pain states suggest that immediate post-translational modifications and longer-term transcriptional changes [4] may further contribute to the plasticity of somatosensory transmission in persistent pain states, both inflammatory or neuropathic in origin. A better understanding of the changes in contribution of receptors/channels underlying these conditions will be essential for the improved treatment regime of chronic pain states.

Models of neuropathic pain states

Clinical symptoms of chronic pain states are not replicated in animal models of acute pain, and therefore models of neuropathic pain based on peripheral nerve damage are essential. Established models of neuropathy include chronic constriction injury, partial sciatic nerve ligation and selective spinal nerve ligation. These models

Mechanisms and Mediators of Neuropathic Pain, edited by Annika B. Malmberg and Sandra R. Chaplan

replicate some aspects of clinical neuropathic pain in particular, mechanical allodynia, cold allodynia and thermal hyperalgesia. There is evidence that the aberrations in somatosensory processing which follow partial nerve injury are the culmination of a number of changes in the peripheral nervous system and the generation of ectopic activity that then impinges upon the spinal cord. Thus, the ongoing activity in peripheral sensory neurones may maintain central states of hyperexcitability, initiated by the afferent barrage producd by nerve damage. Clinically, ectopic activity has been shown to contribute to ongoing neuropathic pain [5]. Animal studies have demonstrated that ectopic activity manifests in both injured and uninjured afferent fibres [6]. Following nerve injury, marked changes in sodium channel transcripts and currents in small dorsal root ganglia have been reported [7] and these changes may contribute to the associated ectopic activity. In this study, transcripts for TTX-resistant (TTX-R) sodium channels (NaN and SNS) were reduced and transcripts for the TTX-sensitive (TTX-S) α-III channel were increased. Importantly these changes in levels of transcripts accompanied a functional reduction of TTX-R currents and the appearance of rapidly repriming TTX-S sodium currents [7]. The profile of ectopic activity and behavioural allodynia are similar and therefore it has been suggested that ectopic activity in A-fibres sensitises the central nervous system leading to the development and maintenance of evoked aberrant behavioural responses [6]. Changes in phenotype of cell bodies corresponding to A-fibres, including the expression of substance P and structural reorganisation of termination sites of large fibre (Aβ) in the spinal cord have been reported [8]. However, may of these latter changes occur after the onset of behavioural allodynia and so are unlikely to be causal factors. Plasticity of spinal nociceptive pathways has also been reported. Abnormal responses of spinal neurones including spontaneous activity and increased after discharges have been reported following chronic constriction injury (see refs in [9]). Following selective spinal nerve ligation, neurones with spontaneous activity, lowered thresholds to thermal and mechanical stimuli coupled with decreased magnitudes of responses and enlarged mechanical receptive fields have been reported [9]. However only subtle changes in the responses of spinal neurones have been reported following partial injury [10]. How the reported changes in neuronal responses translate into the resultant aberrant pain response remains unclear although a number of hypotheses have been proposed [8].

Analgesic targets for persistent pain states

Elucidation of the key substrates underlying aberrant somatosensory transmission in neuropathic and inflammatory pain states will provide a number of additional potential targets for novel analgesics that may have improved efficacy and reduced side effect profile. In the long-term transcriptional changes in receptors and channels will overtly compromise targets of choice. Putative analgesic targets for persistent pain states can be identified by cell biology methods and comparative *in vivo* pharmacology.

Sodium channel blockers

There is good evidence that substrates contributing to neuropathic pain states are sensitive to membrane stablizing drugs such as the anticonvulsant carbamazepine [11, 12]. The use of this drug as the main line treatment of trigeminal neuralgia is established from clinical systematic reviews. Although carbamazepine may also be considered as a primary agent for other types of nerve injury pains its use is limited by side effects. Local anaesthetics such as lignocaine and the Ib antiarrhythmic sodium channel blocker mexiletine are also effective in neuropathic pain states. The sites of action of systemically administered sodium channel blockers such as lignocaine and mexiletine in relieving clinical neuropathic pain may include both peripheral and central sites [13].

A number of studies have evaluated the effects of sodium channel blockers in animal models of neuropathic pain. We have demonstrated with an *in vivo* electrophysiological approach that systemic carbamazepine reduces spontaneous activity of spinal neurones [14] as well as having a novel inhibitory effect on innocuous and noxious evoked responses of spinal neurones of spinal nerve ligated rats [14]. Although a behavioural report indicates carbamazepine does not influence the behavioural consequences of nerve injury [15], here the effects of carbamazepine may have been masked by overt vehicle effect. Systemic lidocaine also reduces both mechanical allodynia [16] and thermal hyperalgesia [17] in animal models of neuropathic pain. Similarly, systemic mexiletine reduces touch evoked pain behaviour [18], mechanical hyperalgesia [19] and cold induced allodynia [20] and evoked responses of spinal neurones [21] in models of neuropathic pain.

Gabapentin

Recent evidence suggests that newer anticonvulsant drugs, such as gabapentin (Neurontin) may also be effective in neuropathic pain states and may offer an alternative therapeutic approach. Gabapentin is an antiepileptic drug reported to have analgesic activity in neuropathic pain states from varying origins (see refs in [14]). A randomised double blind placebo controlled trial of gabapentin in postherpetic neuralgia patients showed gabapentin to be effective therapy and improved quality of life [22]. Similarly, another large trial showed marked effectiveness in diabetic neuropathy patients [23]. Furthermore, gabapentin has been reported to be effective in pain due to peripheral nerve injury and central lesions, with particular effectiveness on paroxysmal pain and allodynia [24].

In keeping with clinical reports, systemic and intrathecal gabapentin reduces heat hyperalgesia and mechano- and cold allodynia in the chronic constriction injury [15, 25] and spinal nerve ligation [26] models of neuropathy and spinally injured rats [27]. Gabapentin has also been shown to block both static and dynamic mechanical allodynia in a model of diabetic neuropathy [28] and anti-GD2 associated allodynia [29].

We have demonstrated that gabapentin has inhibitory effects on innocuous and noxious evoked responses of spinal neurones of rats with spinal nerve ligation, with effects comparable in magnitude to those of carbamazepine [14]. In contrast to carbamazepine, gabapentin produced similar inhibitory effects on neuronal responses of sham operated rats. These electrophysiological studies suggest that the effectiveness of gabapentin is not determined as a direct consequence of spinal nerve ligation but by general invasive surgery and/or tissue damage. A recent study demonstrating that gabapentin is anti-allodynic in a rat model of postoperative pain [30] further corroborates this idea.

Recently, the inhibitory effects of gabapentin on glutamatergic excitatory transmission in the rat dorsal horn have been attributed to a pre-synaptic site of action [31]. Evidence that gabapentin binds to the $\alpha 2\delta$ subunit of voltage-gated Ca^{2+} channels [32] may provide a basis for this action *via* interactions with presynaptic Ca^{2+} channels. Although effects of gabapentin on voltage-gated Ca^{2+} channel currents are inconsistent, this may result from gabapentin only interacting with a specific subpopulation of voltage-gated Ca^{2+} channels [33] and requiring particular circumstances for effective block of these channels.

Ca^{2+} channel blockers

There is considerable evidence that neuronal N type calcium channels, located in the superficial laminae of the spinal cord, play a pivotal role in spinal nociceptive processing. N-type calcium channel blockers, such as ziconotide (SNX-111) have been shown to attenuate thermal hyperalgesia [34] and mechanical allodynia [35–37] in animal models of neuropathic pain. Recently, ziconotide has been shown to inhibit mechanical allodynia and thermal hyperalgesia in a rat model of postoperative pain [38]. Following nerve injury there is a reported increase in $\alpha 1\beta$ Ca^{2+} channel subunit expression in small DRG neurones and the superficial laminae of the spinal cord [39]. This finding correlates well with the evidence that N-type calcium channel blockers are more effective in reducing neuronal responses to sensory stimuli after nerve injury [40]. To date there have been few studies of the clinical analgesic effect of N-type calcium channel blockade. Spinal ziconotide has been reported to provide pain relief in a patient with intractable brachial plexus avulsion pain [41]. However, recent evidence suggests that severe adverse effects are associated with spinal ziconotide [42].

Cannabinoid receptor agonists

Currently there is a heightened interest in the analgesic potential of cannabinoid receptor agonists such as Δ^9-tetrahydrocannabinol (Δ^9-THC) and synthetic agonists [43]. Behavioural [44] and electrophysiological [45] studies provide evidence for a role of spinal cannabinoid (CB) receptors in the modulation of acute nociceptive transmission. Spinal cannabinoid receptors have been identified [46]. Studies with the selective cannabinoid CB_1 receptor (CB_1 receptor) antagonist SR141716A suggest the antinociceptive effects of spinally administered cannabinoids are mediated by CB_1 receptors [47]. The majority of CB_1 receptor messenger RNA in dorsal root ganglia is in medium and large cell bodies [48], raising the possibility of functional pre-synaptic CB_1 receptors on afferent A-fibres. Recent studies have provided evidence for cannabinoid binding sites both pre-synaptic on primary afferents and post-synaptic on dorsal horn neurons, with pre-synaptic receptors residing on both C-fibre and A-fibre primary afferents.

We have recently demonstrated that under normal conditions spinal CB_1 receptors predominantly modulate C-fibre transmission, in particular the post-discharge response, with a lesser effect on Aδ-fibre evoked responses and a minor effect on Aβ-fibre transmission at high concentrations [49]. These antinociceptive effects of the synthetic cannabinoid agonist HU210 were blocked by selective CB_1 receptor antagonism with SR141716A. Similarly, systemic administration of the cannabinoid agonist WIN 55,212-2 has been shown to predominantly inhibit C-fibre responses of spinal neurons [50].

The analgesic potential of the cannabinoids for the treatment of sustained pain states associated with hyperalgesia or allodynia is of clinical interest [51]. There are few basic reports on the effect of cannabinoids in models of persistent/chronic pain states. In the complete Freund's adjuant model of chronic inflammation low doses of the cannabinoid agonist WIN55,212-2 reverse mechanical allodynia of the inflamed paw, whereas higher doses have analgesic effects in both paws [52]. Interestingly in this study blockade of the spinal cannabinoid CB_1 receptors with the selective antagonist SR141716A did not influence thresholds in inflamed rats, suggestive of a loss of spinal endogenous cannabinoid tone following chronic inflammation [52].

The potential of cannabinoid agonists as analgesics for neuropathic pain states is largely unknown. WIN55,212-2 has been shown to reverse mechanical allodynia and mechanical and thermal hyperalgesia in the chronic constriction injury model of neuropathy [53]. In this study administration of the CB_1 receptor antagonist exacerbated hyperalgesia and allodynia, suggesting that the endogenous cannabinoid receptor system remains intact under these conditions [53].

Recently, we have compared the effects of the cannabinoid agonist, HU210, on evoked responses of spinal neurones in rats with spinal nerve ligation and sham operated controls. Following spinal nerve ligation, rats exhibited reproducible mechanical allodynia and 14 days post-surgery the effect of spinal *versus* systemic administration of HU210 on A- and C-fibre evoked responses of spinal neurones was studied [54]. Spinal and systemic HU210 consistently and significantly reduced C-fibre evoked responses of neurones in sham operated rats but not nerve injured rats. However, spinal HU210 significantly reduced late Aδ-fibre evoked responses of spinal neurones in both sham and nerve injured rats. As discussed earlier, there is evidence for CB_1 receptor mRNA in medium sized DRG cell bodies [48], which probably correspond to Aδ-fibres. Our results suggest that there is plasticity of the spinal cannabinoid receptor system following peripheral nerve injury and that cannabinoid receptor activation can markedly modulate Aδ-fibre evoked responses of neurones under these conditions. In view of the evidence that A-fibres play a pivotal role in allodynia, the reported effects of cannabinoid agonists on mechanical allodynia [53] and Aδ-fibre evoked responses [54] in models of neuropathic pain suggest that cannabinoid agonists may be clinically effective at reducing allodynia.

References

1 Casey KL (1999) Forebrain mechanisms of nociception and pain: analysis through imaging. *Proc Natl Acad Sci USA* 96: 7668

2 Tolle TR, Kaufmann T, Siessmeier T, Berthele A et al (1999) Region-specific encoding of sensory and affective components of pain in the human brain: a PET correlation analysis. *Annals Neurology* 45: 40

3 Iadarola MJ, Berman KF, Zeffiro TA, Byas-Smith MG et al (1998) Neural activation during acute capsaicin-evoked pain and allodynia assessed with PET. *Brain* 121: 931

4 Woolf CJ, Costigan M (1999) Transcriptional and posttranslational plasticity and the generation of inflammatory pain. *Proc Natl Acad Sci USA* 96: 7723

5 Gracely RH, Lynch SA, Bennett GJ (1992) Painful neuropathy: altered central processing maintained dynamically by peripheral input. *Pain* 51 (2): 175–194

6 Gold MS (2000) Spinal nerve ligation: what to blame for the pain and why. *Pain* 84: 117–120

7 Dib-Hajj SD, Fjell J, Cummins TR, Zheng Z, Fried K, LaMotte R, Black JA, Waxman SG (1999) Plasticity of sodium channel expression in DRG neurones in the chronic constriction injury model of neuropathic pain. *Pain* 83: 591–600

8 Woolf CJ, Mannion RJ (1999) Neuropathic Pain: aetiology, symptoms, mechanisms, and management. *Lancet* 353: 1959

9 Chapman V, Suzuki R, Dickenson AH (1998) Electrophysiological characterisation of spinal neuronal response properties of anaesthetised rats after ligation of spinal nerves L5-6. *J Physiol* 507.3: 881–894

10 Takaishi K, Eisele JH, Carstens E (1996) Behavioural and electrophysiological assessment of hyperalgesia and changes in dorsal horn responses following partial sciatic nerve ligation in rats. *Pain* 66: 297–306

11 McQuay H, Carroll D, Jasdad AR, Wiffen P, Moore A (1995) Anticonvulsant drugs for management of pain: a systemic review. *Br Med J* 311: 1047–1052
12 Fields HL, Rowbotham MC, Devor M (1997) Excitability blockers: antconvulasants and low concentration local anesthetics in the treatment of chronic pain. In: AH Dickenson, JM Besson (eds): *The pharmacology of pain, vol 130, Handbook of experimental pharmacology*. Springer, Heidelberg, 93–111
13 Wallace MS, Dyck JB, Rossi SS, Yaksh TL (1996) Computer-controlled lidocaine infusion for the evaluation of neuropathic pain after peripheral nerve injury. *Pain* 66: 69–77
14 Chapman V, Suzuki R, Chamarette HLC, Rygh LJ, Dickenson AH (1998) An electrophysiological study of the inhibitory effects of systemic carbamazepine and gabapentin on spinal neuronal responses in spinal nerve ligated rats. *Pain* 75: 261–273
15 Hunter JC, Gogas KR, Hedley LR, Jacobson LO, Kassotakis L, Thompson J, Fontana DJ (1997) The effect of novel anti-epileptic drugs in rat experimental models of acute and chronic pain. *Eur J Pharmacol* 324: 153–160
16 Chaplan SR, Bach FW, Shafer SL, Yaksh TL (1995) Prolonged alleviation of tactile allodynia by intravenous lidocaine in neuropathic rats. *Anesthesiology* 83: 775–785
17 Abram SE, Yaksh TL (1994) Systemic lidocaine blocks nerve injury induced hyperalgesia and nociceptor-driven sensitization in the rat. *Anesthesiology* 80: 383–391
18 Jett MF, McCuirk J, Waligora D, Hunter JC (1997) The effects of mexiletine, desipramine and fluoxetine in rat models involving central sensitization. *Pain* 69: 161–169
19 Koch BD, Faurot GF, Mc Guirk JR, Clarke DE, Hunter JC (1996) Modulation of mechano-hyperalgesia by clinically effective analgesics in rats with a peripheral mononeuropathy. *Analgesia* 2: 157–164
20 Hedley LR, Martin B, Waterbury LD, Clarke DE, Hunter JC (1995) A comparison of the action of mexiletine and morphine in rodent models of acute and chronic pain. *Proc West Pharmacol Soc* 38: 103–104
21 Chapman V, Ng J, Dickenson AH (1998) A novel spinal action of mexiletine in spinal somatosensory transmission of nerve injured rats. *Pain* 77: 289–296
22 Rowbotham M, Harden N, Stacey B, Podolnick P, Magnus-Miller L (1998) Gabapentin for treatment of post-herpetic neuralgia: a randomised controlled trial. *JAMA* 280: 1837–1842
23 Backonja M, Beydoun A, Edwards KR, Schwartz SL, Fonseca V, Hes M, LaMoreaux L, Garofalo E (1998) Gabapentin for the symptomatic treatment of painful neuropathy in patients with diabetes mellitus: a randomized controlled trial. JAMA 280: 1831–1836
24 Attal N, Brasseur L, Parker F, Chauvin M, Bouhassira D (1998) Effects of gabapentin on the different components of peripheral and central neuropathic pain syndromes: a pilot study. *Eur Neurol* 40: 191–200
25 Xiao WH, Bennett GJ (1995) Gabapentin relieves abnormal pains in a rat model of painful peripheral neuropathy. In: *Pain modulation: Anatomy and physiology. Neuropathic pain* Vol 21, Society for Neuroscience.
26 Chen SR, Eisenach JC, McCaslin PP, Pan HL (2000) Synergistic effect between intrathe-

cal non-NMDA antagonist and gabapentin on allodynia induced by spinal nerve ligation in rats. *Anesthesiology* 92: 500–506
27 Hao JX, Xu XJ, Urban L, Wiesenfeld-Hallin Z (2000) Repeated administration of systemic gabapentin alleviates allodynia-like behaviours in spinally injured rats. *Neurosci Letts* 25: 211–214
28 Field MJ, McCleary S, Hughes J, Singh L (1999) Gabapentin and pregabalin, but not morphine and amitriptyline, block both static and dynamic components of mechanical allodynia induced by streptozocin in the rat. *Pain* 80: 391–398
29 Gillian S, Sorkin LS (1998) Gabapentin reverses the allodynia produced by the administration of anti-GD2 ganglioside, an immunotherapeutic drug. *Anesth Analg* 86: 111–116
30 Cheng JK, Pain HL, Eisenach JC (2000) Antiallodynic effect of intrathecal gabapentin and its interaction with clonidine in a rat model of postoperative pain. *Anesthesiology* 92: 1126–1131
31 Shimoyama M, Shimoyama N, Hori Y (2000) Gabapentin affects glutamatergic excitatory neurotransmission in the rat dorsal horn. *Pain* 85: 405–414
32 Gee NS, Brown JP, Dissanayake VU, Offord J, Thurlow R, Woodruff GN (1996) The novel anticonvulsant drug, gabapentin (Neurontin), binds to the alpha2delta subunit of a calcium channel. *J Biol Chem* 271: 5768–5776
33 Taylor MT, Bonhaus DW (2000) Allosteric modulation of [3H]gabapentin binding by ruthenium red. *Neuropharmacology* 39: 1267–1273
34 Yamamoto T, Sakashita Y (1998) Differential effects of intrathecally administered N- and P-type voltage-sensitive calcium channel blockers upon two models of experimental mononeuropathy in the rat. *Brain Res* 794: 329–332
35 Bowersox SS, Gadbois T, Singh T, Pettus M, Wang YX, Luther RR (1996) Selective N-type neuronal voltage-sensitive calcium channel blocker, SNX-111, produces spinal antinociception in rat models of acute, persistent and neuropathic pain. *J Pharmacol Exp Ther* 279: 1243–1249
36 Chaplan SR, Pogrel JW, Yaksh TL (1994) Role of voltage-dependent calcium channel subtypes in experimental tactile allodynia. *J Pharmacol Exp Ther* 269: 1117–11123
37 Xiao WH, Bennett GJ (1995) Synthetic omega-conopeptides applied to the site of nerve injury suppress neuropathic pains in rats. *J Pharmacol Exp Ther* 274: 666–672
38 Wang Y-X, Pettus M, Gao D, Phillips C, Bowersox SS (2000) Effects of intrathecal administration of ziconotide, a selective neuronal N-type calcium channel blocker, on mechanical allodynia and heat hyperalgesia in a rat model of postoperative pain. *Pain* 84: 151–158
39 Cizkova D, Marsala M, Stauderman K, Yaksh T (1999) Calcium channel α1β subunit in spinal cord/DRG of normal and nerve-injured rats. *9th World Congress on Pain*, Abstract Number 11
40 Matthews EM, Dickenson AH (2001) Effects of spinally delivered N- and P-type voltage-dependent calcium channel antagonists on dorsal horn neuronal responses in a rat model of neuropathy. *Pain* 92: 235–246

41 Brose WG, Gutlove DP, Luther RR, Bowersox SS, McGuire D (1997) Use of intrathecal SNX-111, a novel, N-type, voltage sensitive, calcium channel blocker, in the management of intractable brachial plexus avulsion pain. *Clin J Pain* 13: 256–259

42 Penn RD, Paice JA (2000) Adverse effects associated with the intrathecal administration of ziconotide. *Pain* 85: 291–296

43 Manzanares J, Corchero J, Romero J, Fernandez-Ruiz JJ, Ramos JA, Fuentes JA (1999) Pharmacological and biochemical interactions between opioids and cannabinoids. *TIPS* 20: 287–294

44 Smith PB, Martin BR (1992) Spinal mechanisms of Δ^9-tetrahydrocannabinol-induced analgesia. *Brain Res* 578: 8–12

45 Hohmann AG, Tsou K, Walker JM (1998) Cannabinoid modulation of wide dynamic range neurons in the lumbar dorsal horn of the rat by spinally administered WIN55,212-2. *Neurosci Letts* 257: 119–122

46 Tsou K, Brown S, Sanudo-Pena MC, Mackie K, Walker JM (1998) Immunohistochemical distribution of cannabinoid CB1 receptors in the rat central nervous system. *Neuroscience* 83: 393–411

47 Welch SP, Huffman JW, Lowe J (1998) Differential blockade of the antinociceptive effects of centrally administered cannabinoids by SR141716A. *JPET* 286: 1301–1308

48 Hohmann AG, Herkenham M (1999) Localization of central cannabinoid CB_1 receptor messenger RNA in neuronal subpopulations of rat dorsal root ganglia: a double-label *in situ* hybridization study. *Neuroscience* 90: 923–931

49 Drew LJ, Harris J, Millns PJ, Kendall DA, Chapman V (2000) Activation of spinal cannabinoid $_1$ receptors inhibits C-fibre driven hyperexcitable neuronal responses and increases $[^{35}S]GTP\gamma S$ binding in the dorsal horn of the spinal cord of non-inflamed and inflamed rats. *Eur J Neurosci* 12: 2079–2086

50 Strangman NM and Walker JM (1999) Cannabinoid WIN55, 212-2 inhibits the activity-dependent facilitation of spinal nociceptive responses. *J Neurophysiol* 81: 472–477

51 Fields HL, Meng ID (1998) Watching the pot boil. *Nature Medicine* 4: 1008–1009

52 Martin WJ, Loo CM, Basbaum AI (1999) Spinal cannabinoids are anti-allodynic in rats with persistent inflammation. *Pain* 82: 199–205

53 Herzberg U, Eliav E, Bennett GJ, Kopin IJ (1997) The analgesic effects of R(+)-WIN 55,212-2 mesylate, a high affinity cannabinoid agonist, in a rat model of neuropathic pain. *Neurosci Letts* 221: 157–160

54 Chapman V (2001) Functional changes in the inhibitory effect of spinal cannabinoid (CB) receptor activation in nerve injured rats. *Neuropharmacology* 41: 870–877

Sprouting and reorganization in the spinal cord after nerve injury

Susan M. Carlton and Richard E. Coggeshall

Department of Anatomy and Neurosciences, Marine Biomedical Institute, University of Texas Medical Branch, Galveston, TX 77555-1069, USA

Introduction

Several different nerve injury models were introduced in the last decades of the 20th century. These include the chronic constriction injury (CCI) or "Bennett" model in which the entire sciatic nerve is loosely ligated with chromic gut suture [1], the partial nerve ligation (PNL) or "Seltzer" model in which 1/3–1/2 of the axons in the sciatic nerve are tightly ligated [2] and the spinal nerve ligation (SNL) or "Chung" model in which the L5 and L6 spinal nerves are tightly ligated [3]. The major advantage these models have over the more commonly used simple sciatic nerve transection is that, unlike sciatic transection, some fibers are still in contact with the periphery within the sciatic distribution. Thus, behavioral testing can be performed in this region of the nerve injured limb to allow assessments of changes in mechanical and thermal sensitivity. The ability to correlate changes in dorsal horn morphology and neurochemistry with behavioral status led to a multitude of studies attempting to obtain insights into changes in spinal cord neurocircuitry that underlie the observed nociceptive behaviors. Although other nerve-injury models exist, the central reorganization of primary afferent fibers [4] or the pain behaviors [5] have not yet been well characterized.

In 1991, an excellent monograph was written by Snow and Wilson, 1991 [6] in which the morphological and neurochemical plasticity occurring in the dorsal horn following peripheral nerve transection was reviewed up to 1990. Therefore, the focus of the present chapter is to review the literature which has been assembled in the past 10 years and the reader is referred to this former work for an in depth analysis of work prior to 1990.

Morphological reorganization

Sprouting of myelinated primary afferent fibers

It is generally accepted that some of the major changes that occur following transection of all or part of the sciatic nerve include transganglionic degeneration of cen-

Mechanisms and Mediators of Neuropathic Pain, edited by Annika B. Malmberg and Sandra R. Chaplan

tral terminals [7, 8], and sprouting of myelinated afferents [9]. Unfortunately evidence supporting the sprouting of unmyelinated afferents is more tenuous. That structural reorganization occurs in the dorsal horn following peripheral nerve injury is quite surprising given the long standing dogma of a "hard-wired" system that tolerated little change other than loss of terminals due to cell death. Evidence that central primary afferents reorganize, however, suggested that this might contribute to the sensory abnormalities that occur following peripheral nerve injury.

Low threshold mechanoreceptors (large myelinated afferents) normally distribute to laminae III and IV [10] while high threshold nociceptors distribute to laminae I, II and V [11–12]. The presence of vacant synaptic sites, resulting either from transganglionic degeneration or axonal withdrawal, in concert with the upregulation of GAP-43 expression [13], a marker associated with axonal growth, in dorsal root ganglion (DRG) cells and afferent fibers in the dorsal horn provided a basis for investigating axonal sprouting in the spinal cord. Evidence for such sprouting came from Woolf and collaborators who demonstrated that following either sciatic or sural nerve section, cholera toxin conjugated to horseradish peroxidase (CT-HRP), which was transported by large myelinated afferents, was distributed throughout laminae I–V by 2 weeks post-injury [9, 14]. This contrasted with the pattern seen in normal animals in which the CT-HRP was confined to lamina I, and lamina III and below, leaving lamina II vacant. Intraxonal filling of Aβ fibers confirmed that in nerve injured animals, myelinated axons sprouted into laminae IIinner (IIi), IIouter (IIo), and sometimes as far dorsally as lamina I. There was a shift not only in the dorsal-ventral arborization of these fibers but also a shift in the rostro-caudal and medial-lateral organizations [9, 15]. Sprouting of Aβ fibers into lamina II was confirmed in the SNL [16] (Figs. 1 and 2) and CCI [17] models of neuropathy. Admittedly, some controversy exists concerning the validity of Aβ sprouting because it has been reported that CT-HRP can be transported by both large and small DRG cells following peripheral nerve injury [18]. Thus, transganglionic transport of CT-HRP by small DRG cells could account for the reaction product that appears in laminae I and II in nerve injured rats. However, Woolf et al. (1995) [14] demonstrated that there was no significant difference in the percentage of labeled myelinated or unmyelinated axons in dorsal roots when comparing nerve injured to normal. Also, intra-axonal filling of single Aβ fibers confirms that their terminal distribution is more widespread in nerve injured compared to normal animals [15]. Finally, electrophysiological studies demonstrate functional synaptic contacts between low threshold mechanoreceptors and neurons in lamina II that normally receive only C fiber input [19]. Thus, the evidence strongly suggests that Aβ fibers do sprout into superficial laminae in nerve injured animals.

Numerous investigators suggest that the presence of myelinated sensory axons in laminae where normally only unmyelinated sensory axons terminate might account for the abnormal processing of sensory information, particularly the touch-evoked allodynia that often arises following nerve injury. Thus a critical question is whether

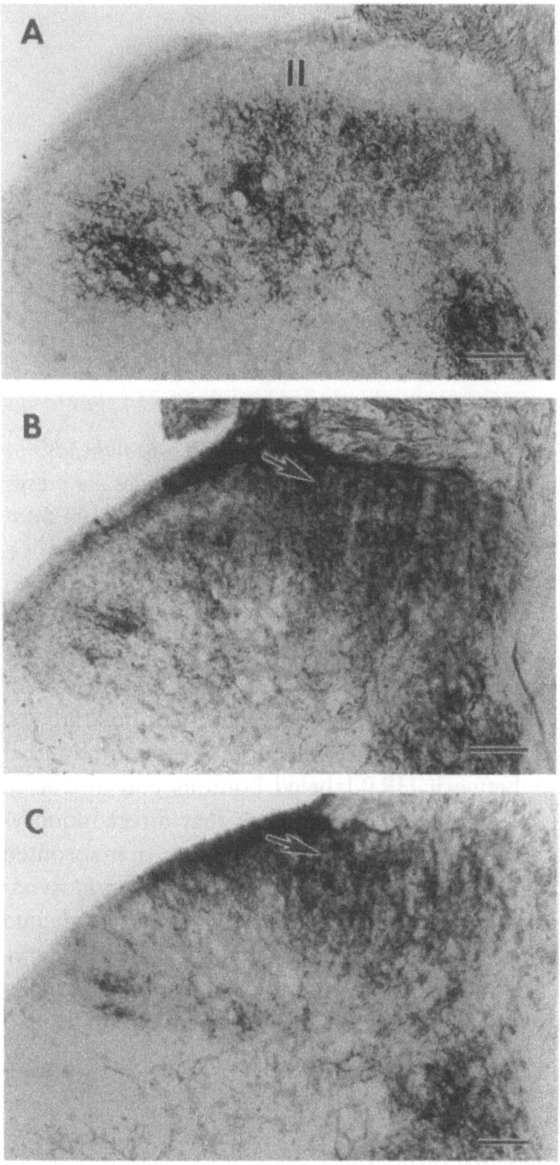

Figure 1
Light micrograph illustrating the extent of CT-HRP labeling in the dorsal horn. Demonstration of the labeling pattern in normal animals where lamina II (II) is clearly devoid of CT-HRP labeling (A). In animals with SNL, CT-HRP label is present in lamina II (arrows) in the ipsilateral dorsal horn at 2 weeks (B) and 4 weeks (C) post-nerve injury. Bar = 100 μm (taken from [16]).

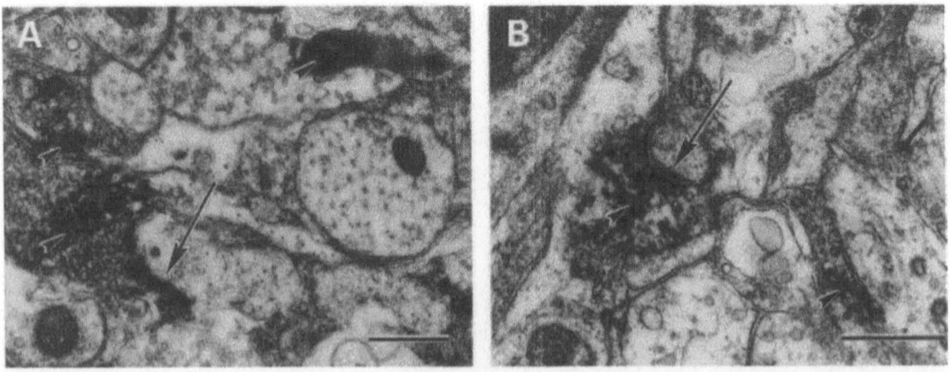

Figure 2
(A) and (B) Electron micrographs illustrating CT-HRP reaction product (arrowheads) in labeled profiles from lamina II in animals with SNL. Some of the labeled profiles are presynaptic terminals with obvious postsynaptic densities (arrows). Bar = 0.5 μm (taken from [16]).

these sprouted afferents make functional contacts in these new territories. Studies where sprouted fibers were intra-axonally filled with HRP [20, 21] or bulk labeled with CT-HRP [14, 16] demonstrate extensive collateralization in laminae I and II and labeled synaptic specializations between HRP-labeled boutons and unlabeled dendrites in laminae I, IIo and IIi, providing strong evidence that direct monosynaptic connections are established (Fig. 2). Another line of evidence that sprouted fibers make functional contacts is derived from demonstrations that the monosynapatic excitatory postsynaptic potentials (EPSPs) in lamina II neurons can be elicited only by high threshold nerve stimulation in normals but also by low threshold nerve stimulation following nerve injury. In the latter case, these EPSPs display properties similar to low-threshold EPSPs caused by large afferent activation in lamina III neurons in control slices [19].

Sprouting of unmyelinated primary afferent fibers

There is little evidence from bulk labeling with wheatgerm agglutinin-horseradish peroxidase (WGA-HRP) that unmyelinated C fibers sprout following peripheral nerve injury [17, 22]. It has been demonstrated, however, that lectin markers such as soybean agglutinin (SBA) and RL29, that presumably label only small diameter DRG cells and their processes, show an expansion of labeled territories in the superficial dorsal horn (Fig. 3) [23, 24], suggesting that unmyelinated afferents sprout into lamina IIi. However, transganglionic tracing studies where SBA or SBA-HRP is injected into the transected sciatic nerve demonstrates that SBA labeling is confined only to laminae I and II [24], providing evidence that unmyelinated axons do not sprout into deeper laminae after axotomy. It is possible that sprouts do not have the ability to transport SBA. However, another possible explanation for the expansion of SBA-labeled territory after nerve injury is that other components in the neuropil e.g. neurons and glia, become labeled by these lectin markers as a result of the injury.

Loss of spinal neurons

Dark "pyknotic" neurons have been described in the dorsal horn following CCI [25, 26] and TUNEL-labeled neurons are seen following sciatic nerve transection in adult animals [27]. It has been suggested that these cells are damaged and perhaps dying due to excess release of glutamate in the cord and the resulting excitotoxicity. Furthermore, it has been hypothesized that these dying cells are inhibitory interneurons and if so, interruption or loss of these neurons should lead to disinhibition in the dorsal horn. This disinhibition may contribute to the abnormal sensory processing and abnormal pain sensations that arise following peripheral nerve injury. However, stereologic cell counts indicate that no cell loss occurs in the dorsal horn following sciatic transection unless stimulation of the transected nerve at A fiber strength follows the transection (Coggeshall, personal communication). Further work is necessary to resolve these discrepancies, but the tentative conclusion is that the above changes are not lethal and so do not progress to actual death of the cells.

Chemical plasticity

Fos expression

C-fos is an immediate early gene whose product, Fos, appears robustly but transiently following nociceptive stimuli. The stimulation-induced Fos appears in nuclei of spinal neurons that are located where nociceptive somatic sensory integration takes place [28]. As regards the expression of Fos in spinal neurons after nerve injury, early studies focused on the fact that nerve injury does lead to Fos expression in dorsal horn neurons [29–32], and that this was not simply a matter of neuronal activity since DRG cells which were also active did not label for Fos [28]. Interestingly when the Fos labeling arose from activity of afferent fibers in a neuroma or from electrical stimulation of fine afferent fibers in a transected nerve, the labeling was very prominent in laminae I and IIo, whereas when a partially denervated limb

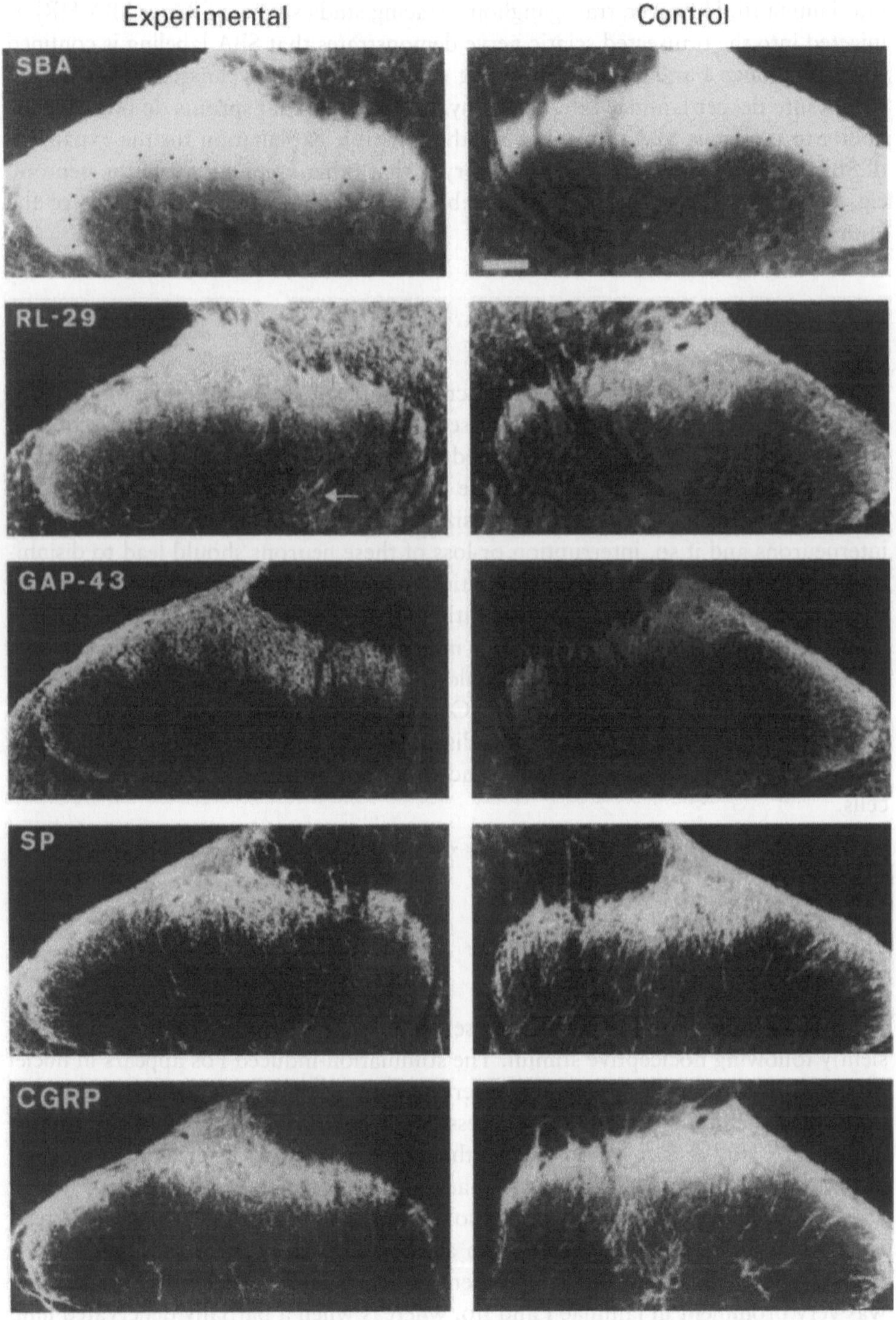
Experimental
Control
SBA
RL-29
GAP-43
SP
CGRP

was stimulated by noxious heat the labeling was in lamina III and below [29]. Notwithstanding differences in laminar localizations, a particular point is that in contrast to brief nociceptive stimuli, the Fos labeling persists for a much longer time following nerve injury, a finding presumably related to ongoing discharge of primary afferent neurons following transection of their peripheral processes [31].

Fos labeling after peripheral nerve injury has been used as evidence of central reorganization and sprouting of injured primary afferent fibers. For example, stimulation of a normal sciatic nerve at a strength that activates only large primary afferents leads to very little Fos expression in the spinal cord [33]. By contrast, stimulating the central stump of a transected sciatic nerve at the same intensity leads to considerable Fos expression in laminae I–V [33]. The Fos expression in laminae III–V presumably indicates an increased responsiveness of spinal neurons to this type of primary afferent stimulation but does not indicate sprouting since these laminae are where these fibers normally terminate. By contrast, the appearance of Fos expressing cells in laminae I–II after injury to a nerve presumably indicates sprouting of primary afferent fibers since these large fibers do not normally end there [9].

Collateral sprouting, which is when new axons and terminals are generated by uninjured fibers in response to denervation of neighboring areas, has also been examined by the Fos technique. Evidence compatible with the presence of collateral sprouting has been obtained by showing that noxious thermal stimuli of the hindpaw or electrical stimulation of the saphenous nerve in animals with a sciatic transection leads to an increase in the numbers of Fos expressing neurons and an expansion of the area containing Fos expressing neurons into part of the sciatic nerve territory [29, 30]. Similarly, Fos neurons appear within an expanded area when the peroneal nerve is stimulated after the neighboring tibial nerve is transected [30, 34, 35]. One reasonable explanation for the appearance of Fos neurons in regions where they are not normally observed is that axons from the non-damaged neighboring nerve have sprouted into the denervated region.

The advantage of the Fos labeling in these types of studies is that it can indicate a "global" response to denervation and thus provide a relatively simple and powerful assay to indicate reorganization of primary afferent fibers following peripheral nerve injury. The limitation of this method is that the role or function of Fos in nociception is not known, but strides are being made towards a better molecular understanding of this immediate early gene in the metabolism of the neurons. As this understanding

Figure 3
Photomicrographs showing labeling in the L4 segment in one CCI animal for each of 5 histochemical markers in the experimental and control dorsal horns. The data indicate that at 28 days post-surgery, there is an increase in the ipsilateral dorsal horn staining density for SBA, RL-29 and GAP-43 and a decrease in SP and CGRP. The boundary between laminae II and III is indicated by the dotted line in the top panel. Bar = 100 μm (taken from [23]).

progresses, Fos localizations will become ever more meaningful in providing insights into the central plastic changes that follow peripheral nerve lesions.

Changes in neurotransmitters

The DRG cells whose peripheral axons are damaged after nerve injury undergo changes in neurotransmitter expression which are reflected not only in the cell body but also in their central and peripheral terminals. Here we focus on the central changes, and both decreases and increases in neurotransmitters are observed in the spinal cord following nerve lesions. Explanations for the disappearance of transmitters from central terminals of primary afferents include (1) depletion/release, (2) dying back of the central process [8] and (3) phenotypic switch in the parent DRG neuron. Evidence can be found supporting each of these phenomena.

Peptides

Jessell et al. (1979) [36] was the first to demonstrate that peripheral axotomy resulted in a decrease in substance P (SP) levels in the dorsal horn. Now, it is generally agreed that in the rat and monkey, injury to peripheral axons because of sciatic transection, SNL, CCI or locally administered neurotoxins results in a down regulation, and is seen as a decrease in immunostaining density or binding for SP and/or calcitonin gene-related peptide (CGRP) (Fig. 3) [23, 25, 37–41] (see, however, [38, 42]) and somatostatin [43]. These changes are attributed to primary afferent sources since peptide staining and mRNA levels also decrease in DRG cells [44–47]. In some studies, the decreases in the dorsal horn were observed as early as 3 days post-injury.

In relation to SP, Marchand et al. (1994) [47] has reported that a population of large diameter DRG cells begin to express this peptide in the CCI model. In a subsequent study it was demonstrated that immunoreactivity for SP increases in DRG neurons that are not axotomized following nerve injury [48]. Thus, depending on cell size and/or the health of the DRG peripheral process, SP expression may increase or decrease following peripheral nerve injury. It has also been demonstrated that a subpopulation of injured DRG cells show increased expression of CGRP [49]. It appears that medium and large diameter DRG cells begin to express mRNA for CGRP and immunohistochemical studies reveal an increase in CGRP staining density in laminae III and IV in the dorsal horn in the CCI. The increased expression of these peptides may contribute to the increased spinal neuronal sensitivity that occurs following nerve injury. Cholecytokinin (CCK) immunostaining is moderately reduced following sciatic transection [50].

Peptides that are up-regulated and thus show an increase in immunostaining density in the spinal cord include galanin (GAL), vasoactive intestinal peptide/peptide histidine isoleucine (VIP/PHI), and neuropeptide Y (NPY) [37, 51–53]. These changes are mainly attributed to primary afferents because peptide and mRNA levels increase in DRG cells after peripheral nerve damage [44, 53]. However, it has also been reported that neither the number of GAL-labeled synapses nor the number or diameter of the dense core vesicles in each GAL-labeled terminal changes after SNL in the primate. Instead, there is a significant increase in the number of GAL-labeled glial cell bodies and processes and this presumably accounts for the increase in staining density on the nerve injured dorsal horn [37]. The authors conclude that the glial processes are "soaking up" the excess GAL that is being released from the primary afferent terminals. However, it has also been reported in the rat that the percentage of GAL-positive large dense core vesicles increases [51] so it is not clear if there is a species difference in the way the galanin system responds to peripheral axotomy. In another study, sciatic transection in the guinea pig results in only a moderate increase in GAL providing further evidence that species differences may exist [54].

Dynorphin (DYN) is present in intrinsic dorsal horn neurons [55] and it and its precursor pre-prodynorphin are up-regulated on the nerve injured side [56–59] particularly in laminae V and VI [25]. A bilateral DYN increase is also observed following ipsilateral cryolesions of peripheral nerves [57]. Another opiate peptide, met-enkephalin (ENK), is also increased in the spinal cord in the CCI model [41]. Evidence that DYN and ENK are involved in neuropathic pain is that increased DYN and ENK staining is present only in animals that demonstrate neuropathic pain behaviors [41, 57, 58]. It has been suggested that increases in spinal DYN outside of the segments where injured primary afferents enter the cord may underlie the development of extraterritorial neuropathic pain [58].

Amino acids

Immunostaining density for glutamate, a major excitatory amino acid, increases in the dorsal horn in the CCI model [60]. In contrast, sciatic nerve transection results in a decrease in the inhibitory transmitter GABA, with a reduction of GABA-immunostained cells in laminae I-III beginning two weeks post-surgery and a reduction in GABA concentration in spinal homogenates [61]. It is hypothesized that the loss of GABA in the dorsal horn could be related to a decrease in primary afferent input. This loss of GABAergic neurons is confirmed in CCI [62, 63] (see, however, [41]), and PNL [64] models. The loss can be reversed by spinal cord transplants of 5HT synthesizing cells [63]. Following biochemical analysis of dorsal horn quadrants in the CCI model, however, an increase in GABA levels is reported [65]. This discrepancy may be due to the differences in techniques since the biochemical analyses sampled both gray and white matter in the cord. Biochemical analysis of dorsal horn quadrants in the CCI model demonstrate an increase in the ipsilateral compared to the contralateral side in another important inhibitory amino acid transmitter, glycine [65]. Intrathecal strychnine enhances the magnitude of the thermal hyperalgesia on the nerve injured hindpaw but not the normal hindpaw suggesting that glycinergic neurons are activated after nerve injury and could modify the transmission of nociceptive input in the cord.

Biogenic amines

Serotonin and noradrenaline are increased in the ipsilateral quadrant of the dorsal horn compared to the contralateral horn in the CCI model [65], suggesting that descending monoaminergic systems possibly play a role in neuropathic pain.

Enzymes

Changes in enzymes in the dorsal horn have been reported and include a decrease in brain nitric oxide synthase (NOS)-labeled cells [66] and an increase in protein kinase Cγ (PKCγ) [67] in the dorsal horn following CCI or tight ligation of the sciatic nerve.

Changes in receptors

Peptides

Numerous studies investigated changes in opioid receptors in the dorsal horn following peripheral nerve injury. Since opioids are relatively ineffective in treating neuropathic pain, it was hoped that understanding the fate of the receptors would elucidate possible underlying mechanisms. The majority of the studies indicate that there is a decrease in binding or immunostaining for the receptors following various nerve injuries [50, 66, 68], however, increases [66, 69] and essentially no change [70] have also been reported.

Sciatic nerve injury results in an up-regulation of the SP receptor NK1 [66, 71] which is consistent with the increase in SP binding that occurs in the dorsal horn [72, 73]. However, this up-regulation is observed not only inside but also outside the normal distribution of nociceptive primary afferents from the sciatic nerve, i.e. in the L1 and S1 spinal segments. The suggestion is that this morphological change may contribute to the secondary hyperalgesia that arises in some nerve injury models. In normal rats, NPY Y1 receptor mRNA is localized in mainly small DRG cells and complicated changes in mRNA levels occur in the DRG following sciatic transection, but no marked changes have been reported in the spinal cord dorsal horn [74]. The NPY Y2 receptor mRNA is localized in medium and large DRG cells and sciatic transection results in an up-regulation of mRNA in DRG cells [75], however, it is unknown if changes occur in the dorsal horn. The CCK_B receptor mRNA expression is very low in normal rat DRG and there is a marked up-regulation in mRNA following sciatic transection [76, 77] and an increase in CCK_B receptor binding in the superficial dorsal horn [76]. The authors suggest that the up-regulation of CCK_B in the dorsal horn is on primary afferent terminals and that this would result in an increased sensitivity of primary afferent terminals to CCK released locally in the cord, possibly reducing the antinociceptive effects of opioids [76].

Amino acids

Changes in ionotropic glutamate receptors occur in the CCI model since there is a marked reduction in the binding of a competitive NMDA receptor antagonist and a decrease in NMDA receptor immunostaining [78]. These decreases are prevented by transplantation of adrenal medullary tissue into the spinal subarachnoid space. It is suggested that transplants interrupt the excitotoxic cascade that is initiated by overactivation of NMDA receptors. In contrast to these findings, a later study reports a bilateral increase in NMDA receptor binding density in the dorsal horn following sciatic nerve crush [73], so there is controversy concerning the fate of NMDA receptors following nerve injury. There is a reported increase in immunostaining density for subunits of the non-NMDA receptor AMPA in the CCI [79] and sciatic transection [80] models. Active zones of glomerular terminals that arise from small diameter primary afferents show enhanced label for the AMPA subunits 2/3 suggesting that AMPA receptors contribute to the increased sensitivity of dorsal horn neurons following peripheral nerve injury. In contrast, a decrease in AMPA receptors is reported following pronase-induced deafferentation [81], but pronase produces a more immediate death of DRG cells as compared to axotomy which could account for the decrease in AMPA receptor staining.

Nerve injury results in a decrease of $GABA_B$ receptor binding (29–33% decrease compared to the contralateral side) but an upregulation of $GABA_A$ binding in lamina II (20–27% increase compared to the contralateral side), 2–4 weeks post-injury [82]. The loss of $GABA_B$ receptors is attributed to disruption/degeneration of the central terminals of primary afferent containing $GABA_B$ receptors [83]. Although $GABA_A$ binding sites are also present on primary afferent terminals, post-synaptic binding sites outnumber presynaptic sites [84], and thus this increase is likely to be associated with non-primary afferent elements in the dorsal horn. Furthermore, the time course of the reduction of GABA-immunostained cells after sciatic transection [61] parallels the up-regulation of $GABA_A$ receptors.

Other

An investigation of α-adrenergic receptors in the dorsal horn following sciatic transection, CCI and SNL demonstrates a decrease in α_{2A} adrenergic receptor immunostaining in all 3 models, but α_{2C} receptor immunoreactivity is increased in only the SNL model [85]. Thus, the authors conclude that neither receptor subtype is likely to play a role in the adrenergic sensitivity that develops after nerve injury. An investigation of $P2X_3$ purinoceptors in the dorsal horn demonstrates that immunostaining density for these receptors increases in both the DRG and the superficial dorsal horn in the CCI, and this is consistent with an up-regulation of the $P2X_3$ receptors in presynaptic terminals of primary afferents [86].

Dorsal horn changes in relation to neuropathic pain

One goal of the present review is to understand the relationship between morphologic changes occurring in the spinal cord and neuropathic pain. Although major insights have been made concerning the sequelae of dorsal horn events following nerve injury, it has remained difficult to find correlations between the anatomical and behavioral changes. For example, it has been suggested that the sprouting of Aβ fibers into inappropriate laminae may underlie the mechanical allodynia which develops following nerve injury. The allodynia, however, can be observed within hours after the injury [3]. It is not likely that myelinated fibers sprout and form functional connections in this short time frame. Therefore, if Aβ sprouting is involved in abnormal sensations, it most likely plays a role in the maintenance and not the induction of mechanical allodynia. In regard to peptides, several extensive time course studies using CCI or sciatic transection have found that peptide mRNA levels in the DRG or peptide staining densities (SP, CGRP, GAL, VIP, NPY) in the dorsal horn have little or no correlation with pain behaviors [42, 44, 87]. Another study compares peptide changes that occur in a nerve injury model associated with neuropathic pain (CCI) with changes that occur in a model that does not develop neuropathic pain (crush injury) [41]. The authors conclude that changes in SP and CGRP are not specific to pain *per se* since decreases are observed in both models. However, an increase in met-enkephalin may be specific for neuropathic pain since there is an increase only in the CCI model and this increase correlates with the resolution of the pain behaviors. Much less information is available concerning receptor changes in the dorsal horn over time, but, peak hyperalgesia in the CCI model parallels the peak increase in dorsal horn NK1 staining density [66]. Furthermore, μ opiate receptor staining density increases as the behavior abates in the CCI model [66]. Thus, with time course studies, we can gain further insight into the significance of the morphological changes observed in the spinal cord dorsal horn and with this knowledge we can begin to design logical therapies and pharmaceuticals that target the malleable features of the dorsal horn following peripheral nerve injury.

Acknowledgements
The authors would like to thank Vicki Wilson for her excellent secretarial help. This work was supported by NS11255, NS27910 and NS40700 to SMC and NS11255 and NS10161 to REC.

Reference list

1 Bennett GJ, Xie Y-K (1988) A peripheral mononeuropathy in rat that produces disorders of pain sensation like those seen in man. *Pain* 33: 87–107

2 Seltzer Z, Dubner R, Shir Y (1990) A novel behavioral model of neuropathic pain disorders produced in rats by partial sciatic nerve injury. *Pain* 43: 205–218

3 Kim SH, Chung JM (1992) An experimental model for peripheral neuropathy produced by segmental spinal nerve ligation in the rat. *Pain* 50: 355–363

4 Decosterd I, Woolf CJ (2000) Spared nerve injury: an animal model of persistent peripheral neuropathic pain. *Pain* 187: 149–158

5 LaMotte CC, Kapadia SE (1987) Deafferentation-induced alterations in the rat dorsal horn: II. Effects of selective poisoning by pronase of the central processes of a peripheral nerve. *J Comp Neurol* 266: 198–208

6 Snow PJ, Wilson P (1991) Plasticity in the somatosensory system of developing and mature mammals – the effects of injury to the central and peripheral nervous system. In: Autrum H, Ottoson D, Perl ER, Schmidt RF, Shimazu H, Willis WD (eds): *Progress in sensory physiology.* Springer-Verlag, Berlin/Heidelberg, 1–482

7 Knyihar-Csillik E, Csillik B (1981) Selective "labelling" by transsynaptic degeneration of substantia gelatinosal cells: an attempt to decipher intrinsic wiring in the rolando substance of primates. *Neurosci Lett* 23: 131–136

8 Liss AG, Wiberg M (1997) Loss of nerve endings in the spinal dorsal horn after a peripheral nerve injury. An anatomical study in Macaca fascicularis monkeys. *Eur J Neurosci* 9: 2187–2192

9 Woolf CJ, Shortland P, Coggeshall RE (1992) Peripheral nerve injury triggers central sprouting of myelinated afferents. *Nature* 355: 75–78

10 Brown AG (1981) *Organization in the spinal cord: the anatomy and physiology of identified neurones.* Springer-Verlag, Heidelberg & New York

11 Sugiura Y, Lee CL, Perl ER (1986) Central projections of identified, unmyelinated (C) afferent fibers innervating mammalian skin. *Science* 234: 358–361

12 Light AR, Perl ER (1979) Spinal termination of functionally identified primary afferent neurons with slowly conducting myelinated fibres. *J Comp Neurol* 186: 133–150

13 Woolf CJ, Reynolds ML, Molander C, O'Brien C, Lindsay RM, Benowitz LI (1990) The growth-associated protein GAP-43 appears in dorsal root ganglion cells and in the dorsal horn of the rat spinal cord following peripheral nerve injury. *Neurosci* 34(2): 465–478

14 Woolf CJ, Shortland P, Reynolds ML, Ridings T, Doubell TP, Coggeshall RE (1995) Reorganization of central terminals of myelinated primary afferents in the rat dorsal horn following peripheral axotomy. *J Comp Neurol* 360: 121–134

15 Shortland P, Woolf CJ (1993) Chronic peripheral nerve section results in a rearrangement of the central axonal arborizations of axotomized A beta primary afferent neurons in the rat spinal cord. *J Comp Neurol* 330: 65–82

16 Lekan HA, Carlton SM, Coggeshall RE (1996) Sprouting of Aβ fibers into lamina II of the rat dorsal horn in peripheral neuropathy. *Neurosci Lett* 208: 147–150
17 Nakamura S, Myers RR (1999) Myelinated afferents sprout into lamina II of L3-5 dorsal horn following chronic constriction nerve injury in rats. *Brain Res* 818: 285–290
18 Tong Y-G, Wang HF, Ju G, Grant G, Hökfelt T, Zhang X (1999) Increased uptake and transport of cholera toxin B-subunit in dorsal root ganglion neurons after peripheral axotomy: possible implications for sensory sprouting. *J Comp Neurol* 404: 143–158
19 Kohama I, Ishikawa K, Kocsis J (2000) Synaptic reorganization in the substania gelatinosa after peripheral nerve neuroma formation: abberrant innervation of lamina II neurons by Aβ afferents. *J Neurosci* 20: 1538–1549
20 Koerber HR, Mirnics K, Brown PB, Mendell LM (1994) Central sprouting and functional plasticity of regenerated primary afferents. *J Neurosci* 14: 3655–3671
21 Koerber HR, Mirnics K, Kavookjian A.M., Light AR (1999) Ultrastructural analysis of ectopic synaptic boutons arising from peripherally regenerated primary afferent fibers. *J Neurophys* 81: 1636–1644
22 Sugimoto T, Gobel S (1982) Primary sensory neurons maintain their central arbors in the spinal dorsal horn following peripheral nerve injury: An anatomical analysis using HRP transganglionic transport. *Brain Res* 248: 377–387
23 Cameron AA, Cliffer KD, Dougherty PM, Willis WD, Carlton SM (1991) Changes in lectin, GAP-43 and neuropeptide staining in the rat superficial dorsal horn following experimental peripheral neuropathy. *Neurosci Lett* 131: 249–252
24 Shortland P, Wang HF, Molander C (1999) Distribution of transganglionically labelled soybean agglutinin primary afferent fibres after nerve injury. *Brain Res* 815: 206–212
25 Bennett GJ, Kajander KC, Sahara Y et al (1989) Neurochemical and anatomical changes in the dorsal horn of rats with an experimental painful peripheral neuropathy. In: Cervero F, Bennett GJ, Headley PM (eds): *Processing of sensory information in the superficial dorsal horn of the spinal cord.* Plenum Press, New York, 463–471
26 Sugimoto T, Bennett GJ, Kajander KC (1990) Transsynaptic degeneration in the superficial dorsal horn after sciatic nerve injury: effects of a chronic constriction injury, transection, and strychnine. *Pain* 42: 205–213
27 Azuke JJ, Zimmermann M, Hsieh T-F, Herdegen T (1998) Peripheral nerve insult induces NMDA receptor-mediated, delayed degeneration in spinal neurons. *Eur J Neurosci* 10: 2204–2206
28 Hunt SP, Pini A, Evan G (1987) Induction of c-fos-like protein in spinal cord neurones following sensory stimulation. *Nature* 328: 632–634
29 Williams S, Evan G, Hunt SP (1991) C-fos induction in the spinal cord after peripheral nerve lesion. *Eur J Neurosci* 3: 887–894
30 Molander C, Hongpaisan J, Shortland P (1998) Somatotopic redistribution of C-fos expressing neurons in the superficial dorsal horn after peripheral nerve injury. *Neurosci* 84: 241–253
31 Chi S-I, Levine JD, Basbaum AI (1993) Peripheral and central contributions to the per-

sistent expression of spinal cord fos-like immunoreactivity produced by sciatic nerve transection in the rat. *Brain Res* 617: 225–237

32 Parpura V, Lui F, Jeftinija K, Haydon PG, Jeftinija S (1995) Neuroligand evoked calcium-dependent release of excitatory amino acids from Schwann Cells. *J Neurosci* 15: 5831–5839

33 Molander C, Hongpaisan J, Persson JKE (1994) Distribution of c-fos expressing dorsal horn neurons after electrical stimulation of low threshold sensory fibers in the chronically injured sciatic nerve. *Brain Res* 644: 74–82

34 Sugimoto T, Yoshida A, Nishijima K, Ichikawa H (1994) c-Fos induction in the rat spinal dorsal horn partially deafferented by dorsal rhizotomy. *Neurosci Lett* 178: 239–242

35 Sugimoto T, Yoshida A, Nishijima K, Ichikawa H (1995) Disruption and restoration of dorsal horn sensory map after peripheral nerve crush and regeneration. *Brain* 63: 49–54

36 Jessell T, Tsunoo A, Kanazaw I, Otsuka M (1979) Substance P: depletion in the dorsal horn of rat spinal cord after section of the peripheral processes of primary sensory neurons. *Brain Res* 168: 247–259

37 Carlton SM, Coggeshall RE (1996) Stereological analysis of galanin and CGRP synapses in the dorsal horn of neuropathic primates. *Brain Res* 711: 16–25

38 Garrison CJ, Dougherty PM, Carlton SM (1993) Quantitative analysis of substance P and calcitonin gene-related peptide immunohistochemical staining in the dorsal horn of neuropathic MK-801-treated rats. *Brain Res* 607: 205–214

39 El-Bohy A, LaMotte CC (1993) Deafferentation-induced changes in neuropeptides of the adult rat dorsal horn following pronase injection of the sciatic nerve. *J Comp Neurol* 336: 545–554

40 Kajander K, Xu J (1995) Quantitative evaluation of calcitonin gene-related peptide and substance P levels in rat spinal cord following peripheral nerve injury. *Neurosci Lett* 186: 184–188

41 Sommer C, Myers RR (1995) Neurotransmitters in the spinal cord dorsal horn in a model of painful neuropathy and in nerve crush. *Acta Neuropathol* 90: 478–485

42 Cameron AA, Cliffer KD, Dougherty PM, Garrison CJ, Willis WD, Carlton SM (1997) Time course of degenerative and regenerative changes in the dorsal horn in a rat model of peripheral neuropathy. *J Comp Neurol* 379: 428–442

43 Hökfelt T, Zhang X, Wiesenfeld-Hallin Z (1994) Messenger plasticity in primary sensory neurons following axotomy and its functional implications. *TINS* 17: 22–30

44 Nahin RL, Ren K, De Leon M, Ruda M (1994) Primary sensory neurons exhibit altered gene expression in a rat model of neuropathic pain. *Pain* 58: 95–108

45 Nielsch V, Bisby MA, Keen P (1987) Effect of cutting or crushing the rat sciatic nerve on synthesis of substance P by isolated L5 dorsal root ganglia. *Neuropeptides* 10: 137–145

46 Noguchi K, Senba E, Morita Y, Sato M, Tohyama M (1990) α-CGRP and β-CGRP mRNAs are differentially regulated in the rat spinal cord and dorsal root ganglion. Mol *Brain Res* 7: 299–304

47 Marchand JE, Wurm WH, Kato T, Kream RM (1994) Altered tachykinin expression by dorsal root ganglion neurons in a rat model of neuropathic pain. *Pain* 58: 219–231

48 Ma W, Bisby MA (1998) Increase of preprotachykinin mRNA and substance P immunoreactivity in spared dorsal root ganglion neurons following partial sciatic nerve injury. *Eur J Neurosci* 10: 2388–2399

49 Miki K, Fukuoka T, Tokunaga A, Noguchi K (1998) Calcitonin gene-related peptide increase in the rat spinal dorsal horn and dorsal column nucleus following peripheral nerve injury: up-regulation in a subpopulation of primary afferent sensory neurons. *Neurosci* 82: 1243–1252

50 Zhang X, de Araujo Lucas G, Elde R, Wiesenfeld-Hallin Z, Hökfelt T (1999) Effect of morpine on cholecystokinin and μ-opioid receptor-like immunoreactivities in rat spinal dorsal horn neurons after peripheral axotomy and inflammation. *Neurosci* 95: 197–207

51 Zhang X, Bean AJ, Wiesenfeld-Hallin Z, Xu X-J, Hökfelt T (1995) Ultrastructural studies on peptides in the dorsal horn of the rat spinal cord – III. Effects of peripheral axotomy with special reference to galanin. *Neurosci* 64: 893–915

52 Zhang X, Bean AJ, Wiesenfeld-Hallin Z, Hökfelt T (1994) Ultrastructural studies on peptpides in the dorsal horn of the rat spinal cord – IV. Effects of peripheral axotomy with special reference to NPY and VIP/NPY. *Neurosci* 64: 917–941

53 Ma W, Bisby MA (1998) Partial and complete sciatic nerve injuries induce similar increases of neuropeptide Y and vasoactive intestinal peptide immunoreactivities in primary sensory neurons and their central projections. *Neurosci* 86: 1217–1234

54 Rydh-Rinder M, Holmberg K, Elfvin L-G, Wiesenfeld-Hallin Z, Hökfelt T (1996) Effects of peripheral axotomy on neuropeptides and nitric oxide synthase in dorsal root ganglia and spinal cord of the guinea pig: an immunohistochemical study. *Brain Res* 707: 180–188

55 Cruz L, Basbaum AI (1985) Multiple opioid peptides and the modulation of pain: immunohistochemical analysis of dynorphin and enkephalin in the trigeminal nucleus caudalis and spinal cord of the cat. *J Comp Neurol* 240: 331–348

56 Kajander KC, Sahara Y, Iadarloa MJ, Bennett GJ (1990) Dynorphin increases in the dorsal spinal cord in rats with a painful peripheral neuropathy. *Peptides* 11: 719–728

57 Wagner R, DeLeo JA, Coombs DW, Willenbring S, Fromm C (1993) Spinal dynorphin immunoreactivity increases bilaterally in a neuropathic pain model. *Brain Res* 629: 323–326

58 Malan TP, Ossipov MH, Gardell LR, Ibrahim M, Bian D, Lai J, Porreca F (2000) Extraterritorial neuropathic pain correlates with multisegmental elevation of spinal dynorphin in nerve-injured rats. *Pain* 86: 185–194

59 Draisci G, Kajander K, Dubner R, Bennett GJ, Iadarloa MJ (1991) Up-regulation of opioid gene expression in spinal cord evoked by experimental nerve injuries and inflammation. *Brain Res* 560: 186–192

60 Al-Ghoul WM, Li Volsi G, Weinberg RJ, Rustioni A (1993) Glutamate immunocytochemistry in the dorsal horn after injury or stimulation of the sciatic nerve of rats. *Brain Res Bull* 30: 453–459

61 Castro-Lopes JM, Tavares I, Coimbra A (1993) GABA decreases in the spinal cord dorsal horn after peripheral neurectomy. *Brain Res* 620: 287–291
62 Ibuki T, Hama AT, Wang X-T, Pappas G, Sagan J (1997) Loss of GABA immunoreactivity in the spinal dorsal horn of rats with peripheral nerve injury and promotion of recovery by adrenal medullary grafts. *Neurosci* 76: 845–858
63 Eaton MJ, Plunkett JA, Karmally S, Martinez MA, Montanez K (1998) Changes in GAD- and GABA-immunoreactivity in the spinal dorsal horn after peripheral nerve injury and promotion of recovery by lumbar transplant of immortalized serotonergic precursors. *J Chem Neuroanat* 16: 57–72
64 Ralston DD, Behbehani MM, Sehlhorst SC et al (1997) Decreased GABA immunoreactivity in rat dorsal horn is correlated with pain behavior: A light and electron microscopic study. In: Jensen TS, Turner JA, Wiesenfeld-Hallin Z (eds): *Proceedings of the 8th World Congress on Pain.* IASP Press, Seattle, 547–560
65 Satoh O, Omote K (1996) Roles of monoaminergic, glycinergic and GABAergic inhibitory systems in the spinal cord in rats with peripheral mononeuropathy. *Brain Res* 728: 27–36
66 Goff JR, Burkey AR, Goff DJ, Jasmin L (1998) Reorganization of the spinal dorsal horn in models of chronic pain: correlation with behaviour. *Neurosci* 82: 559–574
67 Mao J, Price DD, Phillips LL, Lu J, Mayer DJ (1995) Increases in protein kinase C gamma immunoreactivity in the spinal cord dorsal horn of rats with painful mononeuropathy. *Neurosci Letts* 198: 75–78
68 deGroot JF, Coggeshall RE, Carlton SM (1997) The reorganization of μ opioid receptors in the rat dorsal horn following peripheral axotomy. *Neurosci Lett* 233: 113–116
69 Stevens CW, Kajander KC, Bennett GJ, Seybold VS (1991) Bilateral and differential changes in spinal mu, delta and kappa opioid binding in rats with a painful, unilateral neuropathy. *Pain* 46: 315–326
70 Porreca F, Tang QB, Bian D, Riedl M, Elde R, Lai J (1998) Spinal opioid mu receptor expression in lumbar spinal cord of rats following nerve injury. *Brain Res* 795: 197–203
71 Abbadie C, Brown JL, Mantyh PW, Basbaum AI (1996) Spinal cord substance P receptor immunoreactivity increases in both inflammatory and nerve injury models of persistent pain. *Neurosci* 70: 201–209
72 Aanonsen LM, Kajander KC, Bennett GJ, Seybold VS (1992) Autoradiographic analysis of ^{125}I-substance P binding in rat spinal cord following chronic constriction of the sciatic nerve. *Brain Res* 596: 259–268
73 Croul S, Radzievsky A, Sverstiuk A, Murray M (1998) NK1, NMDA, 5HT1a, and 5HT2 receptor binding sites in the rat lumbar spinal cord: modulation following sciatic nerve crush. *Exp Neurol* 154: 66–79
74 Zhang X, Wiesenfeld-Hallin Z, Hökfelt T (1994) Effect of peripheral axotomy on expression of neuropeptide Y receptor mRNA in rat lumbar dorsal root ganglia. *Eur J Neurosci* 6: 43–57
75 Zhang X, Shi T, Holmberg K, Huang W, Xiao H, Ju G, Hökfelt T (1997) Expression

and regulation of the neuropeptide Y Y2 receptor in sensory and autonomic ganglia. *Proc Natl Acad USA* 94: 729–734
76 Bras JMA, Laporte A-M, Benoliel JJ, Bourgoin S, Mauborgne A, Hamon M, Cesselin F, Pohl M (1999) Effects of peripheral axotomy on cholecystokinin neurotransmission in the rat spinal cord. *J Neurochem* 72: 858–867
77 Zhang X, Dagerlind A, Elde RP, Castel M-N, Broberger C, Wiesenfeld-Hallin Z, Hökfelt T (1993) Marked increase in cholecystokinin B receptor messenger RNA levels in rat dorsal root ganglia after peripheral axotomy. *Neurosci* 57: 227–233
78 Hama AT, Unnerstall JR, Siegan JB, Sagan J (1995) Modulation of NMDA receptor expression in the rat spinal cord by peripheral nerve injury and adrenal medullary grafting. *Brain Res* 687: 103–113
79 Harris JA, Corsi M, Quartaroli M, Arban R, Bentivoglio M (1996) Upregulation of spinal glutamate receptors in chronic pain. *Neurosci* 74: 7–12
80 Popratiloff A, Weinberg RJ, Rustioni A (1998) AMPA receptors at primary afferent synapses in substantia geletinosa after sciatic nerve section. *Eur J Neurosci* 10: 3220–3230
81 Helgren ME, Arsenault K, Kapadia SE, LaMotte CC (1999) Deafferentation-induced regulation of AMPA receptors in the spinal cord of the adult rat. *Somatosen & Motor Res* 16: 39–48
82 Castro-Lopes JM, Malcangio M, Pan BH, Bowery NG (1995) Complex changes of $GABA_A$ and $GABA_B$ receptor binding in the spinal cord dorsal horn following peripheral inflammation or neurectomy. *Brain Res* 679: 289–297
83 Price GW, Wilkin GP, Turnbull MJ, Bowery NG (1984) Are baclofen-sensitive $GABA_B$ receptors present on primary afferent terminals of the spinal cord? *Nature* 307: 71–74
84 Singer E, Placheta P (1980) Reduction of [^{3}H]muscimol binding sites in rat dorsal spinal cord after neonatal capsaicin treatment. *Brain Res* 202: 484–487
85 Stone LS, Vulchanova L, Riedl MS, Wang J, Williams FG, Wilcox GL, Elde R (1999) Effects of peripheral nerve injury on alpha-2A and alpha-2C adrenergic receptor immunoreactivity in the rat spinal cord. *Neurosci* 93: 1399–1407
86 Novakovic SD, Kassotakis LC, Oglesby IB, Smith JAM, Eglen RM, Ford APDW, Hunter JC (1999) Immunocytochemical localization of P_{2x3} purinoceptors in sensory neurons in naive rats and following neuropathic injury. *Pain* 80: 273–282
87 Munglani R, Harrison SM, Smith GD, Bountra C, Birch PJ, Elliot PJ, Hunt SP (1996) Neuropeptide changes persist in spinal cord despite resolving hyperalgesia in a rat model of mononeuropathy. *Brain Res* 743: 102–108

Neuropathic pain: preclinical and clinical paradigms

Pre-clinical nerve ligation models: behavior and electrophysiology

Jin Mo Chung and Kyungsoon Chung

Marine Biomedical Institute and Department of Anatomy & Neurosciences, University of Texas Medical Branch, 301 University Boulevard, Galveston, TX 77555-1069, USA

Introduction

The effects of nerve injury have been studied for a long time. However, early studies typically examined the electrophysiological and anatomical effects of complete sectioning of the sciatic nerve and it was assumed that the animal was experiencing pain. In 1988, a more meaningful model for neuropathic pain was introduced by Bennett and Xie [1]. Most importantly, the model used by Bennett and Xie [1] allowed for the quantification of behaviors that are likely associated with neuropathic pain. Determining the effect of various manipulations on behavioral endpoints provides a valuable tool to examine the potential mechanisms involved in neuropathic pain.

Since the development of the chronic constriction injury (CCI) model by Bennett and Xie [1], numerous neuropathic pain models have subsequently been introduced. These can be categorized by which nerve is being injured and by how an injury is produced. As for the injury site, the sciatic nerve [1-3], the segmental spinal nerve [4], the dorsal root [5], and distal branches of the sciatic nerve [6], and the tail nerve [7] have been used. These models also differ in the method of injury: producing axotomy in a nerve [2, 4, 6, 7], placing loose ligatures around the nerve [1, 5], or freezing the nerve [3]. All these models involve a clear injury to a peripheral nerve and therefore represent complex regional pain syndrome type II [8]. Recently, a model was introduced which does not produce a clear injury to a peripheral nerve and hence represents complex regional pain syndrome type I [9]. All these models are different in some aspects but presumably share some common mechanisms that produce neuropathic pain behaviors.

In this chapter, we attempted to describe some of the important aspects of neuropathic pain behaviors in various nerve ligation models, such as types of behavioral tests and factors influencing neuropathic pain behaviors. Rather than providing a detailed description of each model, cross comparisons were often made to highlight the differences of various models. In the second part of this chapter, electrophysiological evidence for abnormal activity of both axotomized and intact afferent neurons is discussed.

Mechanisms and Mediators of Neuropathic Pain, edited by Annika B. Malmberg and Sandra R. Chaplan

Neuropathic pain behaviors

Types of behavioral tests

Behavioral testing is an extremely important process for any animal model for human disease because careful observation of animal behavior is the only way to judge whether or not the animal is experiencing a condition comparable to human disease. For the neuropathic pain model, therefore, it is vital to conduct behavioral tests that are most relevant to the human neuropathic pain condition. Neuropathic pain patients suffer from various types of pain including mechanical allodynia, spontaneous burning pain, cold allodynia, and heat hyperalgesia; and the dominant type may vary in different patients [10]. A good neuropathic pain model should therefore show behaviors representing one or more of these types of pain characteristics and all of the above mentioned models do. However, since behavioral tests rely on animals' reactions, which are frequently simple motor reflexes, and on human interpretation of these reactions, the results of behavioral tests require very careful interpretation. Changes in animals' reactions to stimuli can depend not only on perception (pain) but also expression (motor), and frequently it is hard to tell which is influenced more by experimental manipulations. Furthermore, an animal's reaction in a particular pain state may be quite different than humans. Therefore, results of behavioral tests should be interpreted cautiously.

Mechanical allodynia

Since mechanical allodynia is a major complaint of neuropathic pain patients [10], testing for signs of mechanical allodynia is an important aspect of behavioral tests for neuropathic pain. Mechanical allodynia is often tested by quantifying mechanical sensitivity using von Frey filaments. Mechanical sensitivity is quantified either by measuring response frequency to a series of stimuli [4, 7] or by determining the mechanical threshold [11, 12]. A particularly useful technique is determination of mechanical threshold by the up-down method [11], since it provides an accurate estimate of the threshold. However, one needs to pay close attention to two factors. First, since this method is designed to use a series of von Frey filaments with bending forces of equal logarithmic distance, von Frey selection has to be done accordingly. Second, since these von Frey filaments have bending forces of equal logarithmic distance (based on the theory that sensory perception is proportional to stimuli in logarithmic force scale), the results need to be treated and expressed on a logarithmic force scale (e.g., Fig. 1).

Another important issue of the behavioral test using mechanical stimuli is the determination of whether the resulting behavior represents allodynia or hyperalgesia. Since allodynia is pain elicited by normally innocuous stimuli, one could argue that any pain-like behavior elicited after a nerve injury would be a sign of allody-

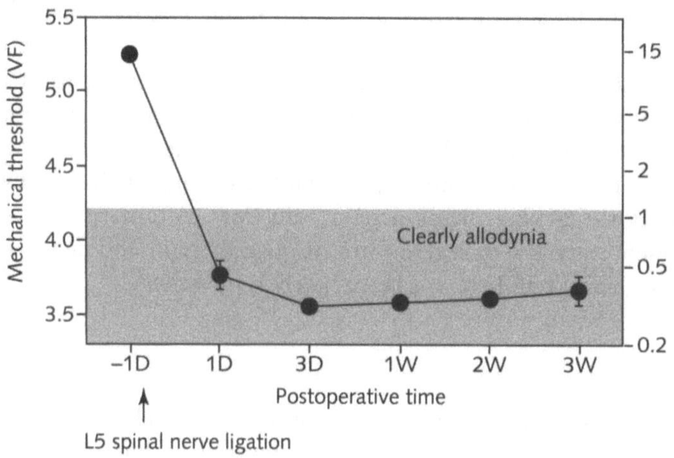

Figure 1
An example of changes in mechanical thresholds in the spinal nerve ligation model of neuropathic pain. In a group of 6 rats, the left L5 spinal nerve was tightly ligated and the mechanical thresholds of the paw (base of the 3rd or 4th toe) for foot withdrawals were measured before and after ligation using the up-down method. The thresholds are expressed in a linear scale of von Frey (VF) filament values on the left side of the graph. For reference, the equivalent gram values are shown in a logarithmic scale on the right side. After the L5 spinal nerve was ligated, mechanical threshold for foot withdrawal declined to around the 0.3 g range by the 3rd day and this level was maintained for weeks. Since thresholds of most nociceptors are higher than 1.5 g, foot withdrawals elicited by lower than this value can be assumed to be mechanical allodynia (shaded area). D: days; W: weeks.

nia. However, nociceptors in normal rats can be activated at quite low thresholds, and hence, it is difficult to judge what stimuli elicit painful sensations. Surveying a large number of cutaneous receptors on the rat foot showed that there was a clear separation of thresholds between low threshold mechanoreceptors and nociceptors, in that the thresholds of most of mechanoreceptors were less than 14 mN whereas those of nociceptors were higher than that value [13]. Therefore, it is safe to assume that a mechanical stimulus at a strength of less than 1.5 g force will activate only low threshold mechanoreceptors in normal conditions and that any pain-like behavior elicited by such stimulus can be assumed to be mechanical allodynia. On the other hand, one cannot be sure whether behaviors elicited by a higher strength (> 1.5 g force) is allodynia or hyperalgesia. An example of changes in mechanical threshold in the L5 spinal nerve ligation model is illustrated in Figure 1.

Ongoing pain

Another important characteristic of neuropathic pain is ongoing pain, usually of a burning type. Therefore, it is important to have behavioral tests representing ongoing pain. The cumulative duration of foot lifts within a given time period in the absence of any obvious stimulus has been used as a sign of ongoing pain in neuropathic rats [1]. However, the magnitude of this behavior is generally small, particularly for models other than the CCI model [14]. This may suggest that many neuropathic pain models are lacking ongoing pain behaviors. Alternatively, foot lift behavior may represent only a very severe form of ongoing pain and hence a large portion of an ongoing pain state may not be properly represented.

Cold allodynia

Cold allodynia is the aspect of neuropathic pain that is best correlated with sympathetically maintained pain [15]. Therefore, testing behaviors that represent cold allodynia in animal models may have important implications. Some studies used the duration of foot lifts on a cold plate as a sign of cold allodynia based on the premise that the animal's avoidance of touching the cold plate was an allodynic reaction. However, a complete denervation of the foot did not change this behavior, making it doubtful that the foot lift behavior is related to allodynia since allodynia would require functioning sensory receptors [16]. Another method that has been used was the observation of foot withdrawals to acetone application [16]. This method has also been used in neuropathic pain patients to test cold allodynia [17].

Heat hyperalgesia

Heat hyperalgesia testing has been used extensively in neuropathic pain models, mostly by use of the Hargreave's heat stimulator [18]. Since the response to a heat stimulus is likely mediated by unmyelinated afferent fibers, this test should be valuable as a test for C fiber function in neuropathic pain.

Factors influencing neuropathic pain behaviors

Local environment of injury site

Ever since the initial development of the neuropathic pain model, various factors have been known to influence pain behaviors. For example, behavioral outcome varies in the CCI model depending on the tightness of loose ligatures placed around the sciatic nerve [1]. Apparently the tightness influences swelling of the nerve and ultimately the number of axotomized fibers as well as the degree of local inflammation. The importance of local inflammation became further evident from a study indicating that the suture material, chromic gut, placed around the sciatic nerve without ligation produced neuropathic pain behaviors [19]. Therefore, both the local environment of the injury site and the number of axotomized afferent fibers are important factors influencing neuropathic pain behaviors.

Anesthesia

NMDA-mediated central sensitization is a critically important mechanism underlying neuropathic pain behaviors in that NMDA blockers applied to the spinal cord greatly reduce pain behaviors [20, 21]. Some anesthetic agents, such as ketamine, have NMDA receptor blocking properties and their use should be avoided when anesthetizing animals during neuropathic surgery.

Diet

Another unexpected but potentially important factor influencing neuropathic pain behaviors is diet. Shir et al. [22] found that rats fed with a high soybean diet produced much diminished levels of neuropathic pain behaviors. Although the underlying mechanisms are not clear, this phenomenon is potentially important for the development of diet based therapy. Furthermore, the results suggest that one should avoid feeding the rat with a high soybean diet if one wants to produce a robust experimental neuropathic pain model.

Genetic factors

Genetic variability is another extremely important factor. It has long been recognized in clinical practice that patients with the same type of injury respond differently in that some patients develop pain while others do not. A similar individual difference may exist in experimental animals. When the same peripheral nerve injury (spinal nerve ligation model) was produced in several in-bred rat strains, there was a big difference in neuropathic pain behaviors among them [23]. Strain difference in pain behaviors was also seen in another model of neuropathic pain [24] as well as in the spinal nerve ligation model of mice [25, 26]. This strain difference is likely due to genetic factors and factors may play an important role in determining neuropathic pain behaviors. Furthermore, a difference in neuropathic pain behaviors was also seen between two groups of Sprague-Dawley rats, a commonly used out-bred strain, obtained from two different venders [23]. This is presumably due to the genetic drift that has taken place during separate breeding by different venders. Because of such genetic variability, it is common to have batch differences even with the same source, particularly in out-bred strains. Therefore, one has to give very serious consideration to genetic factors and to their potential contribution to the variability of neuropathic pain behaviors.

Age

Another factor potentially influencing neuropathic pain behavior is age of the animal. Although neonatal rats do show neuropathic pain behaviors [27], young adult rats show more robust behaviors than old rats. Such age differences can also be seen at various ages of the same animal [28]. However, it is not clear whether such robust behavior in young adult rats is actually due to a stronger sensory experience or whether it is merely a reflection of a more lively motor expression due to the young age of the animal. Such suspicion is strengthened by a report that older rats showed stronger neuropathic pain behaviors in the tail nerve transection model [29]. Therefore, although it is not clear whether older or younger animals have stronger neuropathic pain behaviors, it is clear that the age of the animal is a factor for the variability of neuropathic pain behaviors.

The best neuropathic pain model

A frequently asked question is "which is the best model?" or "which model should I use for my experiment?" The answer is "it depends on one's need." Having so many models provides the opportunity to choose the best model for one's need. For example, if one wants to most closely imitate an accidental traumatic neuropathic injury in a human, either the partial sciatic nerve ligation [2] or the CCI model [1] would be preferable since it is hard to imagine an accident would cause either a complete transection of a selected segmental spinal nerve [4], a freeze-injury to a nerve [3], or an injury to a tail nerve [7]. On the other hand, if knowing the time of injury of individual fibers is an important factor, one would be better off choosing one of the tight ligation models, such as the partial sciatic nerve or segmental spinal nerve ligation model. If the number of injured fibers needs to be consistent between animals or a clear separation between axotomized and intact fibers is an important factor, the model with a whole nerve injury (spinal nerve ligation, tail nerve cut, etc.) would be a better choice.

Another aspect one has to pay attention to when choosing an animal model is the difference in behavioral outcome between animal models. When comparisons were made between the CCI, partial sciatic nerve ligation, and segmental spinal nerve ligation models, behavioral signs of ongoing pain were most obvious in the CCI model while those of mechanical allodynia were strongest in the spinal nerve ligation model [14]. In addition, reversal of behaviors by surgical sympathectomy was most evident in the spinal nerve ligation model. Another difference is morphine sensitivity of behaviors. Although pain behaviors in the CCI model are known to be extremely sensitive to systemically applied morphine [30, 31], the spinal nerve ligation model is quite resistant to morphine administered either intrathecally [32] or intraperitoneally [33].

An obvious factor to be considered would be the species of animal one chose to experiment with. All neuropathic pain models were initially developed using the rat due to the fact that suitable behavioral tests were readily available for this species. Subsequently, some of these models have been extended to species other than the rat. The spinal nerve ligation model has been successfully applied to monkeys [34] as well as mice [25]. The partial sciatic nerve ligation model was extended to [35] and studied in genetically manipulated mice [36, 37]. Extension to the mouse is potentially important since genetic manipulation, such as "knock out" of a gene, is much easier in this species.

Electrophysiology

Sustained input to the spinal cord of abnormal neuronal activity from the periphery is the driving force for central sensitization, which is the critically important mechanism underlying neuropathic pain behaviors. Therefore, examining the characteristics of abnormal activity as well as investigating its generation mechanism is an important step toward the understanding of neuropathic pain. It is now clear that abnormal activity arises from both axotomized and intact afferent neurons after a peripheral nerve injury.

Activity of axotomized afferents – ectopic discharges

The normal impulse generation site for the primary afferent neuron is the sensory receptor. When afferent neurons are disconnected from their receptors by an axotomy, the vast majority of them become totally silent, except for a few unusual afferents which display a very low rate of activity for some time [38, 39]. Some of these silent units gradually become active 12–15 h after spinal nerve ligation, and the activity continues for days and presumably weeks [40–42]. Since the sensory receptor is denervated, the activity must have been generated from a site other than the normal impulse-generating site – ectopic discharges. As an example, Figure 2 shows the time course of development of ectopic discharges after spinal nerve ligation. At different times after spinal nerve ligation, the dorsal root ganglion along with its corresponding dorsal root and spinal nerve was removed, and then placed in an *in vitro* recording chamber. The dorsal root was teased into small fascicles and spontaneously active single units were searched for [40]. During an exhaustive search, no spontaneously active unit was found in the dorsal root recorded at either 3 or 7 h (from 8 preparations each) after spinal nerve ligation. However, we were able to record some units showing ectopic discharges (on average, 1.75 units per preparation) from the dorsal root recorded at later time points (9 and 11 h). Ectopic discharges of all these units, however, were not sustained and activity ceased during the

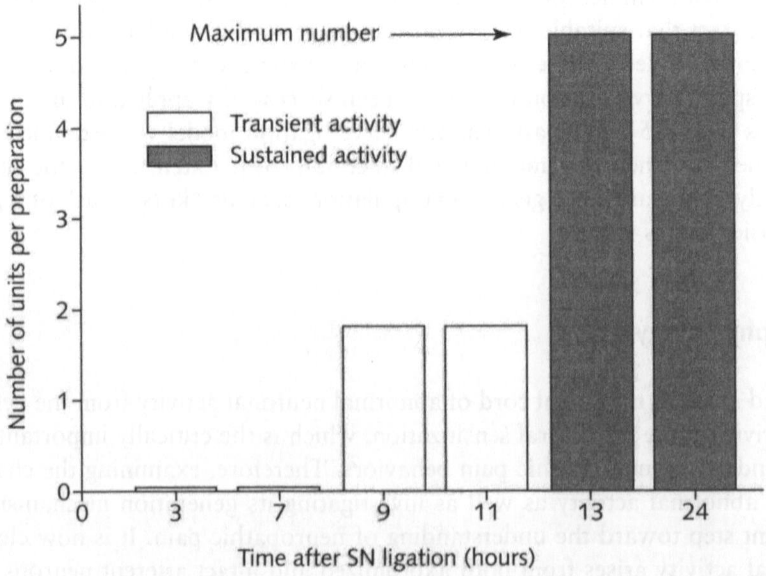

Figure 2
The early time course of development of ectopic discharges after spinal nerve ligation. At different times after spinal nerve ligation, the dorsal root ganglion along with the dorsal root and spinal nerve was removed and placed in an in vitro recording chamber (8 for each time point). Spontaneously active single unit activity was searched for and recorded from teased dorsal root fascicles. No spontaneously active unit was found from the preparations at either 3 or 7 h after spinal nerve ligation. Some units (average 1.75 per preparation) showed spontaneous activity at both 9 and 11 post ligation hours. However, these activities were transient and all of them ceased to fire at some point during the one-hour recording period. At the 13th hour and beyond, sustained spontaneous activity could be recorded from at least 5 units, which was the pre-determined maximum number per preparation. These data suggest that ectopic discharges are generated rather abruptly some time after 11–13 h following spinal nerve ligation.

1-h recording session. The situation became completely different at 13 post-spinal nerve ligation hours or beyond (sampled up to one week). At this point, there were numerous active single units in the dorsal root and we were able to quickly record 5 units per dorsal root, which was our pre-determined maximum number of recorded units. This data suggests that important ectopic discharge generating mechanisms are triggered at between 11 and 13 hours after spinal nerve ligation.

After a peripheral nerve injury, both the injury site and the dorsal root ganglion are known to become ectopic discharge generation sites [43–47]. However, ectopic discharges are generated almost exclusively in the dorsal root ganglion after spinal nerve injury [40]. The pattern of firing of ectopic discharges (regular, irregular, or bursting type) varies a great deal between each individual unit and some units display a very unusual pattern of activity [41, 47–49].

A puzzling phenomenon is that the majority of afferent fibers displaying ectopic discharges are either Aβ or Aδ fibers, but very few C fibers show such discharges [40–42, 48, 49]. This is puzzling because ectopic discharges are thought to enter the spinal cord to set up central sensitization and the C fiber input is known to be critically important in triggering central sensitization [50]. The lack of abundant ectopically active C fibers suggests that either C fiber input need not be big to have an important consequence or ectopic discharges of Aδ/Aβ fibers plays an important role with some unknown mechanisms. It is not clear which is the case at the present time.

Modification of ectopic discharges

Mechanisms of ectopic discharge generation are not clear. An important potential mechanism is the up-regulation of ion channels in the injured sensory neurons, particularly sodium channels. Accumulation of sodium channels at the injury site of sensory neurons has been observed [51, 52] and theoretical grounds for discharge generation by such sodium channel accumulation has been provided [53]. Consistent with this possibility, various ways of blocking sodium channel functions are known to reduce ectopic discharge rate [45, 54–56] or neuropathic pain behavior [57, 58]. However, the exact subtype of sodium channel that plays the most important role in the generation of ectopic discharges is not clear. In addition, since blockers for potassium or calcium channels also influence the ectopic discharge rate [46, 56, 59–61], contributions of these channels should also be considered.

The sympathetic nervous system plays an important role in ectopic discharges. However, it does not seem to function as the discharge generator, but rather serves as a modulator. This is evident because ectopic discharges can still be demonstrated after sympathectomy [41, 49] or in an *in vitro* preparation of isolated dorsal root ganglion with attached dorsal root [40, 62], in which the sympathetic nervous system is obviously denervated and there are no circulating sympathetic ligands. Therefore, the sympathetic nervous system is not essential for the development of ectopic discharges. However, ectopic discharges can be modified by the sympathetic nervous system. Application of adrenergic agonists or electrical stimulation of the sympathetic chains enhances the discharges [40, 43, 47, 62–64] while sympathectomy reduces the activity ([49]; however, see [41]). Although such adrenergic sensitivity has long been recognized, a puzzling phenomenon is that the ongoing rate of ectopic discharges is not influenced by adrenergic receptor blockers. In fact, Wall and Gutnick [43] in their early work described that "If the alpha blocking agent, phentolamine, 0.5 mg/kg was given by itself there was no significant effect. When a combination of phentolamine 0.5 mg/kg with 5 μg/kg noradrenaline was given, the noradrenaline excitation was blocked." The same observation has been made repeatedly since then (e.g. [47]). These data suggest that although ectopic discharges can be evoked by adrenergic ligands, their rate may not necessarily be maintained by adrenergic receptor function. One possible explanation for such an apparent paradox is the possibility of the involvement of the non-adrenergic sympathetic system, particularly the purinergic system. Since sympathetic postganglionic neurons contain not only noradrenaline but also other substances such as ATP, these substances may play an important role in sympathetic function. In fact, one of our recent studies [65] showed that neuropathic pain behaviors were significantly reduced by an injection of a combination of phentolamine and suramin (a purinergic receptor blocker), while phentolamine alone was ineffective. To test the importance of non-adrenergic function in ectopic discharges, effects of the adrenergic receptor blocker and purinergic receptor blocker on ectopic discharges were examined. One week after L5 spinal nerve ligation, the rats (Sprague-Dawley) were anesthetized and single unit recordings were made from filaments of the L5 dorsal root using an *in vivo* recording setup. In each of 4 rats, only one unit was tested due to the long recording time and to avoid the complication of a mixture of drugs. As shown in Figure 3, ectopic discharges recorded from 2 of 4 rats reduced their activity after an injection of suramin (100 mg/kg) along with 5 mg/kg of phentolamine, but not after injection of either agent alone. These data suggest that sympathetic modulation of ectopic discharges is accomplished by concurrent actions of both the adrenergic and non-adrenergic (purinergic) systems. However, it is important to note that this particular experiment was conducted using Sprague-Dawley rats. A lack of effect by phentolamine alone on the rate of ongoing ectopic discharges is consistent with the reports of many previous studies conducted in non-Lewis rats [43, 47]. In contrast, an injection of phentolamine alone reduces not only the rate of ectopic discharges [62] but also pain behaviors [66] in the Lewis strain. Therefore, it seems that strain difference is apparent not only in the amount of pain behaviors after peripheral nerve injury as discussed in the previous section, but also in the mode of action by the sympathetic nervous system, in that Lewis rats are less dependent on non-adrenergic function than Sprague-Dawley rats.

Importance of ectopic discharges

Several lines of evidence suggest that ectopic discharges are important contributors to the generation of neuropathic pain behaviors. The first line of evidence is that the time course of development of ectopic discharges coincides with that of neuropath-

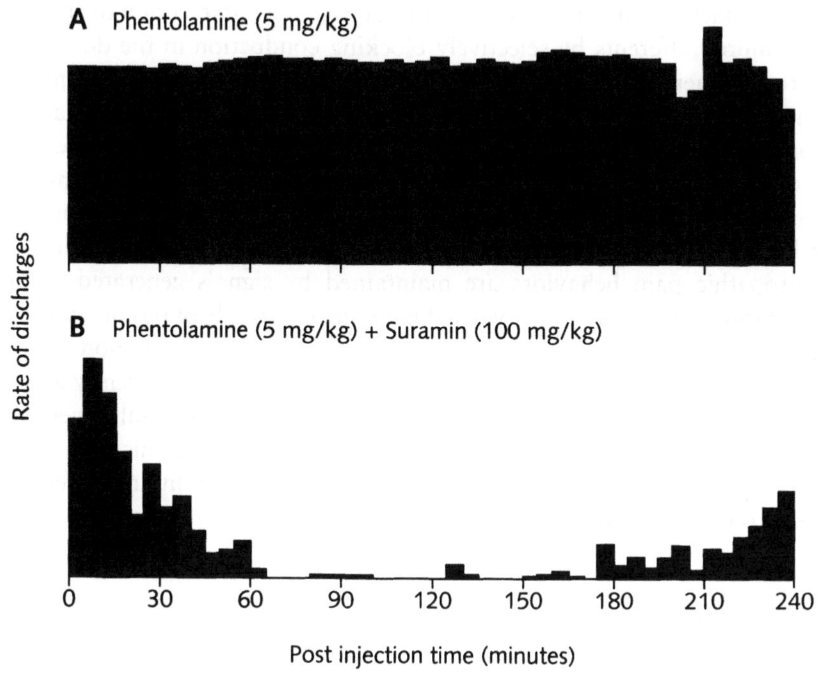

Figure 3
The effects of a combination of drugs on the activity of ectopic discharge of sensory neurons in the neuropathic rat. A spontaneously active single unit was recorded from the L5 dorsal root filament at 2 weeks after L5 spinal nerve ligation using an in vivo *recording setup. (A) and (B) show two units recorded from two different rats. The ongoing activity was recorded for 20 min in each unit before injection of any agent to make sure it maintained a stable level of firing frequency. In (A), intraperitoneal injection of phentolamine (5 mg/kg), an alpha adrenergic receptor blocker, was made at time 0 and the activity did not change. In (B), a combination of phentolamine and suramin (100 mg/kg), a purinergic receptor blocker, was injected and the rate of ectopic discharges greatly reduced. The time course of the reduction of discharge rate was similar to the antiallodynic effect seen after an injection of the same combination of agents in a previous study [65]. These data suggest that a reduction of the ongoing rate of ectopic discharges requires not only the blockage of adrenergic receptors but also purinergic receptors.*

ic pain behaviors [4, 40, 42]. As mentioned above, ectopic discharges first appear about 12–15 h after spinal nerve ligation. This is about the time when signs of mechanical allodynia become apparent as well. More definitive evidence comes from dorsal rhizotomy experiments. In some animal models with clear separation of injured and intact spinal segments, one can examine the contribution of spinal input from the injured afferents by selectively blocking conduction in the dorsal root of the injured segment. In the spinal nerve ligation model, dorsal rhizotomy of the spinal nerve injured segments produced a significant reversal of neuropathic pain behaviors [67, 68]. Furthermore, conduction block by application of local anesthetics on the dorsal root produced a reversible reduction in neuropathic pain behaviors [68]. In addition, transection of the dorsal root also reversed neuropathic pain behaviors in the tail nerve model [69]. The results of these studies suggest that neuropathic pain behaviors are maintained by signals generated in injured afferents. However, this is a controversial issue since a couple of recent studies could not reproduce the effect of dorsal rhizotomy in the spinal nerve ligation model [70, 71]. Since this is an important issue, we repeated the experiment recently and found that dorsal rhizotomy produced not a complete, but a partial reversal of neuropathic pain (unpublished observation). These data suggest that neuropathic pain behaviors depend on spinal inputs from two different sources: one from injured afferents and the other from intact afferents (see below).

Importance of intact afferent input

The fact that dorsal rhizotomy of an injured segment produces only partial, not complete, reversal of neuropathic pain behaviors suggests that factors other than input from injured segments are important contributors to neuropathic pain behaviors. One likely factor would be abnormal activity generated from sensitized nociceptors in the intact nerves. Although intact mechanoreceptors innervating the foot were found, by and large, to be normal after segmental spinal nerve injury [72], some nociceptors are found to be sensitized [73]. Sensitized nociceptors produce spontaneous activity, which enters the spinal cord and presumably contributes to central sensitization. It is not clear what triggers sensitization of intact nociceptors when neighboring fibers are axotomized. One proposed mechanism is that axotomized fibers undergo Wallerian degeneration distal to the injury site and the degeneration process induces an inflammatory response in neighboring intact nociceptors [70]. Although the validity of this proposal remains to be proven, involvement of a type of inflammatory process seems to be likely since inducing inflammation in a nerve (neuritis) produces sensitization of nociceptors [74].

Conclusions

Several animal models have been developed and used for studying the mechanisms underlying various aspects of neuropathic pain. Availability of such a large number of models is good on one hand because it not only provides a large selection when one tries to conduct experiments but also allows investigation of many different aspects of neuropathic pain mechanisms. On the other hand, having many different models can be confusing in terms of choosing the right one for an experiment or interpreting experimental results obtained by using one particular model. To reduce such potential confusion, we attempted to provide several aspects of neuropathic pain behaviors that need to be carefully considered.

It is clear that abnormal activity generated by sensory neurons plays a key role in initiating and maintaining neuropathic pain behaviors. It is also clear that abnormal activities are generated not only from axotomized but also from intact afferent neurons after peripheral nerve injury. However, what the relative contribution of these two sources (axotomized and intact fibers) is to pain behaviors is not clear, nor are the detailed generation mechanisms of each of the two components. Future investigational efforts should be directed toward the clarification of these issues.

References

1 Bennett GJ, Xie Y-K (1988) A peripheral mononeuropathy in rat that produces disorders of pain sensation like those seen in man. *Pain* 33: 87–107

2 Seltzer Z, Dubner R, Shir Y (1990) A novel behavioral model of neuropathic pain disorders produced in rats by partial sciatic nerve injury. *Pain* 43: 205–218

3 DeLeo JA, Coombs DW, Willenbring S, Colburn RW, Fromm C, Wagner R, Twitchell BB (1994) Characterization of a neuropathic pain model: sciatic cryoneurolysis in the rat. *Pain* 56: 9–16

4 Kim SH, Chung JM (1992) An experimental model for peripheral neuropathy produced by segmental spinal nerve ligation in the rat. *Pain* 50: 355–363

5 Kawakami M, Weinstein JN, Spratt KF, Chatani K, Traub RJ, Meller ST, Gebhart GF (1994) Experimental lumbar radiculopathy. Immunohistochemical and quantitative demonstrations of pain induced by lumbary nerve root irritation of the rat. *Spine* 19: 1780–1794

6 Decosterd I, Woolf CJ (2000) Spared nerve injury: an animal model of persistent peripheral neuropathic pain. *Pain* 87: 149–158

7 Na HS, Han JS, Ko KH, Hong SK (1994) A behavioral model for peripheral neuropathy produced in rat's tail by inferior caudal trunk injury. *Neurosci Lett* 177: 50–52

8 Stanton-Hicks M, Jänig W, Hassenbusch S, Haddox JD, Boas R, Wilson P (1995) Reflex sympathetic dystrophy: changing concepts and taxonomy. *Pain* 63: 127–133

9 Vatine JJ, Argov R, Seltzer Z (1998) Brief electrical stimulation of C-fibers in rats produces thermal hyperalgesia lasting weeks. *Neurosci Lett* 245: 1–4

10 Bonica JJ (1990) Causalgia and other reflex sympathetic dystrophies. In: JJ Bonica (ed): *The management of pain.* Lea and Febiger, Philadelphia, 220–243

11 Chaplan SR, Bach FW, Pogrel JW, Chung JM, Yaksh TL (1994) Quantitative assessment of tactile allodynia in the rat paw. *J Neurosci Methods* 53: 55–63
12 Ma Q-P, Woolf CJ (1996) Progressive tactile hypersensitivity: an inflammation-induced incremental increase in the excitability of the spinal cord. *Pain* 67: 97–106
13 Leem JW, Willis WD, Chung JM (1993) Cutaneous sensory receptors in the rat foot. *J Neurophysiol* 69: 1684–1699
14 Kim KJ, Yoon YW, Chung JM (1997) Comparison of three rodent neuropathic pain models. *Exp Brain Res* 113: 200–206
15 Frost SA, Raja SN, Campbell JN, Meyer RA, Khan AA (1988) Does hyperalgesia to cooling stimuli characterize patients with sympathetically maintained pain (reflex sympathetic dystrophy)? In: R Dubner, GF Gebhart, MR Bond (eds): *Proceedings of the Vth world congress on pain*. Elsevier, Amsterdam, 151–156
16 Choi Y, Yoon YW, Na HS, Kim SH, Chung JM (1994) Behavioral signs of ongoing pain and cold allodynia in a rat model of neuropathic pain. *Pain* 59: 369–376
17 Davis KD, Treede RD, Raja SN, Meyer RA, Campbell JN (1991) Topical application of clonidine relieves hyperalgesia in patients with sympathetically maintained pain. *Pain* 47: 309–317
18 Hargreaves K, Dubner R, Brown F, Flores C, Joris J (1988) A new and sensitive method for measuring thermal nociception in cutaneous hyperalgesia. *Pain* 32: 77–88
19 Maves TJ, Pechman PS, Gebhart GF, Meller ST (1993) Possible chemical contribution from chromic gut sutures produces disorders of pain sensation like those seen in man. *Pain* 54: 57–69
20 Mao J, Price DD, Mayer DJ, Lu J, Hayes RL (1992) Intrathecal MK-801 and local nerve anesthesia synergistically reduce nociceptive behaviors in rats with experimental peripheral mononeuropathy. *Brain Res* 576: 254–262
21 Mao J, Price DD, Hayes RL, Lu J, Mayer DJ, Frenk H (1993) Intrathecal treatment with dextrorphan or ketamine potently reduces pain-related behaviors in a rat model of peripheral mononeuropathy. *Brain Res* 605: 164–168
22 Shir Y, Ratner A, Raja SN, Campbell JN, Seltzer Z (1998) Neuropathic pain following partial nerve injury in rats is suppressed by dietary soy. *Neurosci Lett* 240: 73–76
23 Yoon YW, Lee DH, Lee BH, Chung K, Chung JM (1999) Different strains and substrains of rats show different levels of neuropathic pain behaviors. *Exp Brain Res* 129: 167–171
24 Devor M, Raber P (1990) Heritability of symptoms in an experimental model of neuropathic pain. *Pain* 42: 51–67
25 Mogil JS, Wilson SG, Bon K, Lee SE, Chung K, Raber P, Pieper JO, Hain HS, Belknap JK, Hubert L et al (1999) Heritability of nociception I: Responses of 11 inbred mouse strains on 12 measures of nociception. *Pain* 80: 67–82
26 Mogil JS, Wilson SG, Bon K, Lee SE, Chung K, Raber P, Pieper JO, Hain HS, Belknap JK, Hubert L et al (1999) Heritability of nociception II. "Types" of nociception revealed by genetic correlation analysis. *Pain* 80: 83–93

27 Lee DH, Chung JM (1996) Neuropathic pain in neonatal rats. *Neurosci Lett* 209: 140–142
28 Chung JM, Choi Y, Yoon YW, Na HS (1995) Effects of age on behavioral signs of neuropathic pain in an experimental rat model. *Neurosci Lett* 183: 54–57
29 Kim YI, Na HS, Yoon YW, Nahm SH, Ko KH, Hong SK (1995) Mechanical allodynia is more strongly manifested in older rats in an experimental model of peripheral neuropathy. *Neurosci Lett* 199: 158–160
30 Attal N, Chen YL, Kayser V, Guilbaud G (1991) Behavioural evidence that systemic morphine may modulate a phasic pain-related behaviour in a rat model of peripheral mononeuropathy. *Pain* 47: 65–70
31 Jazat F, Guilbaud G (1991) The 'tonic' pain-related behaviour seen in mononeuropathic rats is modulated by morphine and naloxone. *Pain* 44: 97–102
32 Yaksh TL, Pogrel JW, Lee YW, Chaplan SR (1995) Reversal of nerve ligation-induced allodynia by spinal alpha-2 adrenoceptor agonists. *J Pharmacol Exp Ther* 272: 207–214
33 Chung JM, Na HS (1996) Effects of systemic morphine on neuropathic pain behaviors in an experimental rat model. *Analgesia* 2: 151–155
34 Carlton SM, Lekan HA, Kim SH, Chung JM (1994) Behavioral manifestations of an experimental model for peripheral neuropathy produced by spinal nerve ligation in the primate. *Pain* 56: 155–166
35 Malmberg AB, Basbaum AI (1998) Partial sciatic nerve injury in the mouse as a model of neuropathic pain: behavioral and neuroanatomical correlates. *Pain* 76: 215–222
36 Malmberg AB, Brandon EP, Idzerda RL, Liu H, McKnight GS, Basbaum AI (1997) Diminished inflammation and nociceptive pain with preservation of neuropathic pain in mice with a targeted mutation of the type I regulatory subunit of cAMP-dependent protein kinase. *J Neurosci* 17: 7462–7470
37 Malmberg AB, Chen C, Tonegawa S, Basbaum AI (1997) Preserved acute pain and reduced neuropathic pain in mice lacking PKCgamma. *Science* 278: 279–283
38 Baik-Han EJ, Kim KJ, Chung JM (1990) Prolonged ongoing discharges of sensory nerves as recorded in isolated nerves in the rat. *J Neurosci Res* 27: 219–227
39 Chung JM, Leem JW, Kim SH (1992) Somatic afferent fibers which continuously discharge after being isolated from their receptors. *Brain Res* 599: 29–33
40 Liu X, Chung K, Chung JM (1999) Ectopic discharges and adrenergic sensitivity of sensory neurons after spinal nerve injury. *Brain Res* 849: 244–247
41 Liu X, Eschenfelder S, Blenk K-H, Jänig W, Häbler H-J (2000) Spontaneous activity of axotomized afferent neurons after L5 spinal nerve injury in rats. *Pain* 84: 309–318
42 Liu C-N, Wall PD, Ben-Dor E, Michaelis M, Amir R, Devor M (2000) Tactile allodynia in the absence of C-fiber activation: altered firing properties of DRG neurons following spinal nerve injury. *Pain* 85: 503–521
43 Wall PD, Gutnick M (1974) Ongoing activity in peripheral nerves: the physiology and pharmacology of impulses originating from a neuroma. *Exp Neurol* 43: 580–593
44 Scadding JW (1981) Development of ongoing activity, mechanosensitivity, and adrenaline sensitivity in severed peripheral nerve axons. *Exp Neurol* 73: 345–364

45 Devor M, Wall PD, Catalan N (1992) Systemic lidocaine silences ectopic neuroma and DRG discharge without blocking nerve conduction. *Pain* 48: 261–268
46 Kajander KC, Wakisaka S, Bennett GJ (1992) Spontaneous discharge originates in the dorsal root ganglion at the onset of a painful peripheral neuropathy in the rat. *Neurosci Lett* 138: 225–228
47 Chen Y, Michaelis M, Jänig W, Devor M (1996) Adrenoreceptor subtype mediating sympathetic-sensory coupling in injured sensory neurons. *J Neurophysiol* 76: 3721–3730
48 Kajander KC, Bennett GJ (1992) Onset of a painful peripheral neuropathy in rat: A partial and differential deafferentation and spontaneous discharge in Aβ and Aδ primary afferent neurons. *J Neurophysiol* 68: 734–744
49 Han HC, Lee DH, Chung JM (2000) Characteristics of ectopic discharges in a rat neuropathic pain model. *Pain* 84: 253–261
50 Woolf CJ, Thompson SWN (1991) The induction and maintenance of central sensitization is dependent on N-methyl-D-aspartic acid receptor activation; implication for the treatment of post-injury pain hypersensitivity states. *Pain* 44: 293–299
51 Devor M, Keller CH, Deerinck TJ, Levinson SR, Ellisman MH (1989) NA^+ channel accumulation on axolemma of afferent endings in nerve end neuroman in Apteronotus. *Neurosci Lett* 149–154
52 Devor M, Govrin-Lippmann R, Angelides K (1993) Na^+ channel immunolocalization in peripheral mammalian axons and changes following nerve injury and neuroma formation. *J Neurosci* 13: 1976–1992
53 Matzner O, Devor M (1992) Na^+ conductance and the threshold for repetitive neuronal firing. *Brain Res* 597: 92–98
54 Omana-Zapata I, Khabbaz MA, Hunter JC, Clarke DE, Bley KR (1997) Tetrodotoxin inhibits neuropathic ectopic activity in neuromas, dorsal root ganglia and dorsal horn neurons. *Pain* 72: 41–49
55 Omana-Zapata I, Khabbaz MA, Hunter JC, Bley KR (1997) QX-314 inhibits ectopic nerve activity associated with neuropathic pain. *Brain Res* 771: 228–237
56 Matzner O, Devor M (1994) Hyperexcitability at sites of nerve injury depends on voltage-sensitive Na^+ channels. *J Neurophysiol* 72: 349–359
57 Chaplan SR, Bach FW, Shafer SL, Yaksh TL (1995) Prolonged alleviation of tactile allodynia by intravenous lidocaine in neuropathic rats. *Anesthesiology* 83: 775–785
58 Lyu YS, Park SK, Chung K, Chung JM (2000) Low dose of tetrodotoxin reduces neuropathic pain behaviors in an animal model. *Brain Res* 871: 98–103
59 Xie Y-K, Xiao WH, Li HQ (1993) The relationship between new ion channels and ectopic discharges from a region of nerve injury. *Science in China* (Series B) 36: 68–74
60 Xiao WH, Bennett GJ (1995) Synthetic omega-conopeptides applied to the site of nerve injury suppress neuropathic pains in rats. *J Pharmacol Exp Ther* 274(2): 666–672
61 Chaplan SR, Pogrel JW, Yaksh TL (1994) Role of voltage-dependent calcium channel subtypes in experimental tactile allodynia. *J Pharmacol Exp Ther* 269: 1117–1123
62 Lee DH, Liu X, Kim HT, Chung K, Chung JM (1999) Receptor subtype mediating the

adrenergic sensitivity of pain behavior and ectopic discharges in neuropathic Lewis rats. *J Neurophysiol* 81: 2226–2233
63 Korenman EM, Devor M (1981) Ectopic adrenergic sensitivity in damaged peripheral nerve axons in the rat. *Exp Neurol* 72: 63–81
64 McLachlan EM, Jänig W, Devor M, Michaelis M (1993) Peripheral nerve injury triggers noradrenergic sprouting within dorsal root ganglia. *Nature* 363: 543–546
65 Park SK, Chung K, Chung JM (2000) Effects of purinergic and adrenergic antagonists in a rat model of painful peripheral neuropathy. *Pain* 87: 171–179
66 Lee DH, Chung K, Chung JM (1997) Strain differences in adrenergic sensitivity of neuropathic pain behaviors in an experimental rat model. *NeuroReport* 8: 3453–3456
67 Sheen K, Chung JM (1993) Signs of neuropathic pain depend on signals from injured nerve fibers in a rat model. *Brain Res* 610: 62–68
68 Yoon YW, Na HS, Chung JM (1996) Contributions of injured and intact afferents to neuropathic pain in an experimental rat model. *Pain* 64: 27–36
69 Na HS, Ko KH, Back SK, Sung B, Yoo DJ, Hong SK (2000) Role of signals from the dorsal root ganglion in neuropathic pain in a rat model. *Neurosci Lett* 288: 147–150
70 Li Y, Dorsi MJ, Meyer RA, Belzberg AJ (2000) Mechanical hyperalgesia after an L5 spinal nerve lesion in the rat is not dependent on input from injured nerve fibers. *Pain* 85: 493–502
71 Eschenfelder S, Häbler H-J, Jänig W (2000) Dorsal root section elicits signs of neuropathic pain rather than reversing them in rats with L5 spinal nerve injury. *Pain* 87: 213–219
72 Na HS, Leem JW, Chung JM (1993) Abnormalities of mechanoreceptors in a rat model of neuropathic pain: possible involvement in mediating mechanical allodynia. *J Neurophysiol* 70: 522–528
73 Ali Z, Ringkamp M, Hartke TV, Chien HF, Flavahan NA, Campbell JN, Meyer RA (1999) Uninjured C-fiber nociceptors develop spontaneous activity and α-adrenergic sensitivity following L_6 spinal nerve ligation in monkey. *J Neurophysiol* 81: 455–466
74 Sorkin LS, Xiao WH, Wagner R, Myers RR (1997) Tumour necrosis factor-α induces ectopic activity in nociceptive primary afferent fibres. *Neurosci* 81: 255–262

Animal models of toxic and metabolic sensory neuropathies

Nigel A. Calcutt and Jason D. Freshwater

Department of Pathology, 0612, University of California San Diego, La Jolla, CA 92093-0612, USA

Introduction

Peripheral neuropathies, excluding those caused by mechanical damage, are most commonly associated with infection, autoimmune disorders, and exposure to neurotoxins or metabolic disease. The sensory nervous system may be selectively damaged, as occurs in platinum toxicity, or there may be a more generalized sensorimotor neuropathy. Symptoms of sensory dysfunction range from sensory loss to inappropriate or exaggerated pain perception and a variety of sensory disorders may occur together or in chronological sequence. For example, while sensory loss is the predominant symptom of diabetic polyneuropathy, 5–10% of patients exhibit a painful neuropathy that may either precede or co-exist with sensory loss. Pain, paresthesias or dysesthesias are a common feature of many peripheral neuropathies but are difficult to quantify in animal models. It is particularly unfortunate that both leprosy and HIV have proven extremely difficult to model in animals that lend themselves to experimental studies of nociceptive processing. As a result, little is yet known of the mechanisms that produce these most prevalent of neuropathies. In recent years, sensory dysfunction has been described in animal models of chemical neurotoxicity and diabetes mellitus. Such models are of value as they allow investigation of the etiology of neuropathic pain and because they provide test systems in which to study potential therapeutic agents that may prevent or ameliorate neuropathic pain. In this chapter we will briefly consider the animal models of toxic neuropathies, which largely produce sensory loss, and then focus upon data obtained from the widely studied animal models of diabetic neuropathy.

Toxic neuropathies

These may be broadly divided into those resulting from accidental exposure to toxins and those that arise as a side effect of using agents for other therapeutic purposes [1]. The former category includes the peripheral neuropathies that develop after exposure to industrial agents such as acrylamide, hexocarbons and organo-

Mechanisms and Mediators of Neuropathic Pain, edited by Annika B. Malmberg and Sandra R. Chaplan

phospates [2] or to heavy metals. Some heavy metals, including arsenic, gold, lead, mercury and thallium were formerly used as therapeutics but have largely fallen out of use due to side effects that include neurotoxicity. There are other agents, however, that remain in widespread use despite inducing peripheral neuropathy as a dose limiting side effect. These include drugs intended as chemotherapeutic agents such as cisplatin, vinka alkaloids and paclitaxel and a variety of other agents including phenytoin, pyroxidine (vitamin B_6) and thalidomide. Practical difficulties in titrating doses of the toxin that are high enough to induce neuropathy but do not kill the animal have restricted many studies of chemical neurotoxicity to descriptions of neuropathologic changes. Moreover, as many neurotoxins induce a non-selective sensorimotor neuropathy, interpretation of behavioral responses to sensory stimuli can be obstructed by the development of limb weakness.

Environmental neurotoxins

Of the industrial agents that induce neurotoxicity, exposure to acrylamide has been most widely studied. The resulting neuropathy is a dying back axonopathy that afflicts large myelinated motor and sensory fibers, resulting in weakness and loss of vibration sense. Pain and paresthesias are rare in patients and a similar picture is seen in acrylamide intoxicated rats, which show a relative sparing of small fiber function [3]. The underlying cause appears to be disruption of retrograde axonal transport [4] with accumulation of neurofilaments in the distal regions of large diameter axons [5]. Indeed, disruption of cytoskeletal organization appears to be a common theme in many of the neuropathies associated with industrial agents [2].

The neurotoxic effects of heavy metals usually occur in the context of other systemic disorders involving the hematopoietic, gastrointestinal and central nervous systems. A number of the heavy metals were formerly used as drugs, although exposure is now more likely to be accidental, except in the case of platinum derivatives, which remain in use as chemotherapeutic agents (see below). While each metal produces a distinct pattern of disease, peripheral effects of heavy metals generally include limb weakness and peripheral sensory loss, although in the case of arsenic, gold and platinum neuropathy there are reports of initial pains and paresthesias in the extremities.

Chemotherapeutic neurotoxins

Neurotoxicity is a dose limiting side effect of a number of chemotherapeutic agents and the development of animal models has been pursued both to investigate the underlying mechanisms of the neuropathies and to evaluate potential protective agents.

Platinum

Platinum-based compounds such as cisplatin and carboplatin remain in use as adjunct therapies for advanced malignancies of the bladder, testes and ovary and may restrict cell proliferation by binding to DNA. Patients treated with cisplatin develop a predominantly large fiber sensory neuropathy that is dose limiting. Animal models of cisplatin and carboplatin neuropathy also develop a sensory neuropathy with damage to cell bodies in the dorsal root ganglia and axonal atrophy, most notably in large myelinated fibers of the dorsal roots and sciatic nerve. This is accompanied by sensory loss, as indicated by thermal hypoalgesia in the tail flick test [6–8]. Unlike clinical reports of cisplatin neurotoxicity, where patients may complain of early distal paresthesias, no indications of hyperalgesia or allodynia have yet been noted in cisplatin-treated animals.

Vinca alkaloids

Vinblastine and vincristine are antineoplastic agents that interact with microtubules, thereby inhibiting spindle formation and arresting mitosis in tumor cells. However, disruption of microtubule organization also has harmful consequences for the nervous system in which the microtubular component of the axonal cytoskeleton is important for anterograde and retrograde axonal transport. Patients undergoing treatment with vinca alkaloids often complain of neurologic disorders including distal paresthesias and loss of sensorimotor functions. An electrophysiologic study of such patients revealed loss of compound action potential amplitude but normal velocities. This is indicative of fiber loss, which was confirmed by morphologic studies that showed axonal degeneration [9]. Animal models of vincristine neuropathy have been developed in rats [10, 11] rabbits [12] and mice [13]. However, a major confound in these studies has been dose titration to find a range in which there is neuropathy without killing many of the cohort. Most recently, continuous infusion of vincristine using a mini-osmotic pump has been used to avoid bolus delivery of high doses and to establish constant circulating levels of the drug [14].

Behavioral measures of sensory dysfunction in vincristine-treated rats have revealed mechanical hyperalgesia and tactile allodynia to von Frey filaments [10, 11, 14]. Thermal withdrawal latency to a radiant heat source has also been measured, with a report of decreased response latency in rats [10] balanced by others showing thermal hypoalgesia in rats [11] and mice [13] or no change [14]. It is plausible that such discrepancies arise from the degree of structural abnormalities induced by particular treatment regimens. Thus, vinblastine-treated mice that developed thermal hypoalgesia, also had a 30% decrease in both stride length and gait support, a decrease in caudal nerve conduction velocity, and an increase in the number of demyelinating and degenerating axons in the mid-thigh sciatic nerve [13]. In contrast, doses that evoked mechanical and thermal hyperalgesia in rats produced only axonal swelling and mild disorganization of microtubules in unmyelinated and myelinated sensory axons [15, 16], accompanied by NCV slowing in A and C fibers and a increased responsiveness of some C fibers to mechanical and thermal stimuli [17].

Paclitaxel

The use of paclitaxel, a compound derived from the bark of the Western Yew, is based on its ability to bind to β-tubulin and polymerize microtubules [18]. Like the vinca alkaloids discussed above, paclitaxel impedes cell proliferation and is in widespread use against a range of tumors. Paclitaxel also shares the neurotoxic side effects resulting from disruption of the normal microtubule cytoskeleton with patients developing a sensorimotor peripheral neuropathy [19, 20]. A number of animal studies have examined the consequences of paclitaxel treatment on the peripheral nervous system [21, 22] and findings vary depending on the dose and frequency of administration. For example, one study that gave a cumulative dose of 50 mg/kg i.p. of paclitaxel over 10 days produced rats with hindpaw thermal hypoalgesia but normal large fiber MNCV, SNCV and structure, in the sciatic nerve [23]. In contrast, rats given two doses of paclitaxel 3 days apart (15 mg/kg i.v.) exhibited MNCV and SNCV slowing, fiber degeneration in the dorsal but not ventral roots and normal thermal responses in both the tail and hindlimbs [24]. None of the regimes explored to date have noted any indication of allodynia or hyperalgesia in paclitaxel treated animals.

Anti-GD2 antibodies

Treatment of pediatric neuroblastoma and malignant melanoma with anti-ganglioside GD2 antibodies has been under clinical evaluation for nearly a decade. The basis for this approach is that GD2 is highly expressed on tumor cells of neuroectodermal origin. However, it was noted in an early clinical trial of a murine monoclonal antibody to the GD2 that patients experienced severe abdominal and lower limb joint pain after intravenous infusion [25]. Clinical use has continued because the pain can be adequately managed with intravenous morphine, although this is obviously a less than optimal treatment. When anti-GD2 antibodies are given as a bolus injection to rats they quickly (within 15 minutes) develop mechanical allodynia [26]. This is accompanied by ectopic activity in Aδ and C fibers and also a significant decrease in mechanical thresholds of Aδ fibers that can be reversed by administration of intravenous lidocaine [27] or gabapentin [28]. The mechanism of the pain seen in animals or patients treated with anti-GD2 antibodies is currently unclear. GD2 is found on the myelin sheath of peripheral nerve [29, 30], making it a target for anti-GD2 antibodies. The anti-GD2 antibody induces death in tumor cells through both antibody-dependent cellular cytotoxicity and complement-dependent lysis [31], both of which are associated with release of various cytokines, which may directly sensitize sensory afferents. These mechanisms could initiate the rapid onset of hyperalgesia [26] and also prompt demyelination. Demyelination has been observed in a sural nerve biopsy from a patient treated with anti-GD2 antibodies [32] and can initiate ectopic discharges. Given these putative mechanisms, use of the anti-GD2 antibody to initiate pain states in rodents may well serve as a useful model for a variety of neuropathies arising from autoimmune disorders.

Thalidomide

Thalidomide was originally introduced as a sedative but was withdrawn because of teratogenic effects that disrupted limb development. It has been subsequently reintroduced in selected patient populations to treat inflammatory skin diseases, HIV and leprosy based upon its immunomodulatory properties. However, prolonged use induces a peripheral neuropathy with largely sensory features such as distal paresthesias and numbness accompanied by large fiber SNCV slowing and decreased sensory action potentials [33]. Thalidomide-treated rabbits show similar functional changes [34] accompanied by thinning of the myelin sheath [35]. Other than inducing neuropathy, acute thalidomide treatment has also been reported to attenuate tactile allodynia and thermal hyperalgesia caused by chronic constriction injury (CCI) to the sciatic nerve [36]. This effect was suggested to be a consequence of the ability to inhibit release of the pro-inflammatory cytokine TNFα from activated macrophages infiltrating the lesion site. Interestingly, direct injection of TNFα into the sciatic nerve produces behavioral and pathologic changes reminiscent of the CCI model [37]. Subsequent electrophysiologic studies showed that the application of TNFα directly to the exposed sciatic nerve resulted in a dose-dependent increase in background firing frequency in C and Aδ fibers without a decrease in mechanical thresholds in cutaneous receptive fields [38]. One proposed mechanism to explain this ectopic activity is that the TNF trimer is able to form a sodium permeable ion pore by inserting into the axonal membrane, as demonstrated in cultured lymphoma cells and planar lipid bilayers [39]. The intraneural injection of TNFα may thus prove to be a useful model for examining painful inflammatory neuropathies.

Metabolic neuropathies

Metabolic neuropathies include those arising from deficiency of a particular nutrient such as vitamins B_1, B_6, B_{12} or E [40]. The neuropathy associated with alcoholism can present with pain or dysestheias in the extremities and is similar to beriberi neuropathy. Both are associated with thiamine (vitamin B_1) deficiency and this may be the cause of the neuropathy, along with potential direct effects of alcohol on the nervous system. Alcohol neuropathy affects sensory, motor and autonomic nerves and sensory loss in the hands and feet can be accompanied by spontaneous burning pain, tactile allodynia and hyperalgesia. Electrophysiologic and pathologic investigations have demonstrated distal axonal degeneration of all fiber classes and sizes that results in fiber loss. Interestingly, while chronic alcohol consumption by rats does not produce structural changes to peripheral nerves [40] thermal hyperalgesia has been reported although this has not been widely explored [41]. Excess of a nutrient may also induce peripheral neuropathy, as is seen in after excessive dietary intake of pyridoxine (vitamin B_6) by humans [42] or animals [43]. This causes degeneration of large sensory fibers with attendant SNCV slowing and loss of proprioceptive function [44]. Diabetes mellitus, the most common cause of peripheral neuropathy in the developed world may also be included in this category, on the largely-validated assumption that nerve disorders are secondary to hyperglycemia.

Diabetic neuropathy

Diabetes is an increasing clinical problem in developed countries and peripheral neuropathy is the most common of the secondary complications associated with the disease. A significant proportion of diabetics exhibit nerve dysfunction at diagnosis and the prevalence in the diabetic population can be as high as 50%, depending on the diagnostic criteria [45]. Diabetic neuropathy may be broadly divided into focal/multifocal neuropathies and symmetric polyneuropathies [46]. In the former, neuropathy appears related to focal ischemic or compression lesions, whereas the etiology of the symmetric polyneuropathies is considered metabolic in origin. The manifestations of sensory dysfunction are varied, ranging from spontaneous pain and paresthesias to sensory loss. In one study, 46% of patients complained of pain that impaired their quality of life [47] and a range of symptoms may be present in the same patient [48]. Pain may develop at the onset of diabetes (hyperglycemic neuropathy [49]), in established diabetics following an acute period of poor glycemic control (acute painful neuropathy [50]) or insidiously over a period of time (chronic painful neuropathy [51]). This pain may be spontaneous and described as burning or lancinating or it may occur in response to either mild or innocuous stimuli.

At present, there is little understanding of the mechanisms that initiate or maintain painful diabetic neuropathy. Pain occurs in both insulin-deficient and insulin-replete diabetics, indicating that it is unlikely to arise from insulin deficiency *per se* and prompting speculation that it may result from either hyperglycemia or its metabolic consequences. Hyperalgesia can be induced in normal subjects by infusions of glucose [52]. However, this was not reproduced in diabetic patients and painful episodes in diabetics are not associated with rapid changes in blood glucose levels [53]. It is therefore unlikely that pain in diabetes is caused by acute hyperglycemic episodes. The prevalence of diabetic neuropathy and availability of animal models of diabetes have prompted a large number of experimental studies into the etiology of nerve dysfunction during prolonged hyperglycemia.

Animal models

Rodents have provided the most widely studied animal models of diabetic neuropathy. A number of rodent strains have been bred that develop spontaneous diabetes arising from autoimmune attack of the pancreas (the BB Wistar rat and NOD mouse) or from peripheral insulin resistance (sand rats, db/db mice and ob/ob mice). Use of these models is somewhat limited because of their restricted availability and also difficulties in maintaining and breeding the colonies. Moreover, genetically diabetic rodents exhibit a relatively early and unpredictable onset of hyperglycemia, which can superimpose maturational impairments on the model when diabetes develops in immature animals. Because of these concerns, and the general assumption that nerve dysfunction is secondary to prolonged hyperglycemia irrespective of its genesis, the majority of published studies have used rodents made diabetic with agents that destroy pancreatic β cells.

Hyperglycemia secondary to injection of the pancreatic β cell toxins streptozotocin (STZ) or alloxan has been extensively characterized in rodents over the last 30 years [54] and offers the experimental advantage of producing reproducible insulin-deficient hyperglycemia at the age of choice. These toxins enter the pancreatic β cell *via* the GLUT 2 transporter [55], which is also present in certain brain areas [56]. The possibility of direct neurotoxic actions must therefore be considered as an alternative to insulin deficiency or subsequent hyperglycemia when investigating the etiology of nerve disorders in STZ and alloxan-diabetic animals. However, the toxins are rapidly cleared from circulation and the ability of insulin replacement therapy to reverse any identified nerve dysfunction can provide a strong argument against a role for direct neurotoxicity.

An alternative and mechanistically selective model of diabetes has been established by inducing galactose intoxication. Animals fed diets containing a high proportion (typically 20–40% by weight) of the hexose sugar galactose develop a number of peripheral nerve disorders despite normal circulating insulin and glucose levels [57, 58]. Use of selective inhibitors suggests that many of the nerve disorders seen in these animals may be related to metabolism of galactose by the enzyme aldose reductase [59, 60]. As aldose reductase also acts on excess glucose during hyperglycemia, galactose intoxication models metabolic flux through aldose reductase as a pathogenic mechanism for diabetic neuropathy in the absence of the systemic and local metabolic consequences of insulinopenia and hyperglycemia.

Sensory neuropathology

Before detailing the behavioral evidence of sensory dysfunction in hyperglycemic animals it is pertinent to consider briefly the structural context in which it occurs. Insulin-deficient rodents may survive for periods of weeks to months in the untreated state, exhibiting marked hyperglycemia (20–40 mmol/L) and progressive weight loss due to muscle wasting. Nevertheless, there is little evidence of overt peripheral neuropathology in these animals [61]. The hallmarks of progressive neuropathy in diabetic patients, such as fiber loss and attempted regeneration are not observed until after a year or more of hyperglycemia in rats. While insulin supplementation to relieve catabolic dominance can extend the survival time for many months [62], practical reasons have dictated that most behavioral studies are performed in animals after 1–3 months of hyperglycemia. In such animals, gross structural changes are restricted to atrophy of the largest myelinated fibers in the sural and sciatic nerves [63, 64]. There is also reduced neuronal cell volume in the dorsal root ganglia [65] and axonal atrophy and myelin splitting in the dorsal roots [60, 66]. These subtle structural changes are accompanied by impaired amino acid uptake into cell bodies of the dorsal root ganglia [67] and retarded slow axonal transport of cytoskeletal components [68]. Such metabolic and functional disorders have the potential to contribute to the axonal atrophy, restricted capacity to respond to trauma [69, 70] and impaired sensory nerve regeneration [71] seen in diabetic rodents.

While the lack of overt pathology clearly limits the utility of diabetic rodents as models of late stage diabetic neuropathy, the absence of degenerative changes provides an opportunity to examine behavioral indices of sensory dysfunction and potential neurochemical mechanisms without the confound of ongoing fiber loss.

Behavioral responses

A body of work has accumulated describing the behavioral responses of diabetic rats to a range of sensory stimuli. Some studies have focused on responses to acute high intensity nociceptive stimuli, which most closely equate to the quantitative sensory tests performed in diabetic patients. Others have sought to model the allodynia seen in some diabetics or to investigate how hyperglycemia alters modulation of nociceptive processing.

Responses to thermal stimuli

Measuring the time to withdrawal of the tail from a noxious thermal stimulus (the "tail flick" test) is probably the least complex thermal sensory test, involving peripheral sensory and motor functions linked *via* a spinal reflex arc. Unfortunately, there is no clear picture of the effects of diabetes on this response. Diabetic mice have been variously described as having increased [72] or unchanged [73–77] response times. In rats, hyperglycemia has been reported to cause reduced response times to noxious heat [78–81] and cold [80], although hypoalgesia has also been noted [82]. The (presumably) methodological reasons for these disparities have yet to be addressed.

Measuring the time to withdrawal of the paw from a heat source applied to the plantar surface represents a more complex test that incorporates supraspinal sensory processing. The heat may come from a hot plate or a focused light source. Most reports to date have examined responses in STZ-diabetic rats and thermal hypoalgesia is widely described [83–90] although there is also a report of thermal hyperalgesia [91]. A time course study in male Sprague-Dawley rats revealed a transient period of mild thermal hyperalgesia before progression to hypoalgesia [92]. The progressive development of thermal hypoalgesia in diabetic rodents is consistent with clinical observations, where patients with diabetic neuropathy show loss of thermal pain sensation [48].

Responses to high threshold mechanical stimuli

Paw withdrawal from incremental surface pressure has been investigated in diabetic rats, which show decreased withdrawal latencies indicative of responses to lower applied pressures [80, 81, 90, 93–100]. An alternative assay has used the response time to uniform pressure applied to the tail with forceps. There is one report of an increased response time in this test in diabetic rats [72] but the majority report decreased withdrawal latencies in both diabetic rats and mice [73–75, 101, 102]. In general, it appears that diabetic rodents exhibit mechanical hyperalgesia in response to high threshold stimuli, although there is no obvious correlate in clinical quantitative sensory tests applied to diabetic patients.

Responses to low threshold mechanical stimuli

A significant proportion of patients with diabetic neuropathy develop tactile allodynia. This phenomenon is replicated in STZ-diabetic rats, which respond to the light touch of von Frey filaments [103] and to light stroking of the paw [104] within weeks of the onset of diabetes. Tactile allodynia can be prevented or reversed in diabetic rats by instituting prolonged normoglycemia using insulin and is of similar magnitude to that induced by spinal nerve ligation [103]. A number of recent studies have used tactile allodynia as a means of illustrating the efficacy of potential therapeutic agents directed at the acute alleviation of painful diabetic neuropathy, as this test is amenable to frequently repeated measurements [105–109].

Noxious chemical stimuli

Injection of formalin into the paw has been widely used as a model of pain-associated behavior that is triggered and maintained by primary afferent activity and modulated at the spinal cord level. While this model may not have an obvious correlate in the sensory tests employed when evaluating diabetic patients, the formalin test allows examination of the effects of diabetes on the mechanisms of spinal nociceptive processing beyond primary afferent input to the spinal cord.

Diabetic mice show reduced formalin-evoked responses [110–113] and this has been associated with changes in regulation of spinal nociceptive processing by endogenous opioid systems [114]. In contrast, diabetic rats exhibit an increased frequency of responses following paw formalin injection [80, 115], with the difference becoming more exaggerated as lower doses of formalin are used [116]. The neurochemical or other reasons for this clear species difference between mice and rats are not yet known.

Etiology of sensory dysfunction

Short-term diabetic rodents show minimal pathologic changes in their peripheral nerves. This allows examination of potential etiologic mechanisms without interpretation being clouded by superimposition of the consequences of fiber loss.

Underlying metabolic etiology

As mentioned above, it is important to exclude direct neurotoxicity of diabetogenic agents as an etiologic mechanism causing behavioral disorders. Insulin replacement therapy that provided extended periods of normoglycemia has prevented thermal hypoalgesia in paw withdrawal tests [83], thermal hyperalgesia in tail flick tests [79], tactile allodynia to von Frey filaments [103], mechanical hyperalgesia [97] and hyperalgesia in the formalin test [103]. It is therefore likely that these disorders are initiated by insulin deficiency and subsequent periods of hyperglycemia. Further evidence suggesting involvement of hyperglycemia and its sequale is provided by studies showing that hyperalgesia during the formalin test [85, 116] and paw thermal hypoalgesia (Fig. 1) are both prevented by prolonged treatment with agents that block metabolism of glucose by the polyol pathway enzyme, aldose reductase. Flux through the polyol pathway has been implicated in many aspects of diabetic neuropathy [117] and metabolic products of excessive polyol pathway flux accumulate in both peripheral nerves and the spinal cord [103]. The presence of mechanical [93] and formalin [85] hyperalgesia in rats fed a high galactose diet also supports involvement of exaggerated polyol pathway flux in the underlying etiology, as this model of hyperglycemia selectively provides hexose substrate for aldose reductase in the context of normal circulating insulin and glucose levels. At present, it is not clear how increased polyol pathway flux causes nerve dysfunction and the mechanisms linking increased sugar metabolism by aldose reductase to behavioral indices of sensory dysfunction remain obscure. This prompts examination of the functional and neurochemical properties of sensory nerves from diabetic animals that may provide further clues to the mechanisms that precipitate sensory dysfunction.

Electrophysiologic dysfunction

A number of experimental studies have evaluated the electrophysiologic function of peripheral sensory nerves and there is no strong evidence to suggest the presence of

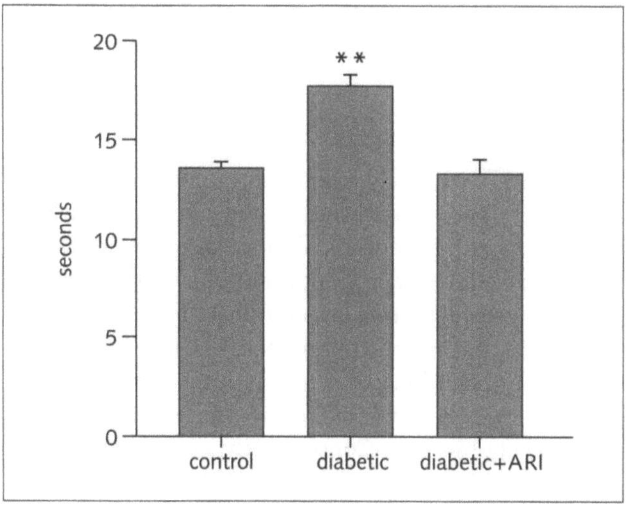

Figure 1
*Thermal hypoalgesia after 8 weeks of STZ diabetes in female Sprague-Dawley rats is prevented by treatment with the aldose reductase inhibitor ICI 222155 (20 mg/kg/day p.o.) for the duration of diabetes. Data are mean ± SEM of N = 13–14/group. ** = $p < 0.01$ vs. controls by ANOVA with Dunnett's post hoc test.*

either conduction block or increased spontaneous activity [118, 119] that could reflect sensory loss or ongoing pain states respectively. Slowing of peripheral large fiber conduction velocities is a hallmark of short-term clinical [120] and experimental [121] diabetes and can be corrected by inhibiting aldose reductase [122]. Large myelinated sensory fibers that mediate light touch and proprioception are widely reported to show a 10–20% reduction in conduction velocity [123], although there is no evidence of any change in the distribution of Aβ receptor types [118]. Thus, it is not clear that peripheral electrophysiologic dysfunction contributes to abnormalities in behavioral tests that reflect Aβ fiber function such as allodynia to light touch.

Small myelinated (Aδ) and unmyelinated (C) fibers transmit nociceptive information and mediate many of the behavioral tests described above. There have been reports of both unchanged [124] and increased [94, 118] conduction velocities for C fibers in diabetic rats. In a detailed series of single fiber studies in the saphenous nerve of STZ-diabetic rats, it was reported that C fibers had mechanical and thermal (hot and cold) thresholds that did not differ from controls. However, increased C fiber afterdischarges following a sustained suprathreshold mechanical stimulus, proportionately fewer silent C fibers and a reduction in Aδ fiber mechanical thresholds were all noted [118, 125] and may contribute to some of the aberrant behavioral responses, such as mechanical hyperalgesia, that are discussed above.

Neurochemical dysfunction

It is becoming clear that adult sensory nerves rely upon continuous neurotrophic support to maintain their phenotype and recent studies indicate that diabetes impairs a variety of neurotrophic support mechanisms. A population of small sensory neurons express the $p75^{NTR}$ and trk A receptors that together confer responsiveness to the neurotrophin NGF. In contrast, large sensory fibers bear the $p75^{NTR}$ and trk C receptors and respond to the neurotrophin NT-3. Diabetes reduces mRNA for both NGF and NT-3 in peripheral target organs, lowers the amount of $p75^{NTR}$ and trk C receptors in sensory nerves and limits retrograde transport of neurotrophins back towards the cell body [126–130]. Consequently, genes regulated by these neurotrophic factors, which include those that determine synthesis of neuropeptides [131], are attenuated in their actions and there is reduced synthesis and axonal transport of both substance P and CGRP [132, 133]. These neuropeptides modulate central nociceptive processing [134–136], and diminished evoked release in diabetes [88, 137, 138] could contribute to abnormal pain perception. In support of this possible association, agents that maintain peripheral nerve levels of substance P in diabetic rats also prevent thermal hypoalgesia in diabetic rodents (Fig. 1 and [82, 87, 132]).

While the reduction of sensory neuropeptides has been investigated extensively in peripheral nerve of diabetic rodents, relatively few studies have examined other neurochemical components potentially involved in peripheral sensory processing. Measurements in primary afferent cell bodies of mRNA for proteins assumed to undergo axonal transport to presynaptic terminals in the spinal cord have shown a 25% decrease in mRNA for the SNS sodium channel [139], the appearance of mRNA for the α_1 adrenoceptor [140] and a small increase in mRNA for nNOS in diabetic rats [141] which is matched by an absence of marked increases in NOS protein or activity [142].

Dysfunction in spinal sensory processing

Although diabetes is considered to produce a largely peripheral neuropathy, it is plausible that disorders beyond primary afferent driven peripheral input contribute to sensory dysfunction. The first site of modulation of primary afferent processing occurs in the spinal cord, where excitatory input can be amplified or suppressed. Until recently, the potential contribution of aberrant spinal nociceptive processing has been relatively ignored in studies of diabetic neuropathy. Spinal tissue content of substance P is reduced [143], as is formalin-evoked release of both substance P [88] and glutamate [144]. For both of these excitatory neurotransmitters there appears to be a compensatory increase in receptor number, as indicated by ligand binding studies for NK-1 [145], NMDA and AMPA [146] receptors. Interestingly, we have recently found that the transient thermal hyperalgesia induced by intrathecal delivery of a fixed amount of substance P is prolonged in diabetic rats [89], which further supports the idea of a spinal modification of nociceptive processing. Pathways involved in local or descending inhibitory pathways have also been investigated. Spinal content of 5-HT is unchanged but centrally evoked release is decreased [147]. There are reports of both decreased [143] and increased [92] levels of met-enkephalin, while a study of nor-adrenaline and its metabolite MHPG found the former elevated and the latter decreased, which was suggested as illustrating decreased noradrenergic activity [148]. Studies have reported normal levels of α_1 adrenoceptor mRNA in the spinal cord of diabetic rats [109] and either normal [109] or increased binding sites [148]. In contrast, α_2 adrenoceptor mRNA and binding sites were reduced [148].

From a neurochemical perspective, the limited evidence to date presents a general picture of depressed input from peripheral primary afferents that is accompanied by a variety of changes to both excitatory and inhibitory mechanisms at the spinal level. A recent electrophysiologic study found increased spontaneous activity of wide dynamic range neurons in the spinal cord of diabetic rats, while none occurred in primary afferents [149], further supporting the concept of modified spinal sensory processing in diabetes. Forming etiologic links between peripheral and spinal neurochemical disorders and the abnormal behavioral responses to sensory stimuli remains a major challenge.

Utility of animal models to pain research

Diabetic neuropathy is a complex syndrome with a range of manifestations extending from sensory loss to spontaneous pain. It is widely acknowledged that hyperglycemic rodents do not model human diabetic neuropathy perfectly. The hyperglycemia is extreme and the duration of diabetes short, as indeed are the nerves themselves compared to human peripheral nerves. These factors may explain why the degenerative neuropathy seen in nerves supplying the hands and feet of many chronically diabetic patients is not replicated in diabetic rodents. At best, the rodent models should be viewed as elaborating the consequences of short-term hyperglycemia upon the metabolism and neurochemistry of peripheral nerve. However, these models may serve to expose the biochemical genesis of peripheral nerve disorders that ultimately progress to neuropathy as experienced by humans and their use should always be viewed in this context.

While the animal models of diabetic neuropathy are not ideal, the same has also been acknowledged of using nocifensive behavioral responses to noxious stimuli as a measure of pain perception in animals. Given the use of indirect measures of pain perception in a less than perfect model of diabetic neuropathy, the utility of these types of studies must be considered carefully before applying findings to the clinical setting. However, within the frame of current knowledge, careful studies can minimize the possibility of erroneous findings. With regard to diabetic rodents, animals exhibiting the signs of behavioral depression [90] and torpor that can accompany extreme hyperglycemia should be excluded from behavioral studies, particularly where hypoalgesia is being examined. Hyperalgesia is less likely to arise from behavioral depression in diabetic rats, but should also be treated with caution as diabetic animals have reduced muscle mass and exhibit changes in their skin and connective tissues that may contribute to altered stimulus transduction properties independent of any changes in nerve function *per se*. Diabetic rats receiving low dose insulin supplementation to maintain normal body weight and behavior in the context of continued hyperglycemia [103] may prove a particularly useful model. In general, extrapolating abnormal behavioral responses in sensory tests to dysfunction of pain processing should be viewed with caution until the disorder can be prevented by agents other than those such as insulin that improve the general metabolic condition of the animal.

In addition to cautious interpretation of data from diabetic rodents, there are also specific experimental considerations that are associated with behavioral testing. Rodents exposed to a variety of repeat-testing regimes tend to modify their behaviors over time and effects may be different between normal and diabetic rats. Further, we have recently found that repeated testing of paw withdrawal from von Frey filaments in diabetic rats did not alter the degree of basal tactile allodynia but obscured the effect of an agent that transiently alleviated the allodynia [89]. It is plausible that the conflicting reports of the effect of diabetes upon a range of sensory tests described earlier in this chapter may derive as much from methodological variations and quirks as from sex, strain and species differences in the rodents being tested.

Screening potential therapeutic agents

Within the limitations considered above, it is hoped that information obtained from animal models may reveal the pathogenesis of sensory disorders associated with diabetes and also predict the efficacy of potential prophylactic or ameliorative treatments. A substantial number of studies have been performed in diabetic rodents using the assays described above as end point indicators of efficacy. A few of these have investigated chronic treatments aimed at preventing onset of sensory dysfunction and therefore suggesting potential pathogenic mechanisms. The efficacy of insulin and aldose reductase inhibitors has been discussed earlier. Given the neurotrophic deficiency experienced by diabetic nerves and subsequent potential neurochemical consequences that may impact sensory functions, the effects of neurotrophic factor replacement therapy or of other neuroactive agents have also been investigated. Thermal hypoalgesia is prevented by treating diabetic rats with NGF

[82] and CNTF, but not BDNF (Calcutt, unpublished observations). The peptide TX14(A), which contains the neuroactive sequence from prosaposin, also prevented onset of thermal hypoalgesia in diabetic rats [87, 89], whereas a peptide derived from ACTH/MSH did not [86]. These data may support an association between protection of neurotrophic support of primary afferent sensory neuropeptide levels and maintenance of thermal nociception. Less is known of the effects of such neuroactive factors on manifestations of hyperalgesia in diabetic rats, although chronic treatment with BDNF or CNTF was without effect on development of tactile allodynia (Calcutt, unpublished observations).

The majority of studies of experimental therapeutics have focused on attempts to rapidly alleviate established sensory disorders in diabetic rodents with the aim of developing analgesics for relief of painful diabetic neuropathy. This approach has the laudable aim of offering pain relief to patients but is unlikely to treat progression of the underlying neuropathy. Early studies examined thermal response latencies and revealed an apparently diminished potency of systemic morphine [150–152]. This may be related to altered distribution and excretion kinetics operational in diabetic animals [153], an issue which will always confound comparisons of systemic drug potency between diabetic and normal animals.

Mechanical hyperalgesia has been used extensively as a screening tool, with temporary efficacy reported after systemic delivery of, amongst others, morphine [90, 152, 153], tricyclic antidepressants [152] clonidine [152], lidocaine [152], mexiletine [154], serotonin [155] the NMDA receptor antagonist MK-801 and the GABA agonist baclofen [90, 100, 156]. Antagonists of the NK-1 receptor were effective in some studies [96] but not others [90, 100], as was local delivery of inhibitors of protein kinase C and cAMP to the paw [94, 95] and delivery of the anti-epileptic lamotrigine, to the spinal cord [99]. Agents that did not alleviate mechanical hyperalgesia included aspirin [152] and L-NAME [90].

More recently, tactile allodynia has also been employed to examine potential agents and rapid, short-lived amelioration has been reported after intrathecal delivery of morphine, the glutamatergic antagonists AP5 and NBQX, the N type calcium channel antagonist SNX-239 and an α_2 adrenoceptor agonist [105]. Systemic delivery of lidocaine [103], morphine [104, 107, 108), gabapentin [104], amitriptyline [104], TX14(A) [89], an adenosine kinase inhibitor [107] and antagonists of the NK-1 [106], endothelin A [108] and α_1 adrenergic [109] receptors also alleviated tactile allodynia. Agents without effect include L-type calcium channel antagonists, L-NAME, ketorolac [105], and an endothelin B antagonist [108].

Hyperalgesia following paw formalin injection in diabetic rats has been less extensively used to date, in part because only one measurement per animal may be made. Agents that reduce both formalin-evoked responses in normal rats and also the exaggerated response of diabetic rats when delivered systemically include a synthetic ganglioside [115], morphine [153] and gabapentin [157]. TX14(A), a peptide fragment of prosaposin is unique to date as it can acutely reduce formalin-evoked responses in diabetic rats without altering the responses of normal rats [89]. This, along with the rapid alleviation of tactile allodynia mentioned above, suggests potential as a therapeutic for painful diabetic neuropathy without the side-effects associated with actions on normal nociceptive processing.

While positive of effects of particular compounds in individual models of diabetic sensory dysfunction may illustrate therapeutic potential, it is only by the successful translation to the clinic that the utility of animal models can be judged. Agents routinely used in the clinical setting such as opiates, tricyclic antidepressants and clonidine show a reasonable correlation between efficacy in humans and against mechanical hyperalgesia in diabetic rats [152]. They are, however, limited by side-effects and selective efficacy [158–160] and in the absence of an approved drug for treating painful diabetic neuropathy there is a considerable need to develop novel therapies. A number of agents have shown promise in clinical trials and off-label use. Of these, gabapentin is currently in widespread off-label use after promising clinical trials [161]. It also showed efficacy against tactile allodynia [104] and formalin-evoked hyperalgesia [157] in diabetic rats. The antiarrythmics mexiletine and infused lidocaine have shown some clinical efficacy against painful diabetic neuropathy [162, 163], and has subsequently been replicated in animal models of mechanical hyperalgesia [152, 154] and tactile allodynia [103]. There is, therefore, some hope that the animal models currently available to us may be imperfect, but useful, tools in the search to find novel treatments of painful diabetic neuropathy and sensory loss.

References

1 Mizisin AP, Powell HC (1995) Toxic neuropathies. *Curr Opin Neurol* 8: 367–371

2 Graham DG (1999) Neurotoxicants and the cytoskeleton. *Curr Opin Neurol* 12: 733–737

3 Edwards PM, Sporel-Ozakat RE, Gispen WH (1991) Peripheral pain fiber function is relatively insensitive to the neurotoxic actions of acrylamide in the rat. *Toxicol Appl Pharmacol* 11: 43–48

4 Miller MS, Spencer PS (1984) Single doses of acrylamide reduce retrograde transport velocity. *J Neurochem* 43: 1401–1408

5 Prineas J (1969) The pathogenesis of dying-back polyneuropathies. II. An ultrastructural study of experimental acrylamide intoxication in the cat. *J Neuropath Exp Neurol* 28: 598–621

6 Whitaker-Azmitia PM, Raio M, Raio D, Borella A (1995) A 5-HT3 receptor antagonist fails to prevent cisplatin-induced toxicity in immature rat spinal cord. *Eur J Pharmacol* 275: 139–143

7 Cavaletti G, Fabbrica D, Minoia C, Frattola L, Tredici G (1998) Carboplatin toxic effects on the peripheral nervous system of the rat. *Ann Onc* 9: 443–447

8 Tredici G, Braga M, Nicolini G, Miloso M, Marmiroli P, Schenone A, Nobbio L, Frattola L, Cavaletti G (1999) Effect of recombinant human nerve growth factor on cisplatin neurotoxicity in rats. *Exp Neurol* 159: 551–558

9 McLeod JG, Penny R (1969) Vincristine neuropathy: an electrophysiological and histological study. *J Neurol, Neurosurg Psych* 32: 297–304

10 Aley KO, Reichling DB, Levine JD (1996) Vincristine hyperalgesia in the rat: a model of painful vincristine neuropathy in humans. *Neurosci* 73: 259–265

11 Authier N,Coudore F, Eschalier A, Fialip J (1999) Pain related behaviour during vincristine-induced neuropathy in rats. *Neuroreport* 10: 965–968

12 Di Gregorio F, Favaro G, Panozzo C, Fiori MG (1990) Efficacy of ganglioside treatment in reducing functional alterations induced by vincristine in rabbit peripheral nerves. *Cancer Chemother Pharmacol* 26: 31–36

13 Contreras PC, Vaught JL, Gruner JA, Brosnan C, Steffler C, Arezzo JC, Lewis ME, Kessler JA, Apfel SC (1997) Insulin-like growth factor-I prevents development of a vincristine neuropathy in mice. *Brain Res* 774: 20–26

14 Nozaki-Taguchi, N, Chaplan SR, Higuera ES, Ajakwe AC, Yaksh TL (2001) Vincristine-induced allodynia in the rat. *Pain* 93: 69–76

15 Tanner KD, Reichling DB, Levine JD (1998) Nociceptor hyper-responsiveness during vincristine-induced painful peripheral neuropathy in the rat. *J Neurosci* 18: 6480–6491

16 Topp KS, Tanner KD, Levine JD (2000) Damage to the cytoskeleton of large diameter sensory neurons and myelinated axons in vincristine-induced painful peripheral neuropathy in the rat. *J Comp Neurol* 424: 563–576

17 Tanner KD, Levine JD, Topp KS (1998) Microtubule disorientation and axonal swelling in unmyelinated sensory axons during vincristine-induced painful neuropathy in rat. *J Comp Neurol* 395: 481–492

18 De Brabander M, Geuens G, Nuydens R, Willebrords R, De Mey J (1981) Taxol induces the assembly of free microtubules in living cells and blocks the organizing capacity of the centrosomes and kinetochores. *Proc Natl Acad Sci USA* 78: 5608–5612

19 Lipton RB, Apfel SC, Dutcher JP, Rosenberg R, Kaplan J, Berger A, Einzig AI, Wiernik P, Schaumburg HH (1989) Taxol produces a predominantly sensory neuropathy. *Neurol* 39: 368–373

20 Rowinsky EK, Eisenhauer EA, Chaudhry V, Arbuck SG, Donehower RC (1993) Clinical toxicities encountered with paclitaxel (Taxol). *Sem Oncol* 20: 1–15

21 Hamers FP, Pette C, Neijt JP, Gispen WH (1993) The ACTH-(4-9) analog, ORG 2766, prevents taxol-induced neuropathy in rats. *Eur J Pharmacol* 233: 177–178

22 Cavaletti G, Tredici G, Braga M, Tazzari S (1995) Experimental peripheral neuropathy induced in adult rats by repeated intraperitoneal administration of taxol. *Exp Neurol* 133: 64–72

23 Campana WM, Eskeland N, Calcutt NA, Misasi R, Myers RR, O'Brien JS (1998) Prosaptide prevents paclitaxel neurotoxicity. *Neurotoxicol* 19: 237–244

24 Cliffer KD, Siuciak JA, Carson SR, Radley HE, Park JS, Lewis DR, Zlotchenko E,

Nguyen T, Garcia K, Tonra JR et al (1998) Physiological characterization of Taxol-induced large-fiber sensory neuropathy in the rat. *Ann Neurol* 43: 46–55

25 Handgretinger R, Baader P, Dopfer R, Klingebiel T, Reuland P, Treuner J, Reisfeld RA, Niethammer D (1992) A phase I study of neuroblastoma with the anti-ganglioside GD2 antibody 14.G2a. *Cancer Immunol Immunother* 35: 199–204

26 Slart R, Yu AL, Yaksh TL, Sorkin LS (1997) An animal model of pain produced by systemic administration of an immunotherapeutic anti-ganglioside antibody. *Pain* 69: 119–125

27 Xiao WH, Yu AL, Sorkin LS (1997) Electrophysiological characteristics of primary afferent fibers after systemic administration of anti-GD2 ganglioside antibody. *Pain* 69: 145–151

28 Gillin S, Sorkin LS (1998) Gabapentin reverses the allodynia produced by the administration of anti-GD2 ganglioside, an immunotherapeutic drug. *Anesth Analgesia* 86: 111–116

29 Svennerholm L, Bostrom K, Fredman P, Jungbjer B, Lekman A, Mansson JE, Rynmark BM. (1994) Gangliosides and allied glycosphingolipids in human peripheral nerve and spinal cord. *Biochim Biophys Acta* 1214: 115–123

30 Yuki N, Yamada M, Tagawa Y, Takahashi H, Handa S (1997) Pathogenesis of the neurotoxicity caused by anti-GD2 antibody therapy. *J Neurol Sci* 149: 127–130

31 Mujoo K, Kipps TJ, Yang HM, Cheresh DA, Wargalla U, Sander DJ, Reisfeld RA (1989) Functional properties and effect on growth suppression of human neuroblastoma tumors by isotype switch variants of monoclonal antiganglioside GD2 antibody 14.18. *Cancer Res* 49: 2857–2861

32 Saleh MN, Khazaeli MB, Wheeler RH, Dropcho E, Liu T, Urist M, Miller DM, Lawson S, Dixon P, Russell CH et al (1992) Phase I trial of the murine monoclonal anti-GD2 antibody 14G2a in metastatic melanoma. *Cancer Res* 52: 4342–4347

33 Fullerton PM, Kremer M (1961) Neuropathy after intake of thalidomide (Distaval). *Brit Med J* 2: 855–858

34 Schwab BW, Arezzo JC, Paldino AM, Flohe L, Matthiessen T, Spencer PS (1984) Rabbit sural nerve responses to chronic treatment with thalidomide and supidimide. *Muscle Nerve* 7: 362–368

35 Schroder JM, Matthiesen T (1985) Experimental thalidomide neuropathy: the morphological correlate of reduced conduction velocity. *Acta Neuropathol* 65: 285–292

36 Sommer C, Marziniak M, Myers RR (1998) The effect of thalidomide treatment on vascular pathology and hyperalgesia caused by chronic constriction injury of rat nerve. *Pain* 74: 83–91

37 Wagner R, Myers RR (1996) Endoneurial injection of TNF-alpha produces neuropathic pain behaviors. *Neuroreport* 7: 2897–2901

38 Sorkin LS, Xiao WH, Wagner R, Myers RR (1997) Tumour necrosis factor-alpha induces ectopic activity in nociceptive primary afferent fibres. *Neurosci* 81: 255–262

39 Kagan BL, Baldwin RL, Munoz D, Wisnieski BJ (1992) Formation of ion-permeable channels by tumor necrosis factor-alpha. *Science* 255: 1427–1430

40 Windebank AJ (1993) Polyneuropathy due to nutritional deficiency and alcoholism. In: PJ Dyck, PK Thomas (eds): *Peripheral neuropathy*. WB Saunders, Philadelphia, 1310–1321

41 Kampov-Polevoy AB, Kasheffskaya OP, Overstreet DH, Rezvani AH, Viglinskaya IV, Badistov BA, Seredenin SB, Halikas JA, Sinclair JD (1996) Pain sensitivity and saccharin intake in alcohol-preferring and nonpreferring rat strains. *Physiol Behav* 59: 683–688

42 Schaumburg H, Kaplan J, Windebank A, Vick N, Rasmus S, Pleasure D, Brown MJ (1983) Sensory neuropathy from pyridoxine abuse. A new megavitamin syndrome. *New Engl J Med* 309: 445–448

43 Xu Y, Sladky JT, Brown MJ (1989) Dose-dependent expression of neuronopathy after experimental pyridoxine intoxication. *Neurol* 39: 1077–1083

44 Helgren ME, Cliffer KD, Torrento K, Cavnor C, Curtis R, DiStefano PS, Wiegand SJ, Lindsay RM. (1997) Neurotrophin-3 administration attenuates deficits of pyridoxine-induced large-fiber sensory neuropathy. *J Neurosci* 17: 372–382

45 Pirart J (1978) Diabetes mellitus and its degenerative complications: a prospective study of 4,400 patients observed between 1947 and 1973. *Diabetes Care* 1: 168–188

46 Thomas PK, Tomlinson DR (1993) Diabetic and hypoglycemic neuropathy. In: PJ Dyck, PK Thomas (eds): *Peripheral neuropathy*. WB Saunders, Philadelphia, 1219–1250

47 Ahroni JH, Boyko EJ, Davignon DR, Pecoraro RE (1994) The health and functional status of veterans with diabetes. *Diabetes Care* 17: 318–321

48 Benbow SJ, Chan AW, Bowsher D, MacFarlane IA, Williams G (1994) A prospective study of painful symptoms, small-fibre function and peripheral vascular disease in chronic painful diabetic neuropathy. *Diabetic Med* 11: 17–21

49 Jordan WR (1936) Neuritic manifestations in diabetes mellitus. *Arch Int Med* 57: 307–311

50 Archer AG, Roberts VC, Watkins PJ (1984) Blood flow patterns in painful diabetic neuropathy. *Diabetologia* 27: 563–571

51 Boulton AJ, Drury J, Clarke B, Ward JD (1982) Continuous subcutaneous insulin infusion in the management of painful diabetic neuropathy. *Diabetes Care* 5: 386–390

52 Morley GK, Mooradian AD, Levine AL, Morley JE (1984) Mechanisms of pain in diabetic peripheral neuropathy: effects of glucose on pain in humans. *Am J Med* 77: 79–86

53 Chan AW, MacFarland IA, Bowsher DR, Wells JCD (1990) Does acute hyperglycaemia influence heat pain threshold? *J Neurosurg* 57: 688–690

54 Rerup CC (1970) Drugs producing diabetes through damage of the insulin secreting cells. *Pharmacol Rev* 22: 485–518

55 Wang Z, Gleichmann H (1998) GLUT2 in pancreatic islets: crucial target molecule in diabetes induced with multiple low doses of streptozotocin in mice. *Diabetes* 47: 50–56

56 Leloup C, Arluison M, Lepetit N, Cartier N, Marfaing-Jallat P, Ferre P, Penicaud L (1994) Glucose transporter 2 (GLUT 2): expression in specific brain nuclei. *Brain Res* 638: 221–226

57 Gabbay KH (1973) Role of sorbitol pathway in neuropathy. *Adv Metab Disord* 2: 417–432
58 Forcier NJ, Mizisin AP, Rimmer MA, Powell HC (1991) Cellular pathology of the nerve microenvironment in galactose intoxication. *J Neuropathol Exp Neurol* 50: 235–255
59 Mizisin AP, Powell HC (1993)Schwann cell injury is attenuated by aldose reductase inhibition in galactose intoxication. *J Neuropathol Exp Neurol* 52: 78–86
60 Mizisin AP, Bache M, DiStefano PS, Acheson A, Lindsay RM, Calcutt NA (1997) BDNF attenuates functional and structural disorders in nerves of galactose-fed rats. *J Neuropathol Exp Neurol* 56: 1290–1301
61 Sharma AK, Thomas PK (1974) Peripheral nerve structure and function in experimental diabetes. *J Neurol Sci* 23: 1–15
62 Willars GB, Calcutt NA, Compton AM, Tomlinson DR, Keen P (1989) Substance P levels in peripheral nerve, skin, atrial myocardium and gastrointestinal tract of rats with long-term diabetes mellitus. Effects of aldose reductase inhibition. *J Neurol Sci* 91: 153–164
63 Jakobsen J (1976) Axonal dwindling in early experimental diabetes. I. A study of cross-sectioned nerves. *Diabetologia* 12: 539–546
64 Nukada H, Dyck PJ, Low PA, Lais AC, Sparks MF (1986) Axonal caliber and neurofilaments are proportionately decreased in galactose neuropathy. *J Neuropathol Exp Neurol* 45: 140–150
65 Sidenius P, Jakobsen J (1980) Reduced perikaryal volume of lower motor and primary sensory neurons in early experimental diabetes. *Diabetes* 29: 182–186
66 Tamura E, Parry GJ (1994) Severe radicular pathology in rats with longstanding diabetes. *J Neurol Sci* 127: 29–35
67 Thomas PK, Wright DW, Tzebelikos E (1984) Amino acid uptake by dorsal root ganglia from streptozotocin-diabetic rats. *J Neurol Neurosurg Psych* 47: 912–916
68 Medori R, Autilio-Gambetti L, Monaco S, Gambetti P (1985) Experimental diabetic neuropathy: impairment of slow transport with changes in axon cross-sectional area. *Proc Natl Acad Sci USA* 82: 7716–7720
69 McLean WG, Chapman JE, Cullum NA (1987) Impaired induction of ornithine decarboxylase activity following nerve crush in the streptozotocin-diabetic rat. *Diabetologia* 30: 963–965
70 Calcutt NA, Mizisin AP, Yaksh TL (1993) Impaired induction of vasoactive intestinal polypeptide after sciatic nerve injury in the streptozotocin-diabetic rat. *J Neurol Sci* 119: 154–161
71 Ekstrom AR; Tomlinson DR (1989) Impaired nerve regeneration in streptozotocin-diabetic rats. Effects of treatment with an aldose reductase inhibitor. *J Neurol Sci* 93: 231–237
72 Levine AS, Morley JE, Wilcox G, Brown DM, Handwerger BS (1982) Tail pinch behavior and analgesia in diabetic mice. *Physiol Behav* 28: 39–43
73 Kamei J, Ohhashi Y, Aoki T, Kasuya Y (1991) Streptozotocin-induced diabetes in mice

reduces the nociceptive threshold, as recognized after application of noxious mechanical stimuli but not of thermal stimuli. *Pharmacol Biochem Behav* 39: 541–544
74 Kamei J, Kawashima N, Kasuya Y (1992) Role of spleen or spleen products in the deficiency in morphine-induced analgesia in diabetic mice. *Brain Res* 576: 139–142
75 Kamei J, Kawashima N, Kasuya Y (1992) Naloxone-induced analgesia in diabetic mice. *Eur J Pharmacol* 210: 339–341
76 Kamei J, Kawashima N, Kasuya Y (1993) Serum glucose level-dependent and independent modulation of mu-opioid agonist-mediated analgesia in diabetic mice. *Life Sci* 52: 53–60
77 Kamei J, Kawashima N, Narita M, Suzuki T, Misawa M, Kasuya Y (1994) Reduction in ATP-sensitive potassium channel-mediated antinociception in diabetic mice. *Psychopharmacol* 113: 318–321
78 Lee JH, McCarty R (1990) Glycemic control of pain threshold in diabetic and control rats. *Physiol Behav* 47: 225–230
79 Lee JH, McCarty R (1992) *Pain* threshold in diabetic rats: effects of good *versus* poor diabetic control. *Pain* 50: 231–236
80 Courteix C, Eschalier A, Lavarenne J (1993) Streptozocin-induced diabetic rats: behavioural evidence for a model of chronic pain. *Pain* 53: 81–88
81 Courteix C, Bourget P, Caussade F, Bardin M, Coudore F, Fialip J, Eschalier A (1998) Is the reduced efficacy of morphine in diabetic rats caused by alterations of opiate receptors or of morphine pharmacokinetics? *J Pharmacol Exp Ther* 285: 63–67
82 Apfel SC, Arezzo JC, Brownlee M, Federoff H, Kessler JA (1994) Nerve growth factor administration protects against experimental diabetic sensory neuropathy. *Brain Res* 634: 7–12
83 Chu PC, Lin MT, Shian LR, Leu SY (1986) Alterations in physiologic functions and in brain monoamine content in streptozocin-diabetic rats. *Diabetes* 35: 481–485
84 Akunne HC, Soliman KF (1987) The role of opioid receptors in diabetes and hyperglycemia-induced changes in pain threshold in the rat. *Psychopharmacol* 93: 167–172
85 Calcutt NA, Malmberg AB, Yamamoto T, Yaksh TL (1994) Tolrestat treatment prevents modification of the formalin test model of prolonged pain in hyperglycemic rats. *Pain* 58: 413–420
86 Calcutt NA, Dines KC, Cesena RM (1998) Effects of the peptide HP228 on nerve disorders in diabetic rats. *Metabolism* 47: 650–656
87 Calcutt NA, Campana WM, Eskeland NL, Mohiuddin L, Dines KC, Mizisin AP, O'Brien JS (1999) Prosaposin gene expression and the efficacy of a prosaposin-derived peptide in preventing structural and functional disorders of peripheral nerve in diabetic rats. *J Neuropathol Exp Neurol* 58: 628–636
88 Calcutt NA, Stiller C, Gustafsson H, Malmberg AB (2000) Elevated substance-P-like immunoreactivity levels in spinal dialysates during the formalin test in normal and diabetic rats. *Brain Res* 856: 20–27
89 Calcutt NA, Freshwater JD, O'Brien JS (2000) Protection of sensory function and anti-

hyperalgesic properties of a prosaposin-derived peptide in diabetic rats. *Anesthesiol* 93: 1271–1278

90 Fox A, Eastwood C, Gentry C, Manning D, Urban L (1999) Critical evaluation of the streptozotocin model of painful diabetic neuropathy in the rat. *Pain* 81: 307–316

91 Forman LJ, Estilow S, Lewis M, Vasilenko P (1986) Streptozocin diabetes alters immunoreactive beta-endorphin levels and pain perception after 8 wk in female rats. *Diabetes* 35: 1309–1313

92 Kolta MG, Ngong JM, Rutledge LP, Pierzchala K, Van Loon GR (1996) Endogenous opioid peptide mediation of hypoalgesic response in long-term diabetic rats. *Neuropeptides* 30: 335–344

93 Wuarin-Bierman L, Zahnd GR, Kaufmann F, Burcklen L, Adler J (1987) Hyperalgesia in spontaneous and experimental models of diabetic neuropathy. *Diabetologia* 30: 653–658

94 Ahlgren SC, Levine JD (1993) Mechanical hyperalgesia in streptozotocin-diabetic rats. *Neurosci* 52: 1049–1055

95 Ahlgren SC, Levine JD (1994) Protein kinase C inhibitors decrease hyperalgesia and C-fiber hyperexcitability in the streptozotocin-diabetic rat. *J Neurophysiol* 72: 684–692

96 Courteix C, Lavarenne J, Eschalier A (1993) RP-67580, a specific tachykinin NK1 receptor antagonist, relieves chronic hyperalgesia in diabetic rats. *Eur J Pharmacol* 241: 267–270

97 Courteix C, Bardin M, Massol J, Fialip J, Lavarenne J, Eschalier A (1996) Daily insulin treatment relieves long-term hyperalgesia in streptozocin-diabetic rats. *Neuroreport* 7: 1922–1924

98 Coudore-Civiale MA, Courteix C, Eschalier A, Fialip J (1998) Effect of tachykinin receptor antagonists in experimental neuropathic pain. *Eur J Pharmacol* 361: 175–184

99 Klamt JG (1998) Effects of intrathecally administered lamotrigine, a glutamate release inhibitor, on short- and long-term models of hyperalgesia in rats. *Anesthesiol* 88: 487–494

100 Malcangio M, Tomlinson DR (1998) A pharmacologic analysis of mechanical hyperalgesia in streptozotocin-diabetic rats. *Pain* 76: 151–157

101 Kamei J, Aoki T, Kasuya Y (1992) Periaqueductal gray matter stimulation-produced analgesia in diabetic rats. *Neurosci Lett* 142: 13–16

102 Kamei J; Saitoh A; Kasuya Y (1995) Involvement of delta 1-opioid receptors in the antinociceptive effects of mexiletine in mice. *Neurosci Lett* 196: 169–172

103 Calcutt NA, Jorge MC, Yaksh TL, Chaplan SR (1996) Tactile allodynia and formalin hyperalgesia in streptozotocin-diabetic rats: effects of insulin, aldose reductase inhibition and lidocaine. *Pain* 68: 293–299

104 Field MJ, McCleary S, Hughes J, Singh L (1999) Gabapentin and pregabalin, but not morphine and amitriptyline, block both static and dynamic components of mechanical allodynia induced by streptozocin in the rat. *Pain* 80: 391–398

105 Calcutt, NA, Chaplan, SR (1997) Spinal pharmacology of tactile allodynia in diabetic rats. *Brit J Pharmacol* 122: 1478–1482

106 Field MJ, McCleary S, Boden P, Suman-Chauhan N, Hughes J, Singh L (1998) Involvement of the central tachykinin NK1 receptor during maintenance of mechanical hypersensitivity induced by diabetes in the rat. *J Pharmacol Exp Ther* 285: 1226–1232

107 Lynch JJ, Jarvis MF, Kowaluk EA (1999) An adenosine kinase inhibitor attenuates tactile allodynia in a rat model of diabetic neuropathic pain. *Eur J Pharmacol* 364: 141–146

108 Jarvis MF, Wessale JL, Zhu CZ, Lynch JJ, Dayton BD, Calzadilla SV, Padley RJ, Opgenorth TJ, Kowaluk EA (2000) ABT-627, an endothelin ET(A) receptor-selective antagonist, attenuates tactile allodynia in a diabetic rat model of neuropathic pain. *Eur J Pharmacol* 388: 29–35

109 Lee YH, Ryu TG, Park SJ, Yang EJ, Jeon BH, Hur GM, Kim KJ (2000) Alpha1-adrenoceptors involvement in painful diabetic neuropathy: a role in allodynia. *Neuroreport* 11: 1417–1420

110 Kamei J, Hitosugi H, Kasuya Y (1993) Formalin-induced nociceptive responses in diabetic mice. *Neurosci Lett* 149: 161–164

111 Takeshita N, Ohkubo Y, Yamaguchi I (1995) Tiapride attenuates pain transmission through an indirect activation of central serotonergic mechanism. *J Pharmacol Exp Ther* 275: 23–30

112 Takeshita N, Yamaguchi I (1997) Insulin attenuates formalin-induced nociceptive response in mice through a mechanism that is deranged by diabetes mellitus. *J Pharmacol Exp Ther* 281: 315–321

113 Takeshita N, Yamaguchi I (1998) Antinociceptive effects of morphine were different between experimental and genetic diabetes. *Pharmacol Biochem Behav* 60: 889–897

114 Kamei J, Kashiwazaki T, Hitosugi H, Nagase H (1997) The role of spinal delta1-opioid receptors in inhibiting the formalin-induced nociceptive response in diabetic mice. *Eur J Pharmacol* 326: 31–36

115 Malmberg AB, Yaksh TL, Calcutt NA (1993) Anti-nociceptive effects of the GM1 ganglioside derivative AGF 44 on the formalin test in normal and streptozotocin-diabetic rats. *Neurosci Lett* 161: 45–48

116 Calcutt NA, Li L, Yaksh TL, Malmberg AB (1995) Different effects of two aldose reductase inhibitors on nociception and prostaglandin E. *Eur J Pharmacol* 285: 189–197

117 Tomlinson DR, Stevens EJ, Diemel LT (1994) Aldose reductase inhibitors and their potential for the treatment of diabetic complications. *TiPS* 15: 293–297

118 Ahlgren SC, White DM, Levine JD (1992) Increased responsiveness of sensory neurons in the saphenous nerve of the streptozotocin-diabetic rat. *J Neurophysiol* 68: 2077–2085

119 Russell LC, Burchiel KJ (1993) Abnormal activity in diabetic rat saphenous nerve. *Diabetes* 42: 814–819

120 Ward JD, Barnes CG, Fisher DJ, Jessop JD, Baker RW (1971) Improvement in nerve conduction following treatment in newly diagnosed diabetics. *Lancet* 1: 428–430

121 Greene DA, De Jesus PV Jr, Winegrad AI (1975) Effects of insulin and dietary myoinos-

itol on impaired peripheral motor nerve conduction velocity in acute streptozotocin diabetes. *J Clin Invest* 55: 1326–1336

122 Tomlinson DR, Holmes PR, Mayer JH (1982) Reversal, by treatment with an aldose reductase inhibitor, of impaired axonal transport and motor nerve conduction velocity in experimental diabetes mellitus. *Neurosci Lett* 31: 189–193

123 Moore SA, Peterson RG, Felten DL, O'Connor BL (1980) A quantitative comparison of motor and sensory conduction velocities in short- and long-term streptozotocin- and alloxan-diabetic rats. *J Neurol Sci* 48: 133–152

124 Julu PO (1988) The correlation between sensory nerve conduction velocities and three metabolic indices in rats treated with streptozotocin. *Diabetologia* 31: 247–253

125 Ahlgren SC, Wang JF, Levine JD (1997) C-fiber mechanical stimulus-response functions are different in inflammatory *versus* neuropathic hyperalgesia in the rat. *Neurosci* 76: 285–290

126 Jakobsen J, Brimijoin S, Skau K, Sidenius P, Wells D (1981) Retrograde axonal transport of transmitter enzymes, fucose-labeled protein, and nerve growth factor in streptozotocin-diabetic rats. *Diabetes* 30: 797–803

127 Rodriguez-Pena A, Botana M, Gonzalez M, Requejo F (1995) Expression of neurotrophins and their receptors in sciatic nerve of experimentally diabetic rats. *Neurosci Lett* 200: 37–40

128 Fernyhough P, Diemel LT, Brewster WJ, Tomlinson DR (1995) Altered neurotrophin mRNA levels in peripheral nerve and skeletal muscle of experimentally diabetic rats. *J Neurochem* 64: 1231–1237

129 Fernyhough P, Diemel LT, Tomlinson DR (1998) Target tissue production and axonal transport of neurotrophin-3 are reduced in streptozotocin-diabetic rats. *Diabetologia* 41: 300–306

130 Delcroix JD, Tomlinson DR, Fernyhough P (1997) Diabetes and axotomy-induced deficits in retrograde axonal transport of nerve growth factor correlate with decreased levels of p75LNTR protein in lumbar dorsal root ganglia. *Brain Res Mol Brain Res* 51: 82–90

131 Lindsay RM, Harmar AJ (1989) Nerve growth factor regulates expression of neuropeptide genes in adult sensory neurons. *Nature* 337: 362–364

132 Robinson JP, Willars GB, Tomlinson DR, Keen P (1987) Axonal transport and tissue contents of substance P in rats with long-term streptozotocin-diabetes. Effects of the aldose reductase inhibitor "statil". *Brain Res* 426: 339–348

133 Diemel LT, Brewster WJ, Fernyhough P, Tomlinson DR (1992) Expression of neuropeptides in experimental diabetes; effects of treatment with nerve growth factor or brain-derived neurotrophic factor. *Brain Res Mol Brain Res* 21: 171–175

134 Hylden JL, Wilcox GL (1981) Intrathecal substance P elicits a caudally-directed biting and scratching behavior in mice. *Brain Res* 217: 212–215

135 Yashpal K, Henry JL (1984) Substance p analogue blocks sp-induced facilitation of a spinal nociceptive reflex. *Brain Res Bull* 13: 597–600

136 Malmberg AB, Yaksh TL (1992) Hyperalgesia mediated by spinal glutamate or substance P receptor blocked by spinal cyclooxygenase inhibition. *Science* 257: 1276–1279
137 Garrett NE, Malcangio M, Dewhurst M, Tomlinson DR (1997) alpha-Lipoic acid corrects neuropeptide deficits in diabetic rats *via* induction of trophic support. *Neurosci Lett* 222: 191–194
138 Calcutt NA, Chen P, Hua XY (1998) Effects of diabetes on tissue content and evoked release of calcitonin gene-related peptide-like immunoreactivity from rat sensory nerves. *Neurosci Lett* 254: 129–132
139 Okuse K, Chaplan SR, McMahon SB, Luo ZD, Calcutt NA, Scott BP, Akopian AN, Wood JN (1997) Regulation of expression of the sensory neuron-specific sodium channel SNS in inflammatory and neuropathic pain. *Mol Cell Neurosci* 10: 196–207
140 Lee YH, Ryu TG, Park SJ, Yang EJ, Jeon BH, Hur GM, Kim KJ (2000) Alpha1-adrenoceptors involvement in painful diabetic neuropathy: a role in allodynia. *Neuroreport* 11: 1417–1420
141 Luo ZD, Chaplan SR, Scott BP, Cizkova D, Calcutt NA, Yaksh TL (1999) Neuronal nitric oxide synthase mRNA upregulation in rat sensory neurons after spinal nerve ligation: lack of a role in allodynia development. *J Neurosci* 19: 9201–9208
142 Zochodne DW, Verge VM, Cheng C, Hoke A, Jolley C, Thomsen K, Rubin I, Lauritzen M (2000) Nitric oxide synthase activity and expression in experimental diabetic neuropathy. *J Neuropath Exp Neurol* 59: 798–807
143 Di Giulio AM, Tenconi B, La Croix R, Mantegazza P, Abbracchio MP, Cattabeni F, Gorio A (1989) Denervation and hyperinnervation in the nervous system of diabetic animals. II. Monoaminergic and peptidergic alterations in the diabetic encephalopathy. *J Neurosci Res* 24: 362–368
144 Calcutt NA, Malmberg AB (1995) Basal and formalin-evoked spinal levels of amino acids in conscious diabetic rats. *Soc Neurosci Abs* 21: 650
145 Kamei J, Ogawa M, Kasuya Y (1990) Development of supersensitivity to substance P in the spinal cord of the streptozotocin-induced diabetic rats. *Pharmacol Biochem Behav* 35: 473–475
146 Li N, Young MM, Bailey CJ, Smith ME (1999) NMDA and AMPA glutamate receptor subtypes in the thoracic spinal cord in lean and obese-diabetic ob/ob mice. *Brain Res* 849: 34–44
147 Suh HW, Song DK, Wie MB, Jung JS, Hong HE, Choi SR, Kim YH (1996) The reduction of antinociceptive effect of morphine administered intraventricularly is correlated with the decrease of serotonin release from the spinal cord in streptozotocin-induced diabetic rats. *Gen Pharmacol* 27: 445–450
148 Bitar MS, Bajic KT, Farook T, Thomas MI, Pilcher CW (1999) Spinal cord noradrenergic dynamics in diabetic and hypercortisolaemic states. *Brain Res* 830: 1–9
149 Pertovaara A, Wei H, Kalmari J, Ruotsalainen M (2001) Pain behavior and response properties of spinal dorsal horn neurons following experimental diabetic neuropathy in the rat: modulation by nitecapone, a COMT inhibitor with antioxidant properties. *Exp Neurol* 167: 425–434

150 Simon GS, Dewey WL (1981) Narcotics and diabetes. I. The effects of streptozotocin-induced diabetes on the antinociceptive potency of morphine. *J Pharmacol Exp Ther* 218: 318–323

151 Raz I, Hasdai D, Seltzer Z, Melmed RN (1988) Effect of hyperglycemia on pain perception and on efficacy of morphine analgesia in rats. *Diabetes* 37: 1253–1259

152 Courteix C, Bardin M, Chantelauze C, Lavarenne J, Eschalier A (1994) Study of the sensitivity of the diabetes-induced pain model in rats to a range of analgesics. *Pain* 57: 153–160

153 Courteix C, Bourget P, Caussade F, Bardin M, Coudore F, Fialip J, Eschalier A (1998) Is the reduced efficacy of morphine in diabetic rats caused by alterations of opiate receptors or of morphine pharmacokinetics? *J Pharmacol Exp Ther* 285: 63–70

154 Kamei J, Hitosugi H, Kawashima N, Aoki T, Ohhashi Y, Kasuya Y (1992) Antinociceptive effect of mexiletine in diabetic mice. *Res Commun Chem Pathol Pharmacol* 77: 245–248

155 Bardin L, Schmidt J, Alloui A, Eschalier A (2000) Effect of intrathecal administration of serotonin in chronic pain models in rats. *Eur J Pharmacol* 409: 37–43

156 Begon S, Pickering G, Eschalier A, Dubray C (2000) Magnesium and MK801 have a similar effect in two experimental models of neuropathic pain. *Brain Res* 887: 436–439

157 Cesena RM, Calcutt NA (1999) Gabapentin prevents hyperalgesia during the formalin test in diabetic rats. *Neurosci Lett* 262: 101–104

158 Max MB, Lynch SA, Muir J, Shoaf SE, Smoller B, Dubner R (1992) Effects of desipramine, amitriptyline, and fluoxetine on pain in diabetic neuropathy. *New Engl J Med* 326: 1250–1256

159 Byas-Smith MG, Max MB, Muir J, Kingman A (1995) Transdermal clonidine compared to placebo in painful diabetic neuropathy using a two-stage "enriched enrollment" design. *Pain* 60: 267–274

160 Harati Y, Gooch C, Swenson M, Edelman SV, Greene D, Raskin P, Donofrio P, Cornblath D, Olson WH, Kamin M (2000) Maintenance of the long-term effectiveness of tramadol in treatment of the pain of diabetic neuropathy. *J Diabetes Comp* 14: 65–70

161 Backonja M, Beydoun A, Edwards KR, Schwartz SL, Fonseca V, Hes M, LaMoreaux L, Garofalo E (1998) Gabapentin for the symptomatic treatment of painful neuropathy in patients with diabetes mellitus: a randomized controlled trial. *JAMA* 280: 1831–1836

162 Dejgard A, Petersen P, Kastrup J (1988) Mexiletine for treatment of chronic painful diabetic neuropathy. *Lancet* 1 8575–6: 9–11

163 Bach FW, Jensen TS, Kastrup J, Stigsby B, Dejgard A (1990) The effect of intravenous lidocaine on nociceptive processing in diabetic neuropathy. *Pain* 40: 29–34

Role of adrenergic transmitters and receptors in nerve and tissue injury related pain

Ralf Baron[1] *and Srinivasa N. Raja*[2]

[1]Klinik für Neurologie, Christian-Albrechts-Universität Kiel, Niemannsweg 147, 24105 Kiel, Germany; [2]Department of Anesthesiology and Critical Care Medicine, The Johns Hopkins Medical Institutions, 600 N. Wolfe St., Osler 292, Baltimore, MD 21287, USA

Introduction

Definition of sympathetically maintained pain (SMP)

Interruption of the sympathetic nerve supply to the affected extremity has been used to treat certain pain syndromes for many years. These syndromes include complex regional pain syndromes (reflex sympathetic dystrophy (RSD) and causalgia), post-traumatic neuralgia, phantom limb pain and to a certain extent acute herpes zoster [1]. Two therapeutical techniques to block sympathetic nerves are currently used: (1) injections of a local anesthetic around sympathetic paravertebral ganglia that project to the affected body part (sympathetic ganglion blocks), and (2) regional intravenous application of guanethidine, bretylium or reserpine (which all deplete noradrenaline in the postganglionic axon) to an isolated extremity blocked with a tourniquet (intravenous regional sympatholysis). However, many authors do not support the concept that the sympathetic nervous system is actively involved in the generation of pain and propose, that the role of the sympathetic nervous system has to be reconsidered or even completely discarded [2, 3]. They claim that interventions that block sympathetic activity lack specificity and argue that the techniques and results of sympathetic blockade have rarely been adequately evaluated and are in most cases not placebo-controlled.

On the basis of experience and recent clinical studies the term "sympathetically maintained pain" was re-defined: Neuropathic pain patients presenting with similar clinical signs and symptoms, can clearly be divided into two groups by the negative or positive effect of selective sympathetic blockade or antagonism of alpha adrenoceptor mechanisms [4, 5]. The pain component that is relieved by specific sympatholytic procedures is considered "sympathetically maintained pain" (SMP). Thus, SMP is now defined to be a symptom or the underlying mechanism in a subset of patients with neuropathic disorders and not a clinical entity. The positive effect of a sympathetic blockade is not essential for the diagnosis. On the other hand, the only possibility to differentiate between SMP and "sympathetically independent pain" (SIP) is the efficacy of a correctly applied sympatholytic intervention [6].

Clinical characteristics of selected pain syndromes with SMP

Complex regional pain syndromes

In recent years, the term reflex sympathetic dystrophy has been inappropriately used to describe a much more extensive range of clinical presentations than originally intended. Moreover, since the pathophysiological mechanisms underlying these syndromes are poorly understood, terms like "reflex" and "sympathetic" can be misleading. Recent studies examined sympathetic neurocirculatory function in patients with RSD and concluded that the majority of patients had an intact anatomical sympathetic innervation, some had a functional sympathetic inhibition in the acute phase and that secondary changes neurovascular transmission are common in chronic patients [7–9]. Therefore, a new terminology was introduced based entirely on elements of history, symptoms and findings on clinical examination with no implied pathophysiological mechanism [6]. According to the IASP (International Association for the Study of Pain) *Classification of Chronic Pain* reflex sympathetic dystrophy and causalgia are now called "complex regional pain syndromes" (CRPS). In CRPS type I (reflex sympathetic dystrophy) minor injuries to a limb or lesions in remote body areas precede the onset of symptoms. CRPS type II (causalgia) develops after injury to a major peripheral nerve [10].

CRPS type I (reflex sympathetic dystrophy)

Paul Sudeck first described this disorder early in the century [11]. These patients develop asymmetrical distal extremity pain and swelling without presenting an overt nerve lesion. Precipitating events include fracture or minor soft tissue trauma, frostbite, burns, as well as stroke and myocardial infarction, and more rarely a low-grade tissue infection. The swelling and pain often develop at a site remote from the inciting injury and there may be no obvious local tissue damaging process at the site of pain and swelling [12].

CRPS I patients often report a burning spontaneous pain [13] felt in the distal part of the affected extremity. Characteristically, the pain is disproportionate in intensity to the inciting event. The pain usually increases when the extremity is in a dependent position. Stimulus-evoked pains are a striking clinical feature; they include mechanical and thermal allodynia and/or hyperalgesia. These sensory abnormalities often appear early, are most pronounced distally, and have no consistent spatial relationship to individual nerve territories or to the site of the inciting lesion [14–16]. Typically the pain can be elicited by movements and pressure at the joints, even if these are not directly affected by the inciting lesion. In addition, somatosensory deficits may be present [17].

Autonomic abnormalities include swelling and changes of sweating and skin blood flow [8, 9, 18–21]. At normal room temperature the skin temperature of the limbs shows a systematic side difference in about 80% of the patients, i.e. the affected extremity is either warmer or colder. In the acute stages of CRPS I the affected limb is more often warmer than the contralateral limb. Sweating abnormalities are present in nearly all CRPS I patients. Either hypohidrosis or, more frequently, hyperhidrosis is present; preferentially at the palmar side of the hand or plantar side of the foot [18, 19, 22, 23].

It is possible that there is an inflammatory component of CRPS I in the acute phase. This is based on the observation that scintigraphic investigations with radiolabelled immunoglobulins show extensive plasma extravasation in patients with acute CRPS I [24]. Furthermore, analysis of joint fluid and synovial biopsies have shown an increase in protein concentration, synovial hypervascularity, and neutrophil infiltration [25]. The acute distal swelling of the affected limb depends very critically on aggravating stimuli. Since it often diminishes after sympathetic blocks it is likely that it is maintained by sympathetic activity [26].

Trophic changes such as abnormal nail growth, increased or decreased hair growth, fibrosis, thin glossy skin and osteoporosis may be present, particularly in chronic stages. Restrictions of passive movement are often present in long-standing cases and may be related to both functional motor disturbances and trophic changes of joints and tendons.

Weakness of all muscles of the affected distal extremity is often present. Small accurate movements are characteristically impaired [26]. Nerve conduction and electromyography studies are normal, except in patients in very chronic and advanced stages. Therefore, the peripheral motor neuron and the neuromuscular junction are unlikely to contribute to the motor dysfunction. About half of the patients have a postural or action tremor that represents an increased physiological tremor [27]. In about 10% of cases dystonia of the affected hand or foot develops [28, 29].

CRPS type II (causalgia)

As defined by the American Civil War physician, S. Weir Mitchell [30], causalgia is burning pain that develops in the distal extremity following traumatic partial peripheral nerve injury. About 10% had a dramatic clinical syndrome which consisted of prominent distal burning pain. In addition to spontaneous pain, patients reported exquisite hypersensitivity of the skin to light mechanical stimulation. Furthermore, movement, loud noises or strong emotions could trigger their pain. In addition to sensory abnormalities, distal extremity swelling, smoothness and mottling of the skin, and in some cases, acute arthritis was present. In most cases the limb was cold and sweaty. The sensory and trophic abnormalities spread beyond the innervation territory of the injured peripheral nerve and often occurred remote from the site of injury. Since all symptoms show many similarities to those of CRPS I, this syndrome is now called CRPS II.

In 1916 Leriche [31] reported that sympathectomy dramatically relieves causalgia. This notion was confirmed in several large clinical series, primarily in wounded soldiers. In his exhaustive 1967 centennial review, Richards describes the clinical features of causalgia and the effect of sympatholytic interventions in hundreds of cases [32]. He was emphatic about the dramatic response of causalgia to sympathetic blockade.

Post-traumatic neuralgia

It is important to recognize that many post-traumatic neuropathy patients have pain but do not have the full clinical picture of causalgia (CRPS II). In these cases, in contrast to causalgia patients, the pain is located largely within the innervation territory of the injured nerve [33]. Although these patients often describe their pain as burning, they exhibit a less complex clinical picture than patients with causalgia and do not show marked swelling or a tendency for progressive spread of symptoms. The cardinal symptoms are spontaneous burning pain, hyperalgesia and mechanical and especially cold allodynia. These sensory symptoms are confined to the territory of the affected periperal nerve although allodynia may extend beyond the border of nerve territories to a certain degree. Spontaneous and evoked pain are felt superficially and not deep inside the extremity and the intensity of both is not dependent on the position of the extremity. The patients occasionally obtain relief with sympatholytic procedures although much less often than those with CRPS.

Following the IASP classification it is possible to use the name "neuralgia" for this type of neuropathic pain (pain within the innervation territory of a lesioned nerve, e.g. post-traumatic neuralgia). However, the new definition of CRPS II includes the statement that symptoms may also be limited to the territory of a single peripheral nerve. Therefore, the term CRPS II provides space to include these localized post-traumatic neuropathies. An inherent weakness of this definition of CRPS II is that different syndromes with different underlying mechanisms are obviously included.

Sympathetic interruption for the treatment of SMP: the clinical evidence

Despite the vast body of uncontrolled clinical evidence, proof of the effectiveness of sympathetic blockade for the treatment of CRPS is scanty [34]. There are several uncontrolled surveys in the literature reviewing the effect of sympathetic interventions, e.g. sympathetic blocks (SB), surgical sympathectomies (SE) and intravenous regional sympathetic blocks (IVRS) in CRPS and post-traumatic neuralgias. In CRPS, about 85% of the patients report a positive acute effect of SE, SB and IVRS, but fewer patients experience long-term relief (60% in SB and 30% in IVRS). In post-traumatic neuralgias the sympatholytic interventions are clearly less effective.

One recent controlled study in patients with CRPS I has shown that SB with local anesthetic has the same immediate effect on pain as a control injection with saline [35]. However, after 24 h patients in the local anesthetic group were much better, indicating that non-specific effects are important initially and that evaluating the efficacy of sympatholytic interventions is best done after 24 h. With these data in mind, the uncontrolled studies mentioned above must be interpreted cautiously. Only 10 out of the 24 studies we reviewed assessed long-term effects. Bonelli et al. [36] compared SB and guanethidine IVRS in a two-arm controlled study (no placebo arm) and could not find a difference between them, in either acute or long-term periods. Five placebo-controlled studies using guanethidine IVRS have been performed [37–41]. Only one demonstrated a significantly better acute effect (1 h after the block) [37]. One potential criticism of these guanethidine data is that four studies used only one IVRS, but multiple blocks may be necessary for treatment of CRPS [42]. Furthermore, the follow-up periods after the block were only 30–60 min in three studies and up to 6 months in two studies. One study demonstrated that IVRS bretylium and lidocaine produce significantly longer pain relief than lidocaine alone [43]. Two studies showed that IVRS reserpine was ineffective [38, 40].

Sympathetically maintained pain (SMP) – the experimental basis for a pathologic interaction of sympathetic and afferent neurons

Animal studies

Under physiological conditions primary afferent nociceptors do not have catecholamine sensitivity and their activity is unaffected by sympathetic outflow [44]. However, under pathologic conditions a sympathetic-afferent interaction may be established, so that the sympathetic efferents can enhance primary afferent nociceptor activity. Two categories of influence of the sympathetic neurons on afferent neurons can be distinguished. This distinction seems to be related to whether the coupling between afferent and sympathetic neurons develops after peripheral tissue inflammation associated with nociceptor sensitization or after traumatic nerve damage. Traumatic nerve injuries may be used as a model for CRPS II and post-traumatic neuralgias in humans whereas animal models of inflammatory pain may share mechanisms with human CRPS I.

Influence of sympathetic activity and catecholamines on primary afferents after tissue inflammation (intact but sensitized primary afferents)

Animal studies demonstrate that catecholamines can activate primary afferent C-nociceptors after thermal [45] and chemical [46, 47] sensitization. The sympathetic-primary afferent interaction is mediated by α_2-adrenoceptors [48]. After chronic sensitization with chloroform, noradrenaline enhances behavioral responses to nociceptive stimulation [49]. Interestingly, surgical sympathetomy prevents the sensitizing effect of injected noradrenaline. The sensitizing effects of noradrenaline may result from a direct action on nociceptors or by an indirect effect on primary afferents *via* the sympathetic postganglionic neuron. Noradrenaline may induce the release of other substances, e.g. prostaglandins (PG), from sympathetic terminals that sensitize the afferents (Fig. 1C).

Sensitization of nociceptive primary afferents by such pro-inflammatory mediators as bradykinin and nerve growth factor (NGF) also depends on an intact sympathetic nervous system [50–52]. Surgical removal of the postganglionic neurons but not interruption of the preganglionic axons (decentralization) decreases the sensitizing effect of these substances [53]. These results indicate that the sensitizing effect of these substances depends on the anatomical integrity of the postganglionic fibers rather than sympathetic activity or noradrenaline release. It has been proposed [53] that bradykinin, like noradrenaline, acts indirectly on nociceptor terminals by inducing the release of prostaglandins from sympathetic fibers (Fig. 1C).

It has to be emphasized that most of the evidence to support these conslusions is based on behavioral experiments and the effects of pharmacological interventions. It is tacitly assumed that the behavioral changes reflect sensitivity changes of nociceptors. A direct neurophysiological demonstration for the proposed presynaptic action on postganglionic fibers by bradykinin or noradrenaline that precipitate the sensitization of nociceptive afferents does not yet exist.

However, this concept is of particular interest because of the evidence that, in its early phase, CRPS I might have an inflammatory component. If this mechanism does contribute significantly to pain then reducing sympathetic postganglionic neural activity would not be effective compared with surgical postganglionic sympathectomy. Furthermore, pain relief should occur only days after surgical (postganglionic) sympathectomy if this mechanism were to operate, because it takes days for the sympathetic terminals to degenerate after removal of their cell bodies. Clinically, however, the pain relief after sympathetic blockade is usually observed within minutes in patients with a significant component of SMP.

Influence of sympathetic activity and catecholamines on primary afferents after nerve damage

After complete nerve injury in experimental animals, surviving cutaneous afferents

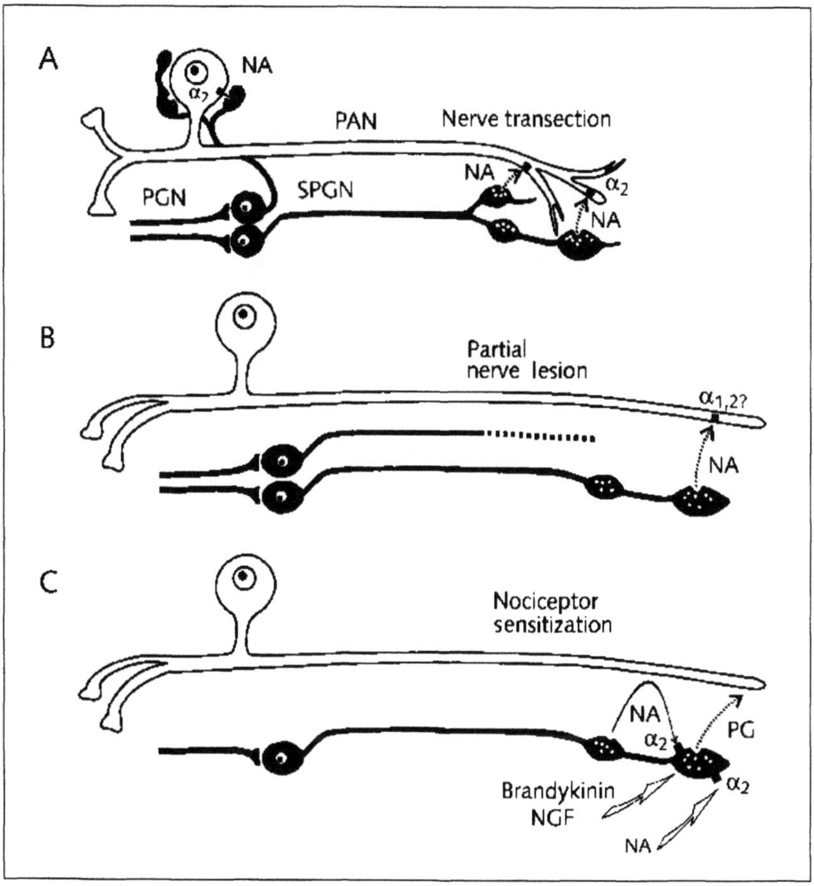

Figure 1

Influence of sympathetic activity and catecholamines on primary afferent neurons (PAN)
(A) Nerve transection. The sympathetic-afferent interaction is located in the neuroma and in the dorsal root ganglion. It is mediated by noradrenaline (NA) released from sympathetic postganglionic neurons (SPGN) and α-adrenoreceptors expressed at the plasma membrane of afferent neurons. PGN, preganglionic neuron.
(B) Partial nerve lesion. Partial nerve injury is followed by a decrease of the sympathetic innervation density (stippled sympathetic postganglionic neuron). This induces an up-regulation of functional α-adrenoceptors at the membrane of intact afferent fibers.
(C) After tissue inflammation intact but sensitized primary afferents acquire noradrenaline sensitivity. Noradrenaline is not acting directly on afferents but induces the release of prostaglandins (PG) from sympathetic terminals that sensitize the afferents. In accordance, bradykinin and nerve growth factor (NGF) induced nociceptor sensitization is also mediated by the release of prostaglandins from sympathetic postganglionic neurons (from [1] with permission).

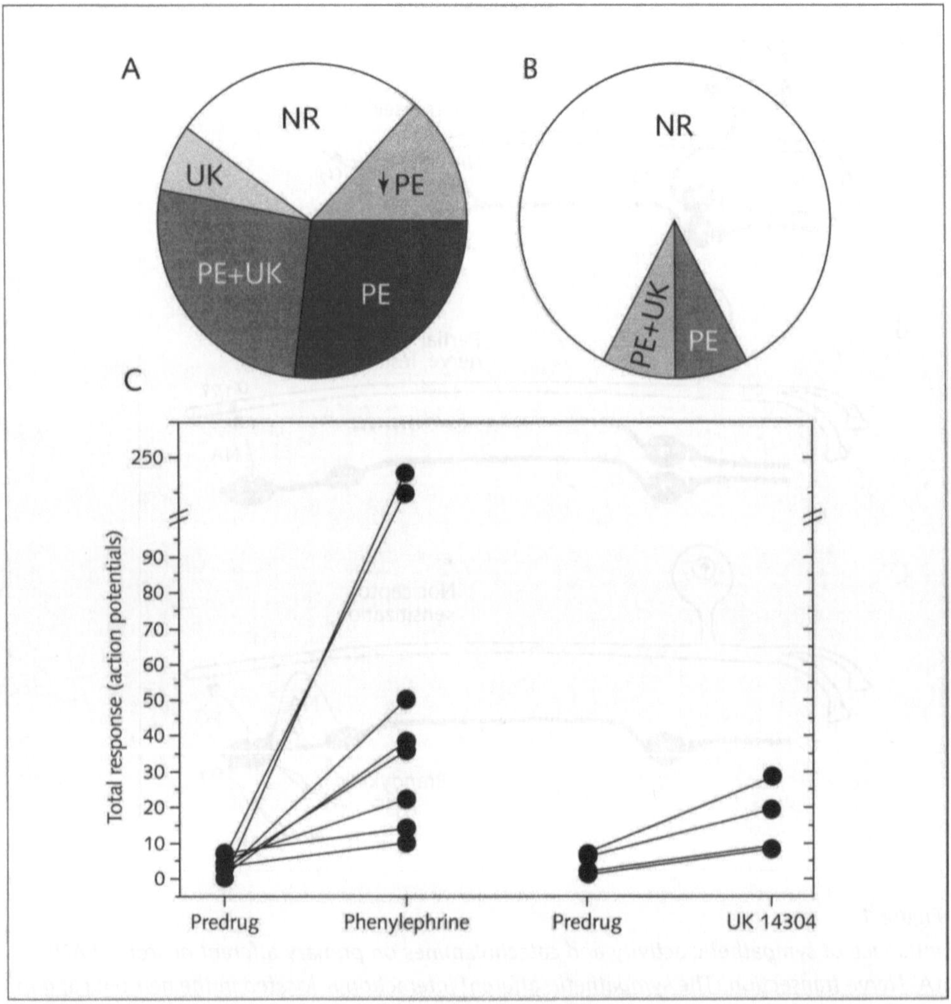

Figure 2

Responses of afferents from the superficial peroneal nerve tested in an in vitro *skin nerve preparation with both phenylephrine and UK14304 (α_1- and α_2-adrenergic agonists)*

(A) Incidence of response in L6-spinal nerve lesioned animals (n = 15).

(B) Incidence of response in control animals (n = 13). PE, phenylephrine-sensitive; UK, UK14304-sensitive; ⇓PE, units whose spontaneous activity decreased during PE administration; NR, nonresponsive.

(C) Peak response to each agonist in the lesioned animals. Peak response and pre-drug response are plotted for the 8 phenylephrine-sensitive fibers (left) and for the 5 UK14304-sensitive fibers (right). Peak response to phenylephrine was significantly greater than the peak response to UK14304 ($p < 0.05$, Wilcoxon matched pairs). From [63] with permission.

may develop noradrenergic sensitivity (Fig. 1A). Myelinated and unmyelinated afferents innervating the stump neuroma can be excited by adrenaline or by stimulation of sympathetic efferents that have regenerated into the neuroma [54–57]. In mature neuromata the catecholamine sensitivity is much less pronounced [56, 58]. However, as late as one year after complete section and re-anastomosis of peripheral nerves electrical stimulation of the sympathetic trunk remains capable of activating regenerated C fibers, probably of nociceptor function [59]. Within two weeks of a partial nerve lesion, electrical stimulation of the sympathetic trunk and injections of catecholamines can activate or sensitize C nociceptors in the partially injured nerve including those in continuity with their peripheral targets [60–62] (Fig. 1B). More recent *in vitro* neurophysiological studies, using a primate skin-nerve preparation, have demonstrated that uninjured cutaneous C-fiber nociceptors that innervate skin partially denervated by a spinal nerve ligation have a higher incidence of adrenergic sensitivity than nociceptive afferents from normal skin [63] (Fig. 2). A recent study reported the development of a novel pattern of sympathetic fiber innervation of the skin after a peripheral nerve injury in the rat [64]. Within a week after bilateral lesions of the mental nerve, dopamine-beta-hydroxylase positive fibers sprout into the upper dermis, a region usually devoid of such fibers. The migration and branching of these fibers occurred gradually reaching a peak at four weeks after the nerve lesion. Ectopic innervation of the superficial dermis by sympathetic fibers may play an important role in the interactions between the sympathetic and regenerating sensory fibers containing substance P.

In addition to the peripheral interaction, coupling of sympathetic and afferent neurons may also occur within the dorsal root ganglion (DRG, Fig. 1A). After a complete nerve lesion, some DRG somata with Aβ-fibers and a few with C-fibers develop ectopic activity. Electrical stimulation of sympathetic efferents innervating the DRG may lead to an α_2-adrenoceptor mediated increase of this spontaneous activity [65–67]. This increased adrenergic sensitivity is paralleled by morphological changes within the DRG (Fig. 1A). After the nerve lesion, sympathetic postganglionic fibers that normally innervate blood vessels within the DRG sprout to form basket-like terminals around large primary afferent somata that project into the injured nerve [68–70]. However, the relationship of this DRG sprouting to pain is uncertain. Sprouting of sympathetic postganglionic terminals is largely around the somata of larger, presumably non-nociceptive primary afferents [68]. Activation of afferent neurons to sympathetic stimulation following sciatic nerve lesion in the rat occurs preferentially in the first 20 day after lesion, but later depression of afferent activity is predominant; yet formation of catecholaminergic baskets around DRG cells in the same nerve lesion model occurs after 30 days [44]. Finally, and most importantly, it has been shown recently that surgical sympathectomy does not alleviate the mechanical allodynia-like behavior which develops in rats following L5 spinal nerve lesion [71].

Adrenoceptors involved in sympathetic-afferent coupling

The excitation and depression of axotomized dorsal root ganglion cells and the excitation of afferent terminals in neuromas generated by activation of the sympathetic innervation are suggested to be mediated by α-adrenoceptors. Any hypothesis about this chemical noradrenaline-mediated coupling between sympathetic and afferent neurons has to explain how functional adrenoceptors appear on the surface of nociceptive and other afferent neurons as a consequence of a nerve lesion. The cellular mechanisms underlying the increased sensitivity are unknown. An obvious explanation is the novel expression or up-regulation of adrenoceptors. Alternatively, the receptors may normally be present on primary nociceptive afferents but not be functional; they may become uncovered and effective during the response to damage. This is consistent with the finding that adrenoceptor mRNA is normally constitutively present in DRG cells [72, 73]. The affinity for ligands or the effectiveness of the subsequent cellular transduction may change in injured nociceptive and non-nociceptive afferent neurons. These possibilities are not mutually exclusive and it is possible that several mechanisms contribute simultaneously to the expression of sympathetically mediated afferent excitation.

Perl and co-workers [60, 74] suggest, on the basis of their experiments on polymodal nociceptors following partial nerve lesion, that the expression or up-regulation of adrenoceptors in primary afferent neurons is related to the sympathetic denervation of the target tissue, analogous to the development of the so-called denervation supersensitivity which is observed in effector tissues following their denervation [75]. This is supported by their recent experimental observation that cutaneous nociceptive C-fibers in the rabbit ear may develop adrenoceptor sensitivity following surgical sympathectomy [76]. However, the percentage of afferents excited and the number of impulses in the responding afferents to close-arterial injection of noradrenaline are small.

Knowledge about the subtypes of α-adrenoreceptor(s) involved in the sympathetic-afferent coupling following nerve trauma is important for an understanding of the underlying neural mechanism, and may be useful in the design of more specific treatment modalities for neuropathic pain conditions involving sympathetic efferent activity. Sato and Perl [60] have shown that excitation and sensitization of polymodal nociceptors in the rabbit ear skin generated by electrical stimulation of the sympathetic supply or by noradrenaline after partial lesion of the auricular nerve, is largely mediated by α_2-adrenoceptors. Investigation of sympathetic-sensory coupling at the site of experimental nerve injury in neuromas and at the cell bodies of axotomized afferent neurons using type-selective agonists and antagonists have shown that this transmission is also largely mediated by α_2-adrenoreceptors [66]. Recordings from primate cutaneous C-fiber nociceptors after an L6 spinal nerve ligation, however, showed a significantly higher incidence of response to both the selective α_1- and α_2-adrenergic agonists, phenylephrine and brimonidine, respectively [63]. In normal (uninjured) DRG neurons mRNAs for α_{2A}- and particularly

α_{2B}-adrenoceptors are present whereas mRNA for α_1-adrenoceptors are either absent or very low in level [73]. After nerve lesion mRNA for α_{2A}-adrenoceptors is clearly up-regulated in DRG neurons [77]. These results were obtained in rats. Caution is required to generalize these results. There may exist species differences as far as the expression of adrenoceptors following trauma is concerned.

Indirect coupling between noradrenergic nerve terminals and nociceptors in the periphery via *the vascular bed*

The possibility for an indirect coupling between sympathetic and afferent nerve terminals should not be discarded. Dramatic changes may occur in vascular perfusion of the micromilieu surrounding the nociceptors after nerve trauma. These changes are related to the consequences of denervation and reinnervation by postganglionic vasoconstrictor neurons and afferent nociceptive neurons and the development of supersensitivity of blood vessels [78, 79]. Whether these changes lead indirectly to sensitization of nociceptors awaits experimental proof [80].

Human experimental studies

Influence of sympathetic activity and catecholamines on primary afferents after tissue inflammation (intact but sensitized primary afferents)

Cutaneous application of the algogenic agent capsaicin causes neurogenic inflammation by activating and sensitizing nociceptors. An adrenergic effect on sensitized cutaneous nociceptors was described in this model in human subjects. The heat hyperalgesia that develops after topical application of capsaicin to the skin of human volunteers was enhanced following iontophoresis of noradrenaline [81]. Phentolamine, a mixed α_1- and α_2-antagonist, inhibited the noradrenaline-induced pain and mechanical hyperalgesia in capsaicin-sensitized skin [82, 83]. However, iontophoresis of the non-adrenergic vasoconstrictors, angiotensin II and vasopressin [84] as well as occlusion of blood flow [85] produced a mild enhancement of hyperalgesia in capsaicin treated skin. This suggests that changes in blood flow may account for some aspects of the enhanced cutaneous sensibility seen in capsaicin-induced hyperalgesic skin. When sympathetic cutaneous vasoconstrictor activity was modulated in physiological ranges by thermoregulatory stress, it did not measurably influence capsaicin-induced pain [86]. Accordingly, activation of skin sympathetic vasoconstrictor neurons does not change the irritant-induced discharge of single microneurographically recorded cutaneous C nociceptors [87]. One controlled study has evaluated the effect of sympathetic blockade on pain after experimental tissue inflammation. Pedersen et al. [88] used a cutaneous heat injury as their pain model and have shown that sympathetic blocks with bupivacaine did not alter spontaneous pain and heat hyperalgesia [89]. In patients with rheumatoid

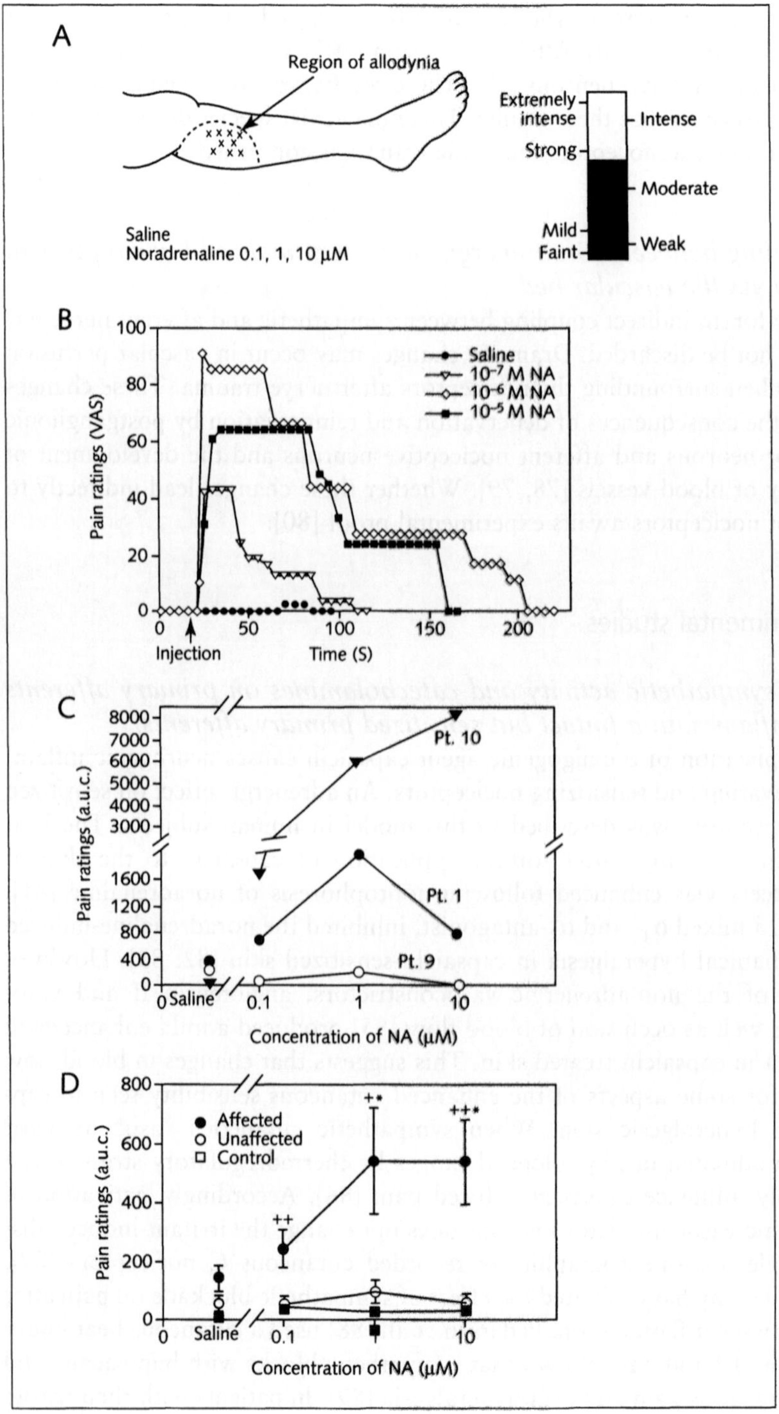
A
Region of allodynia
Extremely intense
Strong
Mild
Faint
Intense
Moderate
Weak
Saline
Noradrenaline 0.1, 1, 10 μM
B
Saline
10⁻⁷ M NA
10⁻⁶ M NA
10⁻⁵ M NA
Pain ratings (VAS)
Injection
Time (S)
C
Pt. 10
Pt. 1
Pt. 9
Pain ratings (a.u.c.)
Saline
Concentration of NA (μM)
D
Affected
Unaffected
Control
Pain ratings (a.u.c.)
Saline
Concentration of NA (μM)

arthritis, however, regional intravenous guanethidine produced a decrease in pain and an increase in pinch strength [90].

Influence of sympathetic activity and catecholamines on primary afferents in patients with nerve damage and CRPS

Clinical studies in humans support the idea that nociceptors develop catecholamine sensitivity after complete or partial nerve lesions. Long after limb amputation, injection of adrenaline around a stump neuroma is reported to be intensely painful [91]. Recent studies also demonstrate that the perineuromal administration of physiological doses of noradrenaline induces more pain than saline injections [92].

Furthermore, intraoperative stimulation of the sympathetic chain induces an increase of spontaneous pain in patients with causalgia (CRPS II) but not in patients with hyperhidrosis [93, 94]. In addition, intravenous phentolamine, but not propranolol, relieves SMP [4, 5]. There is good agreement on the pain relief resulting from both procedures in patients subjected to both a local anesthetic sympathetic ganglion block and a phentolamine infusion [5, 95].

Figure 3
The pain induced by intradermal injections of noradrenaline in normal subjects and patients with sympathetically maintained pain
(A) Experimental paradigm: The region of allodynia was mapped using a camel's hairbrush. Sites for intradermal injections were marked on the skin (x) within the allodynic region, and the mirror-image contralateral extremity, such that they were separated by at least 3 cm. The patient underwent a sympathetic ganglion block with a 0.25% bupivacaine solution. Intradermal injections of saline and incremental concentrations of noradrenaline (NA) were made at different sites first on the unaffected extremity. The patient rated the pain induced by the injection continuously using a computerised visual analog scale with descriptive markers as shown in the figure.
(B) Pain ratings to saline and NA in the affected limb of a patient with SMP. This patient (# 10) experienced no pain to the saline injection but had a dose-related increase in intensity and duration of pain to NA.
(C) Dose-response curves relating the total pain rating (area under the curve) for saline and the varying NA concentrations in the affected limb of 3 different subjects. Patient 10 had a dose-related increase in pain, while patient 1 had the most intense pain with the 1 μM dose. Patient 9 had, in contrast, had no significant pain to the saline or the NA injections.
(D) Mean pain ratings to the different concentrations of NA in control subjects and the affected and unaffected limbs of patients with SMP. Note that the pain ratings to the NA injections are significantly increased in the affected limb compared to the saline injections ($=p<0.05$) and compared to the unaffected limb and control subjects (+ $=p<0.05$; ++ $=p<0.01$, ANOVA). From [98] with permission.*

In CRPS and post-traumatic neuralgias, intracutaneous application of noradrenaline into a symptomatic area rekindles spontaneous pain and dynamic mechanical hyperalgesia that had been relieved by sympathetic blockade, supporting the idea that noradrenergic sensitivity of human nociceptors is present after partial nerve lesion [96]. Also in postherpetic neuralgia, spontaneous pain and mechanical hyperalgesia are enhanced after injection of adrenaline or phenylephrine [97]. A potential criticism of the above mentioned studies where pain was rekindled with exogenous adrenergic agonists is that the doses of noradrenaline (NA) used were much higher than are likely to exist *in vivo*. Therefore, the algesic effects of peripheral administration of NA in physiologically relevant doses was compared in patients with SMP and normal subjects [98]. Intradermal NA, in physiologically relevant doses, was demonstrated to evoke greater pain in the affected regions of patients with SMP, than in the contralateral unaffected limb, and in control subjects (Fig. 3). Moreover, the majority of the patients who had an increase in pain to injection of NA in the affected extremity also reported a decrease in pain to systemic administration of phentolamine. These observations support a role for cutaneous adrenergic receptors in the mechanisms of sympathetically maintained pain. In accordance with this, it could be clearly demontrated in patients with CRPS that spontaneous pain and mechanical hyperalgesia was augmented when sympathetic cutaneous vasoconstrictor neurons were activated physiologically by thermoregulatory stress (Baron, unpublished; Fig. 4).

There are additional lines of evidence to support a peripheral α-adrenergic mechanism in SMP. Topical application of the α_2-adrenergic agonist, clonidine relieves hyperalgesia at the site of application in SMP, but not SIP patients [99, 100]. This effect was considered to be secondary to a reduction in the release of NA *via* activation of the α_2-adrenergic autoreceptor on the sympathetic terminal. Injection of NA and the α_1-adrenergic agonist, phenylephrine, into clonidine-treated area produced marked pain and hyperalgesia [99]. Quantitative autoradiographic studies indicate that the number of α_1-adrenoceptors in hyperalgesic skin of patients with SMP is significantly greater than in the skin of normal subjects [85]. Finally, an increase in venous α-adrenoceptor responsiveness has been observed in patients with CRPS I [101].

Aspects for future investigations

There are several questions with regard to the pathophysiology and the clinical management of CRPS and SMP that remain uncertain and are worthy of future investigations. For example, is CRPS a single clinical entity or does it comprise of multiple subsets of patients with distinct pathophysiological mechanisms? In experimental studies, partial injury to a peripheral nerve or a spinal nerve ligation results in neuropathic pain behavior in the majority of animals. In contrast, only a small pro-

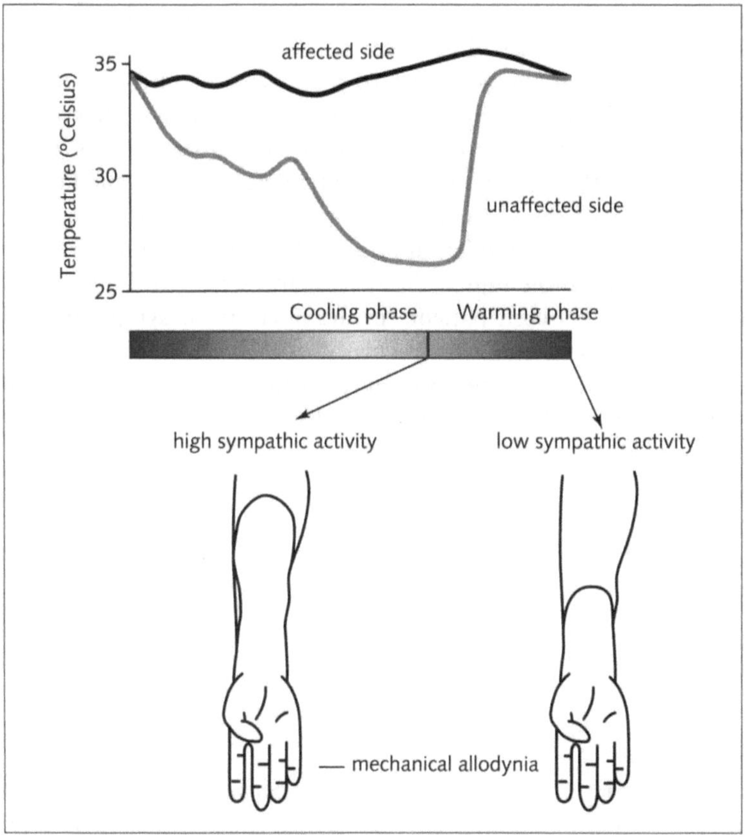

Figure 4

Influence of cutaneous sympathetic vasoconstrictor activity on pain in patients with CRPS. Physiological reflex stimuli were applied that induce a physiological activation of sympathetic activity and noradrenaline release: With the help of a thermal suit, whole body cooling and warming was performed to alter sympathetic skin nerve activity. The subject was lying in a suit supplied by tubes, in which running water of 12 °C and 50 °C respectively (inflow temperature) was used to cool or warm the whole body. Cooling induces a massive tonic activation of these neurons, warming leads to a considerable decrease of activity. By these means sympathetic activity can be switched on and off in a controlled manner during simultaneous measurement of pain sensations. High sympathetic activity during cooling induces a decrease of skin temperature due to vasoconstriction (unaffected side). During the experiment the forearm temperature on the affected side was clamped at 35 °C by a feedback-controlled heat lamp to exclude temperature effects on the sensory receptor level. Activation of sympathetic neurons leads to a considerable increase of the area of allodynia in SMP patients indicating that in CRPS a pathologic coupling between sympathetic and nociceptive neurons does exist.

portion of patients develop persistent neuropathic pain after a peripheral nerve injury. What could explain this species difference? CRPS may result from a wide variety of injuries ranging from a simple sprain to a gun-shot wound to a major peripheral nerve. Are there predictors based on the nature of the injury, the sympathetic activity, or coexistence of stress factors that predispose a given patient to the development of CRPS and SMP? Are there specific inflammatory or tropic factors that trigger the development of the chronic pain state? The sympathetic maintenance of the pain behavior in animal models appears to be dependent on a number of factors including the nature of the injury, the species and strain, and diet [102]. What explains the variability between patients in the intensity of symptoms and signs, the progression of the disease, and the recovery, or lack of, following therapy? Finally, is there a critical window of opportunity during which aggressive pain management and rehabilitation will lead to better results?

Acknowledgements
This work was supported by the Wilhelm Sander-Stiftung and the Deutsche Forschungsgemeinschaft (DFG Ba 1921/1-1) and the National Institute of Health, NS-26363 (SNR).

References

1 Baron R, Levine JD, Fields HL (1999) Causalgia and reflex sympathetic dystrophy: Does the sympathetic nervous system contribute to the generation of pain? *Muscle Nerve* 22: 678–695
2 Schott GD (1994) Visceral afferents: their contribution to 'sympathetic dependent' pain. *Brain* 117: 397–413
3 Verdugo RJ, Ochoa JL (1994) 'Sympathetically maintained pain.' I. Phentolamine block questions the concept. *Neurology* 44: 1003–1010
4 Arnér S (1991) Intravenous phentolamine test: diagnostic and prognostic use in reflex sympathetic dystrophy. *Pain* 46: 17–22
5 Raja SN, Treede RD, Davis KD, Campbell JN (1991) Systemic alpha-adrenergic blockade with phentolamine: a diagnostic test for sympathetically maintained pain. *Anesthesiology* 74: 691–698
6 Stanton-Hicks M, Jänig W, Hassenbusch S, Haddox JD, Boas R, Wilson P (1995) Reflex sympathetic dystrophy: changing concepts and taxonomy. *Pain* 63: 127–133
7 Goldstein DS, Tack C, Li ST (2000) Sympathetic innervation and function in reflex sympathetic dystrophy. *Ann Neurol* 48: 49–59
8 Wasner G, Heckmann K, Maier C, Baron R (1999) Vascular abnormalities in acute reflex sympathetic dystrophy (CRPS I) – complete inhibition of sympathetic nerve activity with recovery. *Arch Neurol* 56: 613–620

9 Wasner G, Backonja MM, Baron R (1998) Traumatic neuralgias: complex regional pain syndromes (reflex sympathetic dystrophy and causalgia): clinical characteristics, pathophysiological mechanisms and therapy. *Neurol Clin* 16: 851–868

10 Merskey H, Bogduk N (1995) *Classification of chronic pain: descriptions of chronic pain syndromes and definition of terms.* IASP Press, Seattle

11 Sudeck P (1902) Über die akute (trophoneurotische) Knochenatrophie nach Entzündungen und Traumen der Extremitäten. *Deut Med Wschr* 28: 336–342

12 Veldman PHJM (1995) *Clinical aspects of reflex sympathetic dystrophy.* Thesis, Nijmegen

13 Galer BS, Jensen MP (1997) Development and preliminary validation of a pain measure specific to neuropathic pain: the Neuropathic Pain Scale. *Neurology* 48: 332–338

14 Price DD, Bennett GJ, Rafii A (1989) Psychophysical observations on patients with neuropathic pain relieved by a sympathetic block. *Pain* 36: 273–288

15 Price DD, Long S, Huitt C (1992) Sensory testing of pathophysiological mechanisms of pain in patients with reflex sympathetic dystrophy. *Pain* 49: 163–173

16 Sieweke N, Birklein F, Riedl B, Neundorfer B, Handwerker HO (1999) Patterns of hyperalgesia in complex regional pain syndrome. *Pain* 80: 171–177

17 Rommel O, Gehling M, Dertwinkel R, Witscher K, Zenz M, Malin JP, Jänig W (1999) Hemisensory impairment in patients with complex regional pain syndrome. *Pain* 80: 95–101

18 Chelimsky TC, Low PA, Naessens JM, Wilson PR, Amadio PC, O'Brien PC (1995) Value of autonomic testing in reflex sympathetic dystrophy. *Mayo Clin Proc* 70: 1029–1040

19 Baron R, Maier C (1996) Reflex sympathetic dystrophy: skin blood flow, sympathetic vasoconstrictor reflexes and pain before and after surgical sympathectomy. *Pain* 67: 317–326

20 Baron R, Blumberg H, Jänig W (1996). Clinical characteristics of patients with complex regional pain syndromes in Germany with special emphasis on vasomotor function. In: W Jänig, M Stanton-Hicks (eds): *Progress in pain research and management, Vol. 6, Reflex sympathetic dystrophy – a reappraisal.* IASP Press, Seattle, 25–48

21 Birklein F, Riedl B, Neundörfer B, Handwerker HO (1998) Sympathetic vasoconstrictor reflex pattern in patients with complex regional pain syndrome. *Pain* 75: 93–100

22 Low PA, Amadio PC, Wilson PR, McManis PG, Willner CL (1994) Laboratory findings in reflex sympathetic dystrophy: a preliminary report. *Clin J Pain* 10: 235–239

23 Birklein F, Sittle R, Spitzer A, Claus D, Neundörfer B, Handwerker HO (1997) Sudomotor function in sympathetic reflex dystrophy. *Pain* 69: 49–54

24 Oyen WJ, Arntz IE, Claessens RM, Van der Meer JW, Corstens FH, Goris RJ (1993) Reflex sympathetic dystrophy of the hand: an excessive inflammatory response? *Pain* 55: 151–157

25 Renier JC, Arlet J, Bregeon C, Basle M, Seret P, Acquaviva P, Schiano A, Serratrice G, Amor B, May V et al (1983) The joint in algodystrophy. Joint fluid, synovium, cartilage. *Rev Rhum Mal Osteoartic* 50: 255–260

26 Blumberg H, Jänig W (1994) Clinical manifestation of reflex sympathetic dystrophy and sympathetically maintained pain. In: PD Wall, R Melzack (eds): *Textbook of pain*, Churchill, Livingstone, 685–697
27 Deuschl G, Blumberg H, Lücking CH (1991) Tremor in reflex sympathetic dystrophy. *Arch Neurol* 48: 1247–1252
28 Bhatia KP, Bhatt MH, Marsden CD (1993) The causalgia-dystonia syndrome. *Brain* 116: 843–851
29 Marsden CD, Obeso JA, Traub MM, Rothwell JC, Kranz H, La Cruz F (1984) Muscle spasms associated with Sudeck's atrophy after injury. *Br Med J (Clin Res Ed)* 288: 173–176
30 Mitchell SW (1865) *Injuries of nerves and their consequences.* Dover, New York
31 Leriche R (1916) De la causalgie envisagée comme une névrite du sympathique et de son traitement par la dénudation et l'excision des plexus nerveux péri-artériels. *Presse Med* 24: 178–180
32 Richards RL (1967) Causalgia. A centennial review. *Arch Neurol* 16: 339–350
33 Baron R, Jänig W (1998) Pain syndromes with causal participation of the sympathetic nervous system. *Anaesthesist* 47: 4–23
34 Kingery WS (1997) A critical review of controlled clinical trials for peripheral neuropathic pain and complex regional pain syndromes. *Pain* 73: 123–139
35 Price DD, Long S, Wilsey B, Rafii A (1998) Analysis of peak magnitude and duration of analgesia produced by local anesthetics injected into sympathetic ganglia of complex regional syndrome patients. *Clin J Pain* 14: 216–226
36 Bonelli S, Conoscente F, Movilia PG, Restelli L, Francucci B, Grossi E (1983) Regional intravenous guanethidine vs. stellate ganglion block in reflex sympathetic dystrophies: a randomized trial. *Pain* 16: 297–307
37 Glynn CJ, Basedow RW, Walsh JA (1981) Pain relief following post-ganglionic sympathetic blockade with I.V. guanethidine. *Br J Anaesth* 53: 1297–1302
38 Rocco AG, Kaul AF, Reisman RM, Gallo JP, Lief PA (1989) A comparison of regional intravenous guanethidine and reserpine in reflex sympathetic dystrophy. A controlled, randomized, double-blind crossover study. *Clin J Pain* 5: 205–209
39 Jadad AR, Carroll D, Glynn CJ, McQuay HJ (1995) Intravenous regional sympathetic blockade for pain relief in reflex sympathetic dystrophy: a systematic review and a randomized, double-blind crossover study. *J Pain Symptom Manage* 10: 13–20
40 Blanchard J, Ramamurthy S, Walsh N, Hoffman J, Schoenfeld L (1990) Intravenous regional sympatholysis: a double-blind comparison of guanethidine, reserpine, and normal saline. *J Pain Symptom Manage* 5: 357–361
41 Ramamurthy S, Hoffman J (1995) Intravenous regional guanethidine in the treatment of reflex sympathetic dystrophy/causalgia: a randomized, double-blind study. Guanethidine Study Group. *Anesth Analg* 81: 718–723
42 Loh L, Nathan PW (1978) Painful peripheral states and sympathetic blocks. *J Neurol Neurosurg Psychiatry* 41: 664–671
43 Hord AH, Rooks MD, Stephens BO, Rogers HG, Fleming LL (1992) Intravenous

regional bretylium and lidocaine for treatment of reflex sympathetic dystrophy: a randomized, double-blind study. *Anesth Analg* 74: 818–821
44 Jänig W, Levine JD, Michaelis M (1996) Interactions of sympathetic and primary afferent neurons following nerve injury and tissue trauma. *Prog Brain Res* 113: 161–184
45 Roberts WJ, Elardo SM (1985) Sympathetic activation of A-delta nociceptors. *Somatosens Res* 3: 33–44
46 Nakamura M, Ferreira SH (1987) A peripheral sympathetic component in inflammatory hyperalgesia. *Eur J Pharmacol* 135: 145–153
47 Hu SJ, Zhu J (1989) Sympathetic facilitation of sustained discharges of polymodal nociceptors. *Pain* 38: 85–90
48 Sato J, Suzuki S, Iseki T, Kumazawa T (1993) Adrenergic excitation of cutaneous nociceptors in chronically inflammed rats. *Neurosci Lett* 164: 225–228
49 Levine JD, Taiwo YO, Collins SD, Tam JK (1986) Noradrenaline hyperalgesia is mediated through interaction with sympathetic postganglionic neurone terminals rather than activation of primary afferent nociceptors. *Nature* 323: 158–160
50 Andreev N, Dimitrieva N, Koltzenburg M, McMahon SB (1995) Peripheral administration of nerve growth factor in the adult rat produces a thermal hyperalgesia that requires the presence of sympathetic post-ganglionic neurones. *Pain* 63: 109–115
51 Woolf CJ, Ma QP, Allchorne A, Poole S (1996) Peripheral cell types contributing to the hyperalgesic action of nerve growth factor in inflammation. *J Neurosci* 16: 2716–2723
52 Khasar SG, Miao FJ, Jänig W, Levine JD (1998) Vagotomy-induced enhancement of mechanical hyperalgesia in the rat is sympathoadrenal-mediated. *J Neurosci* 18: 3043–3049
53 Khasar SG, Miao FJ-P, Jänig W, Levine JD (1998) Modulation of bradykinin-induced mechanical hyperalgesia in the rat skin by activity in the abdominal vagal afferents. *Eur J Neurosci* 10: 435–444
54 Devor M, Jänig W (1981) Activation of myelinated afferents ending in a neuroma by stimulation of the sympathetic supply in the rat. *Neurosci Lett* 24: 43–47
55 Scadding JW (1981) Development of ongoing activity, mechanosensitivity, and adrenaline sensitivity in severed peripheral nerve axons. *Exp Neurol* 73: 345–364
56 Blumberg H, Jänig W (1984) Discharge pattern of afferent fibers from a neuroma. *Pain* 20: 335–353
57 Burchiel KJ (1984) Spontaneous impulse generation in normal and denervated dorsal root ganglia: sensitivity to alpha-adrenergic stimulation and hypoxia. *Exp Neurol* 85: 257–272
58 Jänig W (1990) Activation of afferent fibers ending in an old neuroma by sympathetic stimulation in the rat. *Neurosci Lett* 111: 309–314
59 Habler HJ, Jänig W, Koltzenburg M (1987) Activation of unmyelinated afferents in chronically lesioned nerves by adrenaline and excitation of sympathetic efferents in the cat. *Neurosci Lett* 82: 35–40
60 Sato J, Perl ER (1991) Adrenergic excitation of cutaneous pain receptors induced by peripheral nerve injury. *Science* 251: 1608–1610

61 Shyu BC, Danielsen N, Andersson SA, Dahlin LB (1990) Effects of sympathetic stimulation on C-fibre response after peripheral nerve compression: an experimental study in the rabbit common peroneal nerve. *Acta Physiol Scand* 140: 237–243
62 O'Halloran KD, Perl ER (1997) Effects of partial nerve injury on the responses of C-fiber polymodal nociceptors to adrenergic agonists. *Brain Res* 759: 233–240
63 Ali Z, Ringkamp M, Hartke TV, Chien HF, Flavahan NA, Campbell JN, Meyer RA (1999) Uninjured C-fiber nociceptors develop spontaneous activity and alpha-adrenergic sensitivity following L6 spinal nerve ligation in monkey. *J Neurophysiol* 81: 455–466
64 Ruocco I, Cuello AC, Ribeiro-Da-Silva A (2000) Peripheral nerve injury leads to the establishment of a novel pattern of sympathetic fibre innervation in the rat skin. *J Comp Neurol* 422: 287–296
65 Devor M, Jänig W, Michaelis M (1994) Modulation of activity in dorsal root ganglion neurons by sympathetic activation in nerve-injured rats. *J Neurophysiol* 71: 38–47
66 Chen Y, Michaelis M, Jänig W, Devor M (1996) Adrenoreceptor subtype mediating sympathetic-sensory coupling in injured sensory neurons. *J Neurophysiol* 76: 3721–3730
67 Michaelis M, Devor M, Jänig W (1996) Sympathetic modulation of activity in dorsal root ganglion neurons changes over time following peripheral nerve injury. *J Neurophysiol* 76: 753–763
68 McLachlan EM, Jänig W, Devor M, Michaelis M (1993) Peripheral nerve injury triggers noradrenergic sprouting within dorsal root ganglia. *Nature* 363: 543–546
69 Chung K, Lee BH, Yoon YW, Chung JM (1996) Sympathetic sprouting in the dorsal root ganglia of the injured peripheral nerve in a rat neuropathic pain model. *J Comp Neurol* 376: 241–252
70 Chung K, Yoon YW, Chung JM (1997) Sprouting sympathetic fibers form synaptic varicosities in the dorsal root ganglion of the rat with neuropathic injury. *Brain Res* 751: 275–280
71 Ringkamp M, Eschenfelder S, Grethel EJ, Höbler HJ, Meyer RA, Jänig W, Raja SN (1999) Lumbar sympathectomy failed to reverse mechanical allodynia and hyperalgesia-like behavior in rats with L5 spinal nerve injury. *Pain* 79: 143–153
72 Nicholas AP, Pieribone V, Hökfelt T (1993) Distributions of mRNAs for alpha-2 adrenergic receptor subtypes in rat brain: an *in situ* hybridization study. *J Comp Neurol* 328: 575–594
73 Pieribone VA, Nicholas AP, Dagerlind A, Hökfelt T (1994) Distribution of alpha 1 adrenoceptors in rat brain revealed by *in situ* hybridization experiments utilizing subtype-specific probes. *J Neurosci* 14: 4252–4268
74 Perl ER (1994). A reevaluation of mechanisms leading to sympathetically related pain. In: HL Fields, JC Liebeskind (eds): *Pharmacological approaches to the treatment of chronic pain: new concepts and critical issues. Progress in pain research and management*, IASP Press, Seattle, 129–150
75 Fleming WW, Westfall DP (1988) Adaptive supersensitivity. In: U Trendelenburg, N Weiner (eds): *Handbook of experimental pharmacology, Vol. 90/I. Catecholamines*, Springer Verlag, New York, 509–559

76 Bossut DF, Shea VK, Perl ER (1996) Sympathectomy induces adrenergic excitability of cutaneous C-fiber nociceptors. *J Neurophysiol* 75: 514–517
77 Shi T-JS, Winzer-Serhan U, Leslie F, Hökfeldt T (2000) Distribution and regulation of alpha-2 adrenoceptors in rat dorsal root ganglia. *Pain* 84: 319–330
78 Jobling P, McLachlan EM, Jänig W, Anderson CR (1992) Electrophysiological responses in the rat tail artery during reinnervation following lesions of the sympathetic supply. *J Physiol (Lond)* 454: 107–128
79 Koltzenburg M, Häbler HJ, Jänig W (1995) Functional reinnervation of the vasculature of the adult cat paw pad by axons originally innervating vessels in hairy skin. *Neuroscience* 67: 245–252
80 Jänig W, McLachlan EM (1994) The role of modifications in noradrenergic peripheral pathways after nerve lesions in the generation of pain. In: HL Fields, JC Liebeskind (eds): *Pharmacological approaches to the treatment of chronic pain: new concepts and critical issues. Progress in pain research and management*, IASP Press, Seattle, 101–128
81 Drummond PD (1995) Noradrenaline increases hyperalgesia to heat in skin sensitized by capsaicin. *Pain* 60: 311–315
82 Liu M, Max MB, Parada S, Rowan JS, Bennett GJ (1996) The sympathetic nervous system contributes to capsaicin-evoked mechanical allodynia but not pinprick hyperalgesia in humans. *J Neurosci* 16: 7331–7335
83 Kinnman E, Nygards EB, Hansson P (1997) Peripheral alpha-adrenoreceptors are involved in the development of capsaicin induced ongoing and stimulus evoked pain in humans. *Pain* 69: 79–85
84 Drummond PD (1998) The effect of noradrenaline, angiotensin II and vasopressin on blood flow and sensitivity to heat in capsaicin-treated skin. *Clin Auton Res* 8: 87–93
85 Drummond PD, Skipworth S, Finch PM (1996) alpha 1-adrenoceptors in normal and hyperalgesic human skin. *Clin Sci (Colch)* 91: 73–77
86 Baron R, Wasner GL, Borgstedt R, Hastedt E, Schulte H, Binder A, Kopper F, Rowbotham M, Levine JD, Fields HL (1999) Effect of sympathetic activity on capsaicin evoked spontaneous pain, hyperalgesia and vasodilatation. *Neurology* 23: 923–932
87 Elam M, Skarphedinsson JO, Olhausson B, Wallin BG (1996) No apparent sympathetic modulation of single C-fiber afferent transmission in human volunteers. In: TS Jensen, JA Turner, Z Wiesenfield-Hallin (eds): *Abstracts of the 8th world congress on pain*, IASP Press, Vancouver, 398
88 Pedersen JL, Rung GW, Kehlet H (1997) Effect of sympathetic nerve block on acute inflammatory pain and hyperalgesia. *Anesthesiology* 86: 293–301
89 Raja SN (1995) Role of the sympathetic nervous system in acute pain and inflammation. *Ann Med* 27: 241–246
90 Levine JD, Fye K, Heller P, Basbaum AI, Whiting-O'Keefe Q (1986) Clinical response to regional intravenous guanethidine in patients with rheumatoid arthritis. *J Rheumatol* 13: 1040–1043
91 Chabal C, Jacobson L, Russell LC, Burchiel KJ (1992) Pain response to perineuromal injection of normal saline, adrenaline, and lidocaine in humans. *Pain* 49: 9–12

92 Raja SN, Abatzis V, Frank SM (1998) Role of α-adrenoceptors in neuroma pain in amputees. *Anesthesiology* 89: A1083
93 Walker AE, Nulsen F (1948) Electrical stimulation of the upper thoracic portion of the sympathetic chain in man. *Arch Neurol Psychiatr* 59: 559–560
94 White JC, Sweet WH (1969) *Pain and the neurosurgeon.* Charles C. Thomas, Springfield Illinois
95 Dellemijn PL, Fields HL, Allen RR, McKay WR, Rowbotham MC (1994) The interpretation of pain relief and sensory changes following sympathetic blockade. *Brain* 117: 1475–1487
96 Torebjörk E, Wahren L, Wallin G, Hallin R, Koltzenburg M (1995) Noradrenaline-evoked pain in neuralgia. *Pain* 63: 11–20
97 Choi B, Rowbotham MC (1997) Effect of adrenergic receptor activation on post-herpetic neuralgia pain and sensory disturbances. *Pain* 69: 55–63
98 Ali Z, Raja SN, Wesselmann U, Fuchs P, Meyer RA, Campbell JN (2000) Intradermal injection of noradrenaline evokes pain in patients with sympathetically maintained pain. *Pain* 88: 161–168
99 Davis KD, Treede RD, Raja SN, Meyer RA, Campbell JN (1991) Topical application of clonidine relieves hyperalgesia in patients with sympathetically maintained pain. *Pain* 47: 309–317
100 Wesselmann UI, Ali Z, Meyer RA, Raja SN (1996) Topical clonidine attenuates ongoing pain and hyperalgesia in patients with sympathetically maintained pain (SMP). *Society for Neuroscience Abstracts* 22: 1803
101 Arnold JM, Teasell RW, MacLeod AP, Brown JE, Carruthers SG (1993) Increased venous alpha-adrenoceptor responsiveness in patients with reflex sympathetic dystrophy. *Ann Intern Med* 118: 619–621
102 Shir Y, Ratner A, Raja SN, Campbell JN, Seltzer Z (1998) Neuropathic pain following partial nerve injury in rats is suppressed by dietary soy. *Neurosci Lett* 240: 73–76

Evaluation of neuropathic pain

Karen M. Park[1] *and Richard H. Gracely*[2]

[1]Department of Anesthesiology and Critical Care Medicine, George Washington Medical Center, Washington, DC 20037, USA; [2]Clinical Measurement and Mechanisms Unit, Pain and Neurosensory Mechanisms Branch, National Institute of Dental and Craniofacial Research, National Institutes of Health, Bethesda, MD 20892, USA

Introduction

Normally, the experience of pain results from activation of neural transducers sensitive to actual or impending tissue damage from mechanical, chemical, thermal, and other physical insults. These transducers, termed nociceptors by Sherrington [1], form the end of an extensive wiring harness that transmits information to the brain, leading to the perception of pain. This "nociceptive" pain is associated with physical harm and, in most cases, can be treated by analgesics. In contrast, "neuropathic pain" does not reflect transduction by nociceptors but damage to the wiring system. Neural injury can create the impression of pain in an undamaged region innervated by the nerve. Input from damaged nerves can also initiate processes of central sensitization that result in pain and tenderness in neighboring undamaged regions. Neuropathic pain is generally more difficult to treat than nociceptive pain. Clinicians typically treat all pains initially as nociceptive and only when these treatments fail do they ask the question, "Could the etiology be neuropathic?"

What are the signs and symptoms that differentiate neuropathic from nociceptive pain? Unfortunately nerve biopsies and/or radiological imaging do not necessarily shed light in determining this. Standard neurophysiological tests can be diagnostic but do not characterize underlying pain mechanisms. Neuropathic pain is still largely a clinical diagnosis. A detailed evaluation can aid in the formulation of a diagnosis. Once a diagnosis is made, the detailed evaluation can aid both the clinical application of available treatments as well as research efforts directed at the increased understanding of this difficult group of syndromes. This chapter aims to describe the physiological characteristics of pain transmission in neuropathic conditions and how to perform an evaluation of neuropathic pain in a variety of settings, ranging from the clinician's office to a research facility.

Mechanisms and Mediators of Neuropathic Pain, edited by Annika B. Malmberg and Sandra R. Chaplan

Nerve injuries

Not all neuropathies are painful. In fact a congenital neuropathy produces a selective loss of pain sensation. Others are painless polyneuropathies with selective large fiber loss such as Friedreich's-ataxia and chronic renal failure. Painful neuropathies characterized as polyneuropathies include etiologies such as metabolic, drug or toxin-induced, infection, vitamin deficiency, hypothyroidism, amyloid, myeloma, and acute inflammation. Painful mononeuropathies include post herpetic neuralgia, diabetic mononeuropathies, ischemic neuropathies (polyarteritis nodosa, rheumatoid arthritis and lupus) and carcinomatous neuropathy. Mechanical trauma is a major cause of neuropathic pain and can range from mild, chronic pressure on a nerve trunk to transection by high-velocity projectiles. Common etiological factors range from natural injuries such as mild sprains to surgical transections in which inadvertent nerve damage manifests as a postsurgical neuropathic pain syndrome localized at or near the incision. This type of damage can lead to specific symptoms. Transection can result in focal ectopic discharge from a neuroma, which in turn may be modulated by sympathetic activity *via* up-regulated adrenergic receptors. Mild compression trauma may result in demyelination and increased excitability, leading to syndromes such as trigeminal neuralgia [2, 3].

Nerves can be damaged by inflammation, viral invasion and metabolic insufficiency. Inflammation of the nerve sheath results in a peripheral neuritis that causes pain along the trunk of the nerve and sensory changes in the region innervated by the nerve [2]. Herpes viruses result in a mosaic of peripheral nerve lesions and in many cases the painful neuropathic sequela of post herpetic neuralgia. Diabetic neuropathy includes both polyneuropathies and mononeuropathies, the most common being a symmetrical sensory polyneuropathy that results in a gradual denervation beginning with the longest axons. Thus diabetic neuropathy first affects the feet and slowly progresses up the leg.

The complexity of nerve injury directly impacts assessment. The qualitatively different types of injury require different types of evaluation. Because the pathophysiology of the nerve injury does not follow a final common pathway, there is no common evaluation. Assessment must be tailored to the patient.

Spontaneous and evoked pain. Allodynia and hyperalgesia

Nerve injuries lead to spontaneous pain, which is perceived in the absence of stimulation, and can vary in intensity, quality, locus and extent. Neuropathic pain conditions are also characterized by striking or subtle evoked pain abnormalities in which normally non-painful stimuli produce pain (allodynia) or in which normally painful stimulation produces an increased pain response (hyperalgesia). The classification of allodynia or hyperalgesia can be problematic at levels near the threshold for pain. A decreased pain threshold can be interpreted perceptually as allodynia. However, since nociceptors are activated at stimulus intensities below threshold, the reduced threshold can be defined mechanistically as hyperalgesia. In addition, some authors prefer the historical term "mechanical hyperalgesia" and use it to describe clear cases of allodynia. This chapter follows the IASP definitions of allodynia, "pain due to a stimulus which does not normally provoke pain," and hyperalgesia, as "an increased response to a stimulus which is normally painful" [4]. Allodynia is a particularly useful term for the description of striking symptoms that satisfy both perceptual and mechanistic definitions. An example is the finding of dynamic mechanical allodynia, described below, that is mediated by the large-diameter Aβ touch afferents. Under normal conditions, activations of these fibers by gentle mechanical stimulation produces only touch sensations. In abnormal cases, gentle stroking, flicking a few hairs or even blowing on the skin evokes an intense pain sensation. In these cases the touch afferents have changed character and now function as nociceptors. Allodynia describes this qualitative change, while hyperalgesia is useful for the description of quantitative changes of increased pain sensitivity.

Primary afferent function and sensory modality

Sensory testing infers the functional status of several classes of primary afferent fibers which can be divided into three major classes: the large diameter, thickly myelinated, Aβ primary afferents with conduction velocities of 50–100 m/s; the smaller diameter, thinly myelinated Aδ afferents that usually conduct at 5–30 m/s (extremes of 2 and 65 m/s); and the small diameter, unmyelinated C-fibers which conduct at relatively slow rates of around 1 m/s. Under normal conditions the Aβ fibers mediate touch sensations produced by mechanical stimulation. Gently stroking or rubbing the skin activates Aβ fibers, as does movement of body hair. Regions with dense Aβ innervation, such as near the mouth and on the fingertips, are particularly sensitive to mechanical stimulation.

Information produced by innocuous mechanical stimulation is partly segregated in the nervous system. Aβ afferents mediating fine touch enter the dorsal horn and ascend to thalamus *via* the dorsal columns and medial lemniscus system. However, Aβ fibers also provide input to wide dynamic range (WDR) pain projection neurons. Similar high rates of WDR neuronal activity can result in either tactile or pain sensations, depending on whether the input to the WDR neuron is Aβ or nociceptive. The complexity of the relationship between primary afferent and spinal cord interneurons makes the neuropathic pain evaluation more difficult because we have less certainty as to the specificity of our tests.

Aβ fibers in the skin are activated by gentle mechanical stimuli such as a wisp of cotton or the edge of cotton gauze. Vibration is an effective stimulus. Usually, sensations evoked by these stimuli can be assumed to activate Aβ fibers. However, in certain conditions the receptors to other afferent fibers can become sensitized to mechanical stimulation. In these situations the sensation evoked by gentle stimulation may represent the contribution of several fiber types, or at low levels, the stimulation of only non-Aβ fibers. Because of this lack of specificity of mechanical stimulation, electrical methods are a valuable adjunct for evaluation of Aβ function. Instead of transduction of physical energy at receptors, electrical stimuli produce activity by direct action on the axon. The sensitivity of axons to electrical stimuli is proportional to axon diameter. Thus, the large-diameter Aβ fibers are the most sensitive, and at the low stimulus levels near the detection threshold, electrical stimuli exclusively activate Aβ afferents [5]. Since this mechanism bypasses the receptor, electrical stimuli detection is not influenced by processes such as receptor sensitization.

The thinly myelinated Aδ fibers can be subdivided into populations that serve distinctly different functions. One group of Aδ fibers mediates the sensation of cool. Pain, usually pricking and sharp, is mediated by two distinct types of Aδ fibers. Type I fibers respond best to mechanical and chemical stimulation. Type I fibers also respond to temperatures that are more intense (> 53 °C) than typically presented to human subjects in pain psychophysical experiments. Type II fibers respond preferentially to heat and chemical stimulation and are assumed to mediate heat pricking pain in human studies. These two types also differ in conduction velocity, the mean velocity for Type I fibers (25 m/s) is greater than the mean for Type II fibers (15 m/s). Since electrical stimuli activate fibers in order of diameter, the pain at the level of the electrical pain threshold is mediated by Aδ fibers. Considerably greater electrical currents are required to activate the very small-diameter C fibers.

Several paradigms can assess the functional status of Aδ afferents. Aδ function can be assessed in a non-aversive manner by assessing thresholds for noticeable cooling and cold. The specific function of Aδ nociceptors can be assessed by methods that deliver painful heat or mechanical punctate stimuli. Electrical pain thresholds can be used if the confound of Aβ-mediated pain can be ruled out. In theory, reaction time latencies to electrical stimuli can provide information about conduction velocity that can confirm specific activation of Aδ afferents. [6]

Populations of C fibers mediate the sensation of warmth and pain. There is considerable spatial summation for the sensation of warmth and thus warm stimulators must be of sufficient and constant size. For example, 1-cm diameter circular contact thermodes commonly used to evoke heat pain [6–8] result in higher, more variable warm thresholds, in comparison to those obtained by larger Peltier thermodes of 10 cm^2 and greater stimulating areas. The characteristics of C-fiber mediated heat pain differ from those of Aδ heat pain. Many of these differences are manifested by the classic phenomenon of double pain. Stepping onto a very hot surface produces a sharp, highly localized, pricking pain sensation mediated by the Aδ afferents. After a noticeable delay, there is a second, diffuse, poorly localized burning sensation mediated by C fiber nociceptors. The lag is due to the difference in conduction velocities and proportional to the primary afferent distance. In this example, assum-

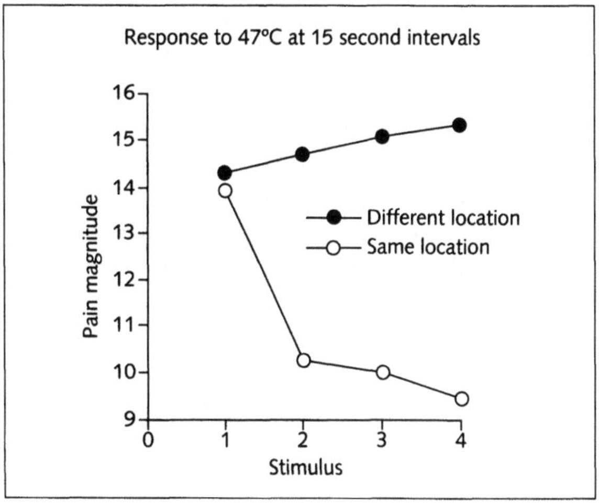

Figure 1
Intensity ratings in the conditions in which the thermal probe remained at the same location for successive stimuli or was moved to a different, adjacent location for each successive stimulus. Responses to 47° C were stable when the probe was moved (ratings = 14.3 to 15.3, mean = 14.8). When the probe remained at a single location for all stimuli, responses to 47° C (stimulus 1 response = 13.9) were significantly decreased by the second stimulus (stimulus 2 response = 10.3, ($F(1.41)$) = 62.6, $P < 0.0001$) and stable thereafter.

ing a distance of 1 meter, the 20 m/s conduction velocity requires 50 ms for the Aδ message to arrive at the spinal cord. In contrast, the 1 m/s conduction velocity of C fibers require a full second to arrive at the spinal level, accounting for the long delay. This delay is approximately halved for stimuli delivered to the hands, and there is practically no delay for stimuli delivered to the face.

Temporal factors

Repeating or prolonging a brief stimulus can significantly influence the evoked sensation. In the example of first and second pain above, the first and second pain sensations can be readily demonstrated by fast thermodes that can transition from neutral or warm baseline temperatures (30–40 °C) to painful levels (49–51 °C) and back to baseline in less than 2 s [8]. If the thermode is not moved and a second stimulus delivered within minutes, the intensity of the first pain is blunted [8, 9]. If the stimuli are delivered repeatedly, the first pain sensation disappears. Figure 1 shows an effect for 4 stimuli delivered to the ulnar distribution of the hand at 15-s intervals in a single subject. Intensity ratings decreases from 13.9 on the first trial to 10.3 on the second trial. Reaction time latencies to the same stimuli decline from 1.657 to 2.738 s. The difference of 1.08 s corrected for the reduction in latency due to reduced stimulus intensity is 0.4 s, which over the asssumed half-meter distance of arm is consistent with a shift from an Aδ conduction velocity of 20 m/s to a C-fiber conduction velocity of 1 m/s [9]. This analysis is consistent with suppression of Aδ afferents. This is one of the most noticeable temporal effects in sensory testing: Aδ-mediated painful heat sensations vanish if the skin is stimulated repeatedly. Thus, evaluation of Aδ heat nociceptors must be made on previously unstimulated skin, which may lead to difficulty if a targeted area is small and repeated measurements must be made.

What happens to the C fiber-mediated second pain sensations? At stimulation intervals of 5 s or greater, intensity ratings of these sensations decline slightly. This rate corresponds to a decline in subsequent stimulus temperatures of 0.13 °C for 47 °C stimuli delivered at 15-s intervals [8, 9]. However, if the stimuli are delivered at intervals of 3 s or less, ratings of the intensity of the C fiber-mediated second pain sensations increase at the same time that the ratings of the Aδ-mediated first pain sensations decrease [8]. The first-pain sensations are generally greater for the first two stimuli while the second-pain sensations are greater thereafter. The mechanisms producing this effect have been demonstrated in human studies and confirmed in animal experiments. The repeated intense stimuli suppress the activity of both Aδ and C-fiber nociceptors. At fast stimulating intervals the effect on C fibers is compensated by a centrally-mediated temporal summation, and effect termed "wind up" in electrophysiological studies. This combination of peripheral suppression and central summation predicts testable effects. For example, moving the stimulating probe to new sites for each stimulus abolishes the first pain suppression, but the central summation is still observed since input from a broad area converges onto the spinal circuitry responsible for the summation [8].

Temporal effects for other modalities are less pronounced. Prolonged warm and cool stimuli result in adaptation that causes the perceived intensity to fade or disappear completely [10]. Repeated stimulation can sensitize as well as suppress nociceptors, although this sensitization is usually observed at stimulus parameters that damage tissue. Repetitive mechanical tapping or squeezing, for example, can quickly produce spontaneous pain and increased evoked pain sensation [11, 12].

The results of sensory evaluation are usually described by terms derived from neurological examinations supplemented by a few additional descriptors.

Terms for diminished sensation

Hypoesthesia describes decreased sensitivity to stimulation, excluding the special senses (i.e. specific for somatic sensations of the skin and viscera), and *anesthesia* describes complete loss of sensation. These could be mixed across modalities, for example during recovery from a diagnostic block with local anesthetic, a patient may experience mechanical hypoesthesia and anesthesia to warmth.

Hypoalgesia describes diminished pain in response to a normally painful stimulus and *analgesia* describes complete loss of pain sensation. Hypoalgesia is commonly referred to as analgesia. These effects can also be modality specific, and a patient can show analgesia without anesthesia, and hypoalgesia without hypoesthesia.

Terms for augmented sensation

Hyperesthesia describes increased sensitivity to stimulation, excluding the special senses. Hyperalgesia is defined by the IASP as "an increased response to a stimulus which is normally painful" [4]. *Hyperalgesia* can either be considered as the painful analog of hyperesthesia or a subset of hyperesthesia. As noted above, hyperalgesia is sometimes used to describe allodynia. Allodynia is used by some authors to refer to an augmented sensation but is classified here as an altered sensation.

Terms for altered sensory quality

Dysesthesia is defined as an unpleasant abnormal sensation, whether spontaneous or evoked (that is not described as painful). *Paresthesia* is defined as an abnormal sensation, whether spontaneous or evoked. *Allodynia* is defined by the IASP as, "pain due to a stimulus which does not normally provoke pain" [4].

Mixed terms

Hyperpathia is a mix of low intensity hypoalgesia and high intensity hyperalgesia, with altered sensation. As stimulus intensity is increased and the pain threshold is reached, the pain is experienced as abnormally intense and disagreeable. The pain usually persists and spreads beyond the point of stimulation. There often may be a delay between the application of the stimulus and the evoked sensation.

Mapping: repeated testing within the limits of the patient

The variety of nerve injuries results in characteristic distributions of spontaneous pain and evoked pain abnormalities which are not appreciated without mapping areas of skin. Diabetic neuropathies are characterized by an irregular "die back" pattern in which the boundary between normal function and denervation anesthesia varies around the circumference of the limb within and between different types of primary afferents. Postherpetic neuralgia results in a micro-patchwork of anesthetic scar tissue and spots of altered sensitivity. Malignant tumors can compress and inflame nerves, resulting in a multi-component neuropathic pain that is difficult to treat. Mechanical nerve injury results in fluctuating sensitivity in the zone of the injury and often a surrounding area of allodynia and punctate hyperalgesia that does not conform to dermatomal divisions and often presents in a glove-stocking distribution.

Characterization of these irregular sensory distributions requires repeated testing to map the spatial pattern of sensitivity for a number of sensory modalities. However, repeatedly evoking pain sensations can quickly overtax the patient. Thus, particular attention should be paid to the patient's endurance and trials should be paced at intervals to minimize discomfort. Repeated testing also can exacerbate spontaneous pain and alter evoked pain sensitivity, leading to erroneous results.

In addition, as described in "temporal factors" repeated testing with heat stimuli can diminish the intensity of Aδ mediated sensations while repeating heat stimulation at 3/s can increase the intensity of sensations mediated by C-fiber stimulation.

The need for repeated testing requires simple methods that can be easily and quickly performed. More extensive and elaborate testing can be reserved for specific identified regions. The discussion of the various methods is organized by complexity rather than modality. A review of simple procedures that can be applied at the bedside will be followed by methods that use specialized equipment and more extensive and time-consuming procedures.

Three stages of evaluation

The three stages below describe a continuum from a brief clinical exam to research laboratory. This organization reflects the realities of medical practice and the distinction between treatment and research settings.

Stage one: patient verbal description of evoked and spontaneous pain with minimal sensory testing

Total equipment: Paper and pencil, optional pain scales.

This stage addresses the common problem of performing evaluations in a busy clinical practice. The evaluation described here could be performed at the patient's bedside or in any physician's office. Many of the patients evaluated in this scenario present with nociceptive pain and a possible neuropathic involvement. A common example is a patient with cancer on conventional opioids for nociceptive pain with suspected tumor compression of a nerve. If a neuropathic component is suspected, adjunctive medications such as anticonvulsants are prescribed empirically. Since many of these newer agents are well tolerated, this tactic implicitly assumes that the consequences of occasional false positive diagnoses of neuropathic involvement are minimal and do not justify a more extensive evaluation. Additional testing may be needed before more invasive treatments such as reducing tumor nerve compression by chemotherapy, radiation therapy or surgical resection are recommended. The use of neurolysis in terminal cases should be based on additional information about the presence and localization of neuropathic involvement.

The first step is to examine the affected region. Acute cases of complex regional pain syndrome are associated with increased temperature, edema, hyperhidrosis (increased sweating), increased nail and/or hair growth on the affected extremity. These signs should be recorded with degree (mild, moderate, excessive) noted. Other observations such as guarding of the affected body part or changes in the skin such as scarring of postherpetic neuralgia can be readily observed. Complex regional pain syndrome is associated with changes in blood flow over time. Initially the affected extremity is much warmer than the contralateral counterpart while in late cases, the affected extremity is cooler. This can easily be assessed by touch.

The second step is to listen carefully as the patients describe their pain in their own words. The clinician should listen for key words such as shooting, electric, episodic or tingling. Neuropathic involvement in the feet in cases of diabetic, chemotherapy or AIDS neuropathy may be signaled by a description of burning pain and limited walking endurance. Neuropathic pain is also indicated by pain distributions that do not follow a dermatomal distribution. The patient's description can be followed by simple questions. Figure 2 shows a simple questionnaire designed to be applied quickly to categorize sensations evoked by electrical brain stimulation during neurosurgical treatments of pain and movement disorders [13]. This questionnaire simply asks the patient if the sensations are natural or unnatural, if they are on or below the skin surface, painful or non-painful, and if they are similar to a simple organization of pain qualities. Unnatural or sensations described as "electric current" are signs of neuropathic involvement. Questions should distinguish between spontaneous pain that the patient feels all of the time and evoked pain that is triggered by movement, contact, or by warm or cold. The effects of ambient temperatures, such as difficulty with winter environments, may also indicate abnormalities of thermal sensation. For example, one can ask if pain is felt when hot/warm water touches the affected area in the shower. Unlike neuropathic pain conditions, chronic nociceptive pain such as osteoarthritis or myofascial pain is more likely to be alleviated with heat application.

These two steps can be followed by more formal self-report assessments. For example, Figure 3 shows a sample page from "The Neuropathic Pain Scale" developed by Galer and Jensen [14]. This validated scale can determine a score which can then be compared between subjects or over time within the same subject.

Location of sensation

[ ] Clearly on the skin surface
[ ] Definitely below the skin surface
[ ] Both

Non-painful

Mechanical	Movement	Temperature	Tingle
[ ] Touch	[ ] Vibration	[ ] Warm	[ ] Electric
[ ] Pressure	[ ] Movement	[ ] Cool	[ ] Tickle/itch
[ ] Sharp			

Painful

Mechanical	Movement	Temperature	Tingle	Affect
[ ] Drilling	[ ] Spreading	[ ] Hot	[ ] Itching	[ ] Frightful
[ ] Stabbing	[ ] Flashing	[ ] Burning	[ ] Electric	[ ] Nauseating
[ ] Sharp	[ ] Flickering	[ ] Cold		[ ] Cruel
[ ] Squeezing	[ ] Throbbing			[ ] Suffocating
[ ] Tugging				[ ] Fatiguing
[ ] Tearing				
[ ] Dull				
[ ] Splitting				

Figure 2
The patient is asked to describe the location of the sensation that he/she feels. If the sensation is non-painful the patient identifies category (i.e., mechanical, movement) from the "non-painful" list. If the sensation is painful the patient identifies the category from the "painful" list. Then the patient would select one or more the descriptors within the category that matched the sensation. Modified from Lenz FA [13].

Stage Two: Fast sensory assessment without specialized equipment or elaborate procedures

Total equipment: marking pen; alcohol preps; cotton applicator or cotton gauze; tuning fork; safety pin or blunt 18 gauge needle; test tube, non-metallic cup or lamp.

This stage focuses on the evaluation of evoked pain abnormalities and general sensory function. Often this assessment must be made within a large area and/or be performed quickly. Results are expressed qualitatively by terms such as those above, or simply classified as normal or abnormal. Regions of abnormality may be marked on the skin for more intensive testing or as a guide to diagnosis or treatment. This evaluation could also be done at a physician's office but requires more time than stage one.

Touch sensitivity can be evaluated by a wisp of cotton or by cotton gauze. Twisting a portion of the cotton away from the stick of a cotton applicator provides a brush useful for evaluation of small areas. Brushing with this or a similar device can be used to distinguish regions of normal sensation from regions of either hypoesthesia or hyperesthesia. The patient can observe the procedure and indicate if the evoked sensation is unusually weak or strong. The presence of allodynia is usually obvious since the sudden evoked pain will often produce appropriate withdrawal reactions. Allodynia is best evoked by a moving stimulus; suspected regions can be examined by multiple brushing of the area. Sensitivity to punctate pressure can be evaluated by a sterile safety pin, blunted 18 gauge needle, the sharp corner of an alcohol prep covering or a broken wooden shaft of a cotton applicator. The evoked pain is unusually intense, characterized by an abnormally painful reaction to a stimulus. The adequate stimulus often involves temporal summation, i.e. repetitive stimuli are usually especially effective in triggering the sensation. There is often a delay between the triggering stimulus and initiation of pain. Neuropathic pain usually radiates and persists after the stimulus is removed. In cases of focal nerve injury, zones of punctate hyperalgesia will often surround zones of mechanical allodynia, or be present without mechanical allodynia.

If the areas of pathology are sufficiently large, they can be mapped by testing along lines extending from outside the pathological center to the assumed center of the affected region. The point at which the quality of the evoked sensation changes from normal is marked on the skin. In the case of allodynia or hyperalgesia, the criteria should be precisely defined, distinguishing between paresthesia, dysesthesia, hyperalgesia and mechanical allodynia.

Patients may exhibit tenderness to blunt pressure, especially in regions near a documented or suspected injury. This sensitivity can be assessed easily by delivering pressure with the thumb. Application of steady pressure on the affected area can, in certain neuropathic conditions such as postherpetic neuralgia, actually relieve spontaneous pain.

Many patients with neuropathic pain present with cold hyperalgesia. Metal at room temperature or slightly cooler evokes a distinct, disagreeable pain sensation. Sensitivity to cool can be evaluated by metallic room temperature objects. The handle of a rubber hammer can be touched to the skin. An excellent method is to warm one counterweight (3–4 cm^2) of a tuning fork in the hand while listening to the initial self report. The warm and cool weights can be alternately applied, evaluating the effect of cool and controlling for mechanical contact. Since patients may have a cool pain threshold below this temperature, standard temperatures cooler than the typical 23 °C room temperature can be used by immersing objects in fresh tap water, or refrigerated water. Stimuli of 19–20 °C provide a good standard probe stimulus. Wiping the wet alcohol prep along the affected area can also test for cold hyperalgesia. Evaporation of alcohol creates a cooling sensation.

Appendix
NEUROPATHY PAIN SCALE

Instructions: There are several different aspects of pain which we are interested in measuring: pain **sharpness, heat/cold, dullness, intensity,** overall **unpleasantness,** and **surface vs. deep** pain.

The distinction between these aspects of pain might be clearer if you think of taste. For example, people might agree on how *sweet* a piece of pie might be (the *intensity* of the sweetness), but some might enjoy it more if it were sweeter while others might prefer it to be less sweet. Similarly, people can judge the loudness of music and agree on what is more quiet and what is louder, but disagree on how it makes them feel. Some prefer quiet music and some prefer it more loud. In short, the *intensity* of a sensation is not the same as how it makes you feel. A sound might be unpleasant and still be quiet (think of someone grating their fingernails along a chalkboard). A sound can be quiet and "dull" or loud and "dull."

Pain is the same. Many people are able to tell the difference between many aspects of their pain: for example, *how much* it hurts and *how unpleasant* or annoying it is. Although often the intensity of pain has a strong influence on how unpleasant the experience of pain is, some people are able to experience more pain than others before they feel very bad about it.

There are scales for measuring different aspects of pain. For one patient, a pain might feel extremely hot, but not at all dull, while another patient may not experience any heat, but feel like their pain is very dull. We expect you to rate very high on some of the scales below and very low on others. We want you to use the measures that follow to tell us exactly what you experience.

1. Please use the scale below to tell us how **intense** your pain is. Place an "X" through the number that best describes the intensity of your pain.

No pain	0	1	2	3	4	5	6	7	8	9	10	The most **intense** pain sensation imaginable

2. Please use the scale below to tell us how **sharp** your pain feels. Words used to describe "sharp" feelings include "like a knife," "like a spike," "jabbing " or "like jolts."

Not sharp	0	1	2	3	4	5	6	7	8	9	10	The most **sharp** sensation imaginable ("like a knife")

3. Please use the scale below to tell us how **hot** your pain feels. Words used to describe very hot pain include "burning" and "on fire."

Not hot	0	1	2	3	4	5	6	7	8	9	10	The most **hot** sensation imaginable ("on fire")

4. Please use the scale below to tell us how **dull** your pain feels. Words used to describe very dull pain include "like a dull toothache," "dull pain," "aching" and "like a bruise."

Not dull	0	1	2	3	4	5	6	7	8	9	10	The most **dull** sensation imaginable

5. Please use the scale below to tell us how **cold** your pain feels. Words used to describe very cold pain include "like ice" and "freezing."

Not cold	0	1	2	3	4	5	6	7	8	9	10	The most **cold** sensation imaginable ("freezing")

Figure 3A
The Neuropathic Pain Scale, first page.

Sensitivity to warm can be evaluated by similar objects placed in warm water baths. These can be simply calibrated by placing a known amount of water in a microwave for a specific amount of time. Reliable testing requires larger areas, at least the size of a tuning fork counter weight. Warmth may also be assessed by radi-

6. Please use the scale below to tell us how **sensitive** your skin is to light touch or clothing. Words used to describe sensitive skin include "like sunburned skin" and "raw skin."

Not sensitive	0	1	2	3	4	5	6	7	8	9	10	The most **sensitive** sensation imaginable ("raw skin")

7. Please use the scale below to tell us how **itchy** your pain feels. Words used to describe itchy pain include "like poison oak" and "like a mosquito bite.".

Not itchy	0	1	2	3	4	5	6	7	8	9	10	The most **itchy** sensation imaginable ("like poison oak")

8. Which of the following best describes the **time** quality of your pain? Please check only one answer.

( ) I feel a background pain all of the time **and** occasional flare-ups (break-through pain) some of the time.

Describe the background pain:____________________

Describe the flare-up (break-through) pain:____________________

( ) I feel a single type of pain all the time. Describe this pain:____________________

( ) I feel a single type of pain only sometimes. Other times, I am pain free.

Describe this occasional pain:____________________

9. Now that you have told us the different physical aspects of your pain, the different types of sensations, we want you to tell us overall how **unpleasant** your pain is to you. Words used to describe very unpleasant pain include "miserable" and "intolerable." Remember, pain can have a low intensity, but still feel extremely unpleasant, and some kinds of pain can have a high intensity but be very tolerable. With this scale, please tell us how **unpleasant** your pain feels.

Not unpleasant	0	1	2	3	4	5	6	7	8	9	10	The most **unpleasant** sensation imaginable ("intolerable")

10. Lastly, we want you to give us an estimate of the severity of your deep versus surface pain. We want you to rate each location of pain separately. We realize that it can be difficult to make these estimates, and most likely it will be a "best guess," but please give us your best estimate.

HOW INTENSE IS YOUR *DEEP* PAIN?

No **deep** pain	0	1	2	3	4	5	6	7	8	9	10	The most **intense deep** pain sensation imaginable

HOW INTENSE IS YOUR *SURFACE* PAIN?

No **surface** pain	0	1	2	3	4	5	6	7	8	9	10	The most **intense surface** pain sensation imaginable

Figure 3B
The Neuropathic Pain Scale, second page.

ant source such as an ordinary lamp, calibrated against the operator. This method requires precaution against excessive heat, especially with halogen lamps.

Vibration sensitivity is lost in diabetic neuropathy. To test for loss of vibration, one can apply the vibrating tuning fork handle to the toe with the operator's hand below the toe and ask the subject if they feel a vibration. If they do, then the operator can ask them to let them know when they no longer feel vibration. As the operator's finger is also feeling the vibration, the time difference between loss of vibration sense of the patient *versus* operator can be recorded. In addition to vibration sensitivity, position sense can be tested by having the subject close their eyes and then the examiner asks the subject whether the toe is up or down. The examiner moves the subject's first toe up or down by only touching the sides of the toe. Many with diabetic neuropathy are not able to tell which way their toe has been moved.

In Stage One we described a neuropathic pain scale based on self report of symptoms. The Leeds Assessment of Neuropathic Symptoms and Signs (LANSS) pain scale (Fig. 4) has been validated as another tool which is based on analysis of sensory description and bedside examination. The LANSS adds a test for allodynia using a cotton wool over the affected area and for pin prick hyperalgesia using a 23 gauge needle [Bennett, 2001 #606].

Stage Three: Detailed quantification of sensory function by specialized stimulators and procedures

Total equipment: Thermal, vibratory and electrical stimulators; calibrated nylon filaments or "electronic" filament; weighted punctate stimulators; vapo-coolant sprays, pen and acrylic sheets, computer software; skin temperature, pulse oximeter with wave form. Regional block tray with epidural needle, regional block needle, local anesthetic.

The devices and personnel required for this testing are available at most tertiary medical centers. The use of nerve blocks in diagnoses could be done by an anesthesiologist in a non-tertiary care setting.

The Quantitative Sensory Testing (QST) uses a variety of methods and devices to define sensory function. The precise assessment provided by QST provides clinical information about diagnosis, treatment choice and evaluation of treatment efficacy. QST research has uncovered mechanisms of neuropathic pain, providing the theoretical base for new, improved treatments. The QST devices include calibrated vibrameters, von Frey filaments, and quantitative measures for temperature using contact thermal probes [5, 15].

Normative values

Often the quantitative result of a test needs to be reduced to qualitative categories of normal or abnormal. One method is to compare the result of each test to standard normative values classified by location, age, gender and other values found to abb.lanss.gif (610x866x16 gif)

THE LANSS PAIN SCALE

Leeds Assessment of Neuropathic Symptoms and Signs

NAME ______________________________ DATE ______________

This pain scale can help to determine whether the nerves that are carrying your pain signals are working normally or not. It is important to find this out in case different treatments are needed to control your pain.

A. PAIN QUESTIONNAIRE

- Think about how your pain has felt over the last week.
- Please say whether any of the descriptions match your pain exactly.

1) Does your pain feel like strange, unpleasant sensations in your skin? Words like pricking, tingling, pins and needles might describe these sensations.

a) NO - My pain doesn't really feel like this.............................. (0)

b) YES - I get these sensations quite a lot.............................. (5)

2) Does your pain make the skin in the painful area look different from normal? Words like mottled or looking more red or pink might describe the appearance.

a) NO - My pain doesn't affect the colour of my skin.................... (0)

b) YES - I've noticed that the pain does make my skin look different from normal .... (5)

3) Does your pain make the affected skin abnormally sensitive to touch? Getting unpleasant sensations when lightly stroking the skin, or getting pain when wearing tight clothes might describe the abnormal sensitivity.

a) NO - My pain doesn't make my skin abnormally sensitive in that area......... (0)

b) YES - My skin seems abnormally sensitive to touch in that area...... .... (3)

4) Does your pain come on suddenly and in bursts for no apparent reason when you're still. Words like electric shocks, jumping and bursting describe these sensations.

a) NO - My pain doesn't really feel like this .... (0)

b) YES - I get these sensations quite a lot .... (2)

5) Does your pain feel as if the skin temperature in the painful area has changed abnormally? Words like hot and burning describe these sensations

a) NO - I don't really get these sensations.. (0)

b) YES - I get these sensations quite a lot .... (1)

Figure 4
The LANSS.

be relevant. Unfortunately, such norms are not yet available. To date such a centralized, normative database has not been developed.

An alternative to normative database is to use patients as their own control. Because sensitivities vary over body regions, the sensitivity of pathological regions can be compared to that of homologous contralateral control regions. This method cannot be used for evaluation of midline structures or for cases in which pathology is suspected in the homologous contralateral region. Modifications such as testing in adjacent unaffected skin are also possible.

Each of these methods has particular advantages. The use of a normative database provides control values for midline structures and is useful for assessment of bilateral pathology in a region. The use of homologous contralateral control values reduces the influence of individual differences in pain sensitivity and in reporting style. This method also allows the evaluation of any lateral structure, while the database method limits testing to groups of standard structures. In addition, the within-patient method provides a greater range of measurement. Kemler et al. [16] directly compared these two methods in a group of patients with suspected neuropathic pain. The results were generally similar with several important differences. Using two standard deviations as the criteria for abnormality, hypoesthesia and hyperalgesia could not be evaluated since the criteria values exceeded the performance or safety limits of the stimulator. In addition, the study showed that the database method provided a different result than the within-patient control method in a substantial number of cases. Based on these findings the authors recommend the use of within-patient controls when possible, resorting to normative values if the within-patient method is not possible.

Quantitative tests of light touch and allodynia

Sensitivity to light touch is commonly evaluated by a set of 20 nylon monofilaments. Each filament is attached to a plastic handle, and calibrated to deliver a specific force calibrated in grams (more appropriately expressed in bars) ranging from 4.5 mg to 447 g. The filament is applied perpendicular to the skin and pressure is applied to maintain a bend in the filament. Care is taken to not drag the tip across the skin or to contact hair. This method quantifies a range of cutaneous sensitivity from very sensitive to severe hypoesthesia. Factors such as the surface area of the tip are ignored, although a jagged tip may produce excessive stimulation as it rotates during the initial bending. In the absence of a set of filaments or for specific applications, a disposable set of filament stimulators can be made from specific lengths of suture material and held by forceps [17]. The forces can be calibrated on a laboratory scale or the clinician can use the published values.

The filaments have been used to test for the presence of allodynia, but they are no more useful for quantification than more simple methods. Allodynia is irregularly evoked by static pressure, the best stimulus is a moving stimulus lightly dragged against the skin like the cotton wisp or described above, or a camel hair brush. LaMotte et al. [18] have used an applicator that delivers a constant pressure by the methods used for the monofilaments. The stimulator is attached to a strip of metal saw blade in turn attached to a handle and the stimulator is moved across the skin at a constant pressure by means of a bend maintained in the metal blade.

Studies with a number of devices show that the intensity of evoked allodynia is a poor measure; allodynia is best quantified by measuring its extent. This is accomplished as described above by applying the applicator from normal skin inward along radial spokes until the patient reports pain. This region is marked and the operation is repeated until a reasonable boundary can be inferred. This boundary is transferred to transparent acrylic sheets and the area determined by computerized methods. The area can also be determined with surprising precision by a method dating to World War II; tracing onto high quality paper, cutting out the area with scissors and weighing on a laboratory scale [19].

Punctate pressure

Sensitivity to punctate stimuli can be assessed by the nylon filaments described above and by calibrated weighted or automatic stimulators [20, 21]. The stimulators should be sterile and not sharp enough to puncture the skin. Hypodermic needles are too sharp. A convenient method is to use a weighted stimulator that has been developed by several groups. One variant of this method cuts the end of a sterile safety pin, places it in a blunt irrigating syringe, crimps it in place, and drops it down the bore of a small syringe. Sterility is easily maintained. Calibrated weights placed in the bore determine the force at the tip when the device is lowered vertically onto the test area. After use the device can be recapped for additional use in the same patient, or discarded in a standard sharps container.

Punctate, or pinprick, hyperalgesia is a common finding in neuropathic pain. Zones of pinprick hyperalgesia usually surround areas of allodynia and are present in the absence of allodynia. The areas appear to be more stable than those of allodynia, which can fluctuate in size and location. Mapping of these areas can be outlined as described in stage two.

Quantitative blunt pressure

The American College of Rheumatology has adapted a method using a force of 4 kg delivered by the examiner's thumb. Pressure may also be delivered by devices termed dolorimeters or algometers that deliver ascending pressure *via* a hard rubber probe tip (e.g. 1 cm^2) and record the pressure delivered at pain threshold.

Thermal sensation

A number of commercially available thermal contact stimulators are based on the thermoelectric Peltier principle developed by Kenshalo and Bergen [22]. The surface areas of the probes are sufficiently large to be used for warmth detection as well as determination of the heat and cold pain threshold and subjective scaling of suprathreshold cold and heat pain. Warmth and cool sensations are influenced by a number of well-known factors, including the most recent thermal experience (e.g. soaking in a hot bath or exposed to cold air temperatures). Warmth, and to a lesser extent cool, are influenced by the area stimulated, and may be influenced by alterations in internal body temperature [23–25].

The methods used to assess thermal sensory and pain thresholds are essentially the same, and are included in software with some of these devices. In the simplest procedure, stimulus temperature is increased from a baseline temperature in a single direction until a button press indicates a subjective level (detection, pain) has been reached, returning the stimulus to baseline or in the opposite direction for a response to the opposite thermal modality. More elaborate procedures use a staircase titration paradigm which provides increased resolution at the expense of increased time [6]. Other, more time intensive psychophysical procedures can detect subtle sensory changes [19]. Choice of method is limited by the time available and determined by the magnitude of the sensory abnormality. Many of the marked changes associated with clinical evaluation of neuropathic pain are easily detected by simple, fast procedures, while more extensive procedures may be most useful for research investigations.

Electrical stimulation

Electrical stimulation of skin can be used to assess the function of Aβ touch fibers (mechanical allodynia, neural inflammation), and temporal summation of nociceptor activation. Because of the precise timing of stimulus onset, electrical stimulation is also used for extensive measures such as the nociceptive spinal reflex, somatosensory evoked cortical potentials and traditional measurement of nerve conduction [26].

Electrical evaluation of Aβ involvement in mechanical allodynia is based on the large diameter of Aβ axons and on the physics of electrical neural activation. Electrical stimuli bypass the receptor, stimulating the axon directly. Thus the method is not influenced by receptor processes such as suppression or sensitization. Sensitivity to electrical axonal stimulation is proportional to axon diameter. As the intensity of electrical current applied to skin is increased from zero, the first axons activated are Aβ touch axons and the first sensation experienced at the threshold for detection is touch sensation. If the stimulus is in a zone of Aβ mediated mechanical allodynia, the sensation experienced at the threshold for detection is also painful.

Electrical stimulation of Aβ afferents can also detect suspected cases of primary afferent neural inflammation. Such inflammation leads to the phenomenon of "progressive tactile hypersensitivity", which is reflected by lowered electrical detection thresholds that indicate Aβ hypersensitivity [17]. This hypersensitivity and the presence of mechanical allodynia can be evaluated with mechanical stimulation. The mechanical methods, however, are not as specific as electrical stimulation. In cases of mechanical hyperalgesia, gentle mechanical stimulation may produce pain due to lowered nociceptor thresholds, while the threshold for electrical detection is unaltered and results in only a touch sensation.

Electrical tests in research settings typically employ precise constant current stimulators that deliver precisely controlled stimuli in the micro-ampere to milli-ampere range. This degree of sophistication is probably unnecessary for clinical evaluation. Simple stimulators such as transcutaneous electrical nerve stimulator (TENS) devices can be used to detect both mechanical allodynia (pain at detection threshold) and Aβ hypersensitivity (lower detection threshold in affected region compared to homologous contralateral or adjacent control site).

Sudomotor tests

Sudomotor function refers to sweat response testing. The sudomotor tests available include the sympathetic skin response and resting sweat output, and quantitative sudomotor axon reflex test (QSART). QSART has shown higher sensitivity in detecting small fiber neuropathy (i.e., C fiber) compared to QST and cardiovascular autonomic testing [27]. Reidel et al designed a new capsule for the distal performance of the QSART. Because many neuropathies show a length-dependent loss of nerve function, they postulated that measurement of the fingers and toes was necessary. The forearm/finger ratios in patients with insulin dependent diabetes and normal sensory nerve conduction studies were a more sensitive indicator of distal-onset neuropathies [28].

Vasomotor tests

The vasomotor system is tested in a variety of ways, the simplest being measuring skin temperature. With vasodilation, the skin temperature rises. Earlier we mentioned the changes in blood flow that occur over time in complex regional pain syndrome, type one. This is secondary to vasodilation or vasoconstriction. The early stage of complex regional pain syndrome involves increased blood flow while the late stages are characterized with decreased blood flow. Gross measures of blood flow include skin temperature and thermographic comparisons with the contralateral body part. Quantification of blood flow is attempted by studies such as Laser Doppler which measures capillary flow in locations such as skin or nail beds.

Unfortunately for the patient, the evaluation of neuropathic pain described thus far causes no pain or induces pain. Clinicians and patients prefer tests that can actually reduce pain. Local anesthetic nerve blocks can be useful in this way.

Differential neural blockade

Local anesthetics block neural transmission in a selective fashion. This can be taken advantage of in two ways. One way is to do a sympathetic nerve block. The sympathetic ganglia for the upper extremity can typically be blocked by injecting local anesthetic into the ganglion which is anatomically close to the transverse process of the sixth cervical vertebra. An injection here can produce sympathetic blockade of the entire upper extremity. Likewise the lower extremity sympathetic ganglia lies along the anterolateral border of the second and third lumbar vertebral bodies. Thus, one could perform this type of procedure and then see if pain is reduced. The caveat of using sympathetic blockade as a diagnostic test is that confirming needle placement on the sympathetic ganglia can be problematic. Injection of contrast dye with fluoroscopy can aid accurate placement but even so, overflow of local anesthetic to nerve roots can not absolutely be ruled out. Because sympathetic blockade increases blood flow and decreases the sweat response, we can test for the presence of sympathetic blockade, i.e., pulse plethysmography, Laser doppler, QSART. Reduced pain after sympathetic block would indicate sympathetically mediated pain.

Local anesthetics decrease conduction of transmission through nerve fibers in a selective fashion. The sensitivity of nerve fibers to this type of block is as follows (from most to least): sympathetic, C, Aδ, Aβ, and so on to the thickest myelinated fibers. The epidural space runs from the base of the skull to the sacrum just outside the intrathecal space. Local anesthetic placed here blocks neural transmission at the dorsal root ganglion at several levels, depending on the volume injected. If an epidural catheter is placed and dosed with a short-acting local anesthetic, the return of nerve fiber function can temporally be correlated with the return of pain and so help determine what fibers are involved. For example, in a patient who has possible neuropathic pain in the foot, after dosing the epidural with local anesthetic at an appropriate dose and volume, the patient will have no motor or sensory function of the foot and leg. In time however, the motor function will return first, followed by return of light touch detection, then pinprick, then temperature. If the patient has no pain until after all fiber function has returned, then the clinician could postulate sympathetically mediated pain as the mechanism for the patient's pain.

Reversible ischemia can be induced on nerve fibers of the extremities by use of a tourniquet. The sensitivity of nerve fibers to ischemia is opposite to local anesthetic, i.e., the nerve fibers most sensitive to ischemia are the large myelinated fibers. The tourniquet should be placed proximal from the site of pain. Approximately 25 min of ischemia will cause loss of sensation to light touch in the area distal to the tourniquet. Ischemic blocks have been used to determine Aβ fiber involvement in neuropathic pain conditions [29].

Combining the diagnostic nerve block with electrical and temperature detection threshold testing would increase the power of the results because non-specific factors can be ruled out. For example, recall that one problem with sympathetic ganglion blocks is that the nerve root could be blocked as well as the sympathetic ganglion. If we determined sensory function of nerve roots was preserved, we would be more certain that we had a pure sympathetic block. For the differential epidural block, if we could more specifically time the return of light touch, heat, pinprick compared to the return of pain, we could be more certain which fibers were involved in the transmission of pain.

The stages of evaluation of neuropathic pain described in this chapter have formed a foundation for subcategorizing the signs and symptoms presented by an individual patient. Characterizing the neuropathic pain states of an individual patient with elaborate testing is justified if the obtained data guides choice of treatments such as choice of initial medications. Currently, treatments for neuropathic pain are tried in random fashion, with long trial periods to determine effectiveness. Clinical studies matching pain testing results to trials of medications with differing mechanisms of action will be the next step in defining paradigms for the treatment of neuropathic pain.

References

1 Sherrington CS (1906) *The integrative action of the nervous system.* Scribner, New York.

2 Eliav E, Gracely RH (1999) Trigeminal neuralgia. In: AR Block, E Kremer, E Ferdandez (eds): *Handbook of chronic pain syndromes*, Erlbaum and Associates, London, 435–453

3 Loeser JD (1994) Tic douloureux and atypical facial pain. In: PD Wall, R Melzack (eds): *Textbook of pain*, 2nd edn, Churchill Livingston, London, 699–710

4 Merskey H, Bogduk N (1994) *Classification of chronic pain: Description of chronic pain syndromes and definition of pain terms*, 2nd edn., IASP Press, Seattle

5 Gracely RH, Price DD, Roberts WJ, Bennett GJ (1996) Quantitative sensory testing in patients with CRPS-I&II. In: W Jänig, M Stanton-Hicks M (eds): *Reflex sympathetic dystrophy – a reappraisal*, IASP Press, Seattle, 151–172

6 Gracely RH, Lota L, Walter DJ, Dubner R (1988) A multiple random staircase method of psychophysical pain assessment. *Pain* 32: 55–63

7 Coghill RC, Sang CN, Maisog JM, Iadarola MJ (1999) Pain intensity processing within the human brain: a bilateral, distributed mechanism. *J Neurophysiol* 82: 1934–1943

8 Price DD, Hu JW, Dubner R, Gracely RH (1977) Peripheral suppression of first pain and central summation of second pain evoked by noxious heat pulses. *Pain* 3: 57–68

9 Gracely RH, Kenshalo DRJ (1996) Human A-delta and C-fiber nociceptor responses to low intensity painful heat. *Society of Neuroscience Abstracts* 22: 113

10 Kenshalo DR, Scott HHJ (1966) Temporal course of thermal adaptation. *Science* 151: 1095–1096

11 Farrell MJ, Petzke F, Grant MAB, Clauw DJ, Gracely RH (2000) Continuous and fluctuating pressure stimuli provoke pain of equivalent intensity and disparate unpleasantness. *American Pain Society 19th Annual Meeting* 147

12 Kohlloffel LU, Koltzenburg M, Handwerker HO (1991) A novel technique for the evaluation of mechanical pain and hyperalgesia. *Pain* 46: 81–87

13 Lenz FA, Gracely RH, Romanoski AJ, Hope EJ, Rowland LH, Dougherty PM (1995) Stimulation in the human somatosensory thalamus can reproduce both the affective and sensory dimensions of previously experienced pain. *Nat Med* 1: 910–913

14 Galer BS, Jensen MP (1977) Development and preliminary validation of a pain measure specific to neuropathic pain. *Neurology* 48: 332–338

15 Boivie J (1988) Projections from the dorsal column nuclei and the spinal cord to the red nucleus in cat. *Behav Brain Res* 28: 75–79

16 Kemler MA, Schouten HJ, Gracely RH (2000) Diagnosing sensory abnormalities with either normal values or values from contralateral skin: comparison of two approaches in complex regional pain syndrome I. *Anesthesiology* 93: 718–727

17 Eliav E, Gracely RH (1998) Sensory changes in the territory of the lingual and inferior alveolar nerves following lower third molar extraction. *Pain* 77: 191–199

18 LaMotte RH, Shain CN, Simone DA, Tsai EF (1991) Neurogenic hyperalgesia: psychophysical studies of underlying mechanisms. *J Neurophysiol* 66: 190–211

19 Gracely RH, Naliboff BD (1996) Measurement of pain sensation. In: L Kruger (ed): *Pain and touch*, Academic Press, San Diego, 243–313

20 Greenspan JD, McGillis SL (1991) Stimulus features relevant to the perception of sharpness and mechanically evoked cutaneous pain. *Somatosens Mot Res* 8: 137–147

21 Greenspan JD, Thomadaki M, McGillis SL (1997) Spatial summation of perceived pressure, sharpness and mechanically evoked cutaneous pain. *Somatosens Mot Res* 14: 107–112

22 Kenshalo DR, Bergen DC (1975) A device to measure cutaneous temperature sensitivity in humans and subhuman species. *J Appl Physiol* 39: 1038-1040

23 Marks LE, Stevens JC (1973) Spatial summation of warmth: influence of duration and configuration of the stimulus. *Am J Psychol* 86: 251–267

24 Marks LE (1971) Spatial summation in relation to the dynamics of warmth sensation. *Int J Biometeorol* 15: 106–110

25 Stevens JC, Marks LE (1979) Spatial summation of cold. *Physiol Behav* 22: 541–547

26 Gracely RH (1999) Normal people. In: PD Wall, R Melzack (eds): *Textbook of pain*, 4th edn, Churchill Livingstone, St. Louis, MO, USA, 385–407

27 Tobin K, Giuliani MJ, Lacomis D (1999) Comparison of different modalities for detection of small fiber neuropathy. *Clin Neurophysiol* 110: 1909–1912

28 Riedel A, Braune S, Kerum G, Schulte-Monting J, Lucking CH (1999) Quantitative sudomotor axon reflex test (QSART): a new approach for testing distal sites. *Muscle Nerve* 22: 1257–1264

29 Byas-Smith MG, Bennett GJ, Gracely RH, Max MB, Robinovitz E, Dubner R (1999) Tourniquet constriction exacerbates hyperalgesia-related pain induced by intradermal capsaicin injection. *Anesthesiology* 91: 617–625

New treatment strategies for neuropathic pain

Glutamate receptor antagonists and neuropathic pain

Per Kristian Eide

Department of Neurosurgery, The National Hospital, University of Oslo, Sognvannsveien 20, N-0027 Oslo, Norway

Glutamate receptors in central nervous system disorders

Glutamate is the quantitatively most important excitatory neurotransmitter in the central nervous system with three main receptor subtypes (metabotropic, non-NMDA and NMDA receptor subtypes). The various glutamate receptor subtypes play an important role in the pathophysiology of numerous central nervous system diseases, including acute excitotoxic neurodegeneration (e.g. cerebral ischaemia, stroke and brain injury), chronic excitotoxic neurodegeneration (e.g. dementia of the Alzheimer type, Parkinsonism and Huntington's disease), epilepsy and psychiatric disorders (e.g. anxiety, depression and schizophrenia) [1].

During the last decade, major attention has been given to the role of glutamate receptors in pathophysiological mechanisms of neuropathic pain. Glutamate receptors (in particular N-methyl-D-aspartic acid (NMDA) receptors) play an important role in so-called central sensitization [2, 3]. Central sensitization may be an important mechanism behind several of the characteristics of neuropathic pain such as spontaneous ongoing pain and abnormally evoked pain (e.g. allodynia and hyperalgesia) [4]. Central sensitization may be induced by primary afferent hyperactivity after nerve injury, and may be manifested as increased spontaneous activity and hyperexcitability of central wide dynamic range (WDR) neurons. This change in neuronal excitability is related to flow of Ca^{2+} ions into the cell following NMDA receptor channel activation. Increased intracellular Ca^{2+} triggers a number of intracellular biochemical processes that produce long-term changes in nerve cell function.

The glutamate receptor antagonists that have been used in clinical trials on neuropathic pain include ketamine, dextromethorphan/dextrorphan and amantadine/memantine. Ketamine has been used as an anaesthetic for more than 30 years

Mechanisms and Mediators of Neuropathic Pain, edited by Annika B. Malmberg and Sandra R. Chaplan

and is a non-competitive NMDA receptor antagonist acting at the PCP binding site in the NMDA receptor channel [5]. The effects of ketamine on neuropathic pain reported in clinical trials appear to be related to blockade of NMDA receptors, not interactions with other receptor types [6]. Dextromethorphan has been used as an antitussive and more recently has been shown to exhibit NMDA receptor antagonist properties [7]. Amantadine/memantine has been used as anti-viral and anti-Parkinsonian drugs and have been shown to act as non-competitive NMDA receptor antagonists [8].

Animal models of neuropathic pain

In various animal models of neuropathic pain, NMDA receptor antagonists have been found to inhibit different types of evoked pain (e.g. mechanical allodynia-like behavior). In an animal model of allodynia induced by inhibition of spinal inhibitory tone, NMDA receptor antagonists inhibited a tactile-evoked pain response [9]. In another model with loose ligation of the common sciatic nerve (chronic constriction injury), systemic or intrathecal administration of NMDA receptor antagonists were shown to inhibit heat-evoked hyperalgesia [10–14], spontaneous pain-related behavior [11, 12], and mechanical hyperalgesia [15]. After tight ligation of spinal nerves, systemic ketamine relieved various symptoms, including mechanical allodynia and hyperalgesia, cold allodynia and spontaneous pain-related behavior [16]. Other NMDA receptor antagonists also inhibited mechanical allodynia after tight nerve ligation [17]. In an animal model of central neuropathic pain after vascular damage of the spinal cord, an allodynia-like behavior is assessed as a painful reaction to light tactile stimulation of the denervated skin area (mechanical allodynia). Acute mechanical allodynia appearing 24–48 h after ischaemic spinal cord injury was inhibited by a non-NMDA excitatory amino acid receptor antagonist though not by an NMDA antagonist [18]. Chronic mechanical allodynia occurring several weeks after ischaemic spinal cord injury was relieved by NMDA receptor blockade though not by non-NMDA receptor blockade [19].

Furthermore, data from various animal models of neuropathic pain suggest that NMDA receptor antagonists may enhance the effects of opiods on mechanical allodynia [20, 21].

Clinical studies on neuropathic pain

Controlled clinical trials including patients with peripheral and/or central neuropathic pain have shown that glutamate receptor antagonists modulate different pain characteristics (Tab. 1). In particular, NMDA receptor antagonists have been found to inhibit different pain characteristics such as spontaneous ongoing pain, allodynia and abnormal temporal summation of pain (often referred to as wind-up like pain).

Spontaneous ongoing pain

A number of case reports have shown that NMDA receptor antagonists, in particular ketamine, may inhibit spontaneous ongoing pain in patients with chronic neuropathic pain. Ketamine given *via* various administration routes (i.v., i.m., s.c., p.o), inhibited ongoing pain in patients with peripheral and/or central neuropathic pain [22–29]. In a patient with neuropathic pain, intrathecal (i.th.) injection of the NMDA receptor antagonist CPP failed to affect spontaneous ongoing pain or allodynia, but inhibited the spread of pain outside the territory of nerve injury and pain that lasted after the end of mechanical stimulation [30]. Long-lasting effects of short trials of NMDA receptor antagonists have been reported. In a case that reported neuropathic pain (complex regional pain syndrome, type II) after sciatic nerve injury for some weeks, epidural infusion of a low dose of ketamine (25 µg/kg/h) for 10 days completely abolished pain for the observation period of 8 months [31]. Eisenberg and Pud [32] reported three patients with peripheral neuropathic pain that became pain free for several months after the end of a short course of amantadine treatment. Among 26 patients with trigeminal neuropathic pain, ketamine (0.4 mg/kg, i.m.) was found to relieve pain for up to 24 h after treatment in eight of the patients [33]. The authors argued that in some patients, the chronic pain might be particularly dependent on NMDA receptor mechanisms.

As indicated in Table 1, several randomized, double-blind and placebo-controlled studies have shown that short-term infusions of ketamine significantly inhibits spontaneous ongoing pain in patients with peripheral [33–37] and central [38] neuropathic pain. One controlled study showed that amantadine inhibits spontaneous ongoing pain in patients with neuropathic cancer pain [39]. Unfortunately, the analgesic effects of short-term infusions of NMDA receptor antagonists are associated with bothersome side-effects.

As indicated in Table 1, less promising results have been reported after long-term treatment with NMDA receptor antagonists. Double-blind and randomized placebo-controlled studies demonstrated no significant relief of spontaneous ongoing pain after treatment with dextromethorphan in patients with neuropathic pain [40, 41], though spontaneous pain was significantly reduced in patients with diabetic neuropathic pain [41]. In another double-blind and randomized placebo-controlled trial, long-term treatment with memantine (daily doses of 10 or 20 mg) failed to significantly inhibit spontaneous ongoing pain in patients with postherpetic neuralgia [42]. Furthermore, we found in a controlled trial that a low dose of ketamine (2 and 4 mg/kg/day) failed to relieve spontaneous ongoing pain in patients with postherpetic neuralgia [43]. In this study, pain relief was enhanced

Table 1 - Results of randomised, double blind, placebo-controlled studies on the analgesic effect of NMDA receptor antagonists in neuropatic pain

Drug	Dose/administration route	N	Pain syndrome	Pain characteristics inhibited	Ref.
Short-term treatments					
Ketamine	0.15 mg/kg, i.v.	8	Post-herpetic neuralgia	↓ Intensity of ongoing pain ↓ Wind-up like pain ↓ Allodynia	[34]
Ketamine	0.75 mg/kg/2 h, i.v.	8	Posttraumatic pain	↓ Intensity of ongoing pain ↓ Allodynia	[36]
Ketamine	5 µg/kg/min after 0.2 mg/kg bolus, i.v.	10	Peripheral neuropathic pain	↓ Intensity of ongoing pain ↓ Area of allodynia	[35]
Ketamine	6 µg/kg/min after 60 µg/kg bolus, i.v.	9	Central neuropathic pain after spinal cord injury	↓ Intensity of ongoing pain ÷ Wind-up like pain ↓ Allodynia	[38]
Ketamine	7 µg/kg/min after 0.1 mg/kg bolus, i.v.	11	Stump and phantom limb pain	↓ Intensity of ongoing pain ↓ Wind-up like pain	[37]
Amantadine	200 mg/3 h, i.v.	15	Neuropathic cancer pain	↓ Intensity of ongoing pain	[39]
Ketamine	0.4 mg/kg, i.m.	8	Trigeminal neuropathic pain	↓ Intensity of ongoing pain	[33]
Long-term treatments					
Dextromethorphan	40.5 or 81 mg/day, p.o.	19	Peripheral or central neuropathic pain	÷ Intensity of ongoing pain	[40]
Dextromethorphan	~ 381 mg/day, p.o.	14	Diabetic neuropathy	↓ Intensity of ongoing pain	[41]
Dextromethorphan	~ 381 mg/day, p.o.	18	Postherpetic neuralgia	÷ Intensity of ongoing pain	[41]
Memantine	10–20 mg/day, p.o.	9	Postherpetic neuralgia	÷ Intensity of ongoing pain ÷ Mechanical allodynia ÷ Wind-up like pain	[42]
Ketamine	2 and 4 mg/kg/day, p.o.	12	Postherpetic neuralgia	÷ Intensity of ongoing pain ÷ Allodynia ↓ Wind-up like pain	[43]

Symbols: ↓ (significantly reduced); ÷ (not significantly changed) by the combined administration of ketamine and morphine, suggesting a possible beneficial effect of combining opioids with NMDA receptor antagonists.

Allodynia

Several case studies reported that allodynia in patients with neuropathic pain was inhibited by acute infusions of ketamine [22, 23, 28].

The effect of NMDA receptor antagonists on allodynia may last for a longer time than the pharmacological action. In a case with mechanical allodynia surrounding an infected surgical wound, i.v. administration of ketamine (0.2–0.5 mg/kg/2 min) during wound dressing produced acute short-term pain relief that was not obtained with opioids [44]. In this case, mechanical allodynia diminished considerably during the 3 months period of intermittent ketamine administration 1–2 times daily.

As indicated in Table 1, several controlled studies showed that acute i.v. injection of ketamine inhibits allodynia in patients with central or peripheral neuropathic pain, and that this effect was associated with inhibition of spontaneous ongoing pain [34–36, 38]. When ketamine was given in an oral dose that failed to relieve spontaneous pain, allodynia was neither affected [43]. The studies assessed dynamic mechanical allodynia, either evoked by applying an electric tooth brush [34, 38, 43] or a cotton swab [35, 36] to the skin. Oral treatment with memantine, on the other hand, failed to inhibit mechanical allodynia in patients with postherpetic neuralgia [42].

Abnormal temporal summation of pain (wind-up like pain)

So-called wind-up like pain was first described in patients with PHN [34]. Repetitive pricking of the skin with a von Frey filament (2741 mN) at a rate of 3/s was initially non-painful but after some seconds led to markedly enhanced pain, with radiation of pain outside the stimulated area and sensation of pain after the end of stimulation. This response was different from temporal summation found in normal individuals or in skin areas without persistent pain. This type of abnormal temporal summation of repetitive mechanical stimuli is considered a pathophysiological activation of the physiological response.

As indicated in Table 1, controlled studies have shown that ketamine inhibits wind-up like pain in patients with peripheral and central neuropathic pain [34, 37, 38, 43]. This effect may be associated with inhibition of ongoing pain [34, 37]. Wind-up like pain in patients with neuropathic cancer pain also was inhibited by amantadine [39]. Oral treatment with memantine, on the other hand, did not relieve wind-up like pain in patients with postherpetic neuralgia [42].

Side-effects of NMDA receptor antagonists

NMDA receptors are ubiquitously present in the human brain. Therefore, NMDA receptor antagonists at analgesic doses interfere not only with pain perception, but also with sensory perception in general. Psychotomimetic side-effects of NMDA receptor antagonists represent the major problem [45]. Thus, the main reason limiting the use of NMDA receptor antagonists in the clinic is the frequency and severity of side effects. Clinical experience suggests that both dextromethorphan and amantadine/memantine have a better side-effect profile than ketamine. However, when memantine was given in a dose with few side-effects, no effects on pain was found in patients with postherpetic neuralgia [42].

Presently efforts are made to reduce the side effect profile of glutamate receptor antagonists, for instance by identifying new receptor subunits and development of drugs interacting with other regulatory sites in the NMDA receptor operated channel (e.g. Mg^{2+}, Zn^{2+}, glycine binding sites) [1].

Conclusions

In various animal models of neuropathic pain, glutamate receptor antagonists consistently inhibit different types of evoked pain behavior such as mechanical allodynia-like behavior. Controlled trials show that short-term treatments with NMDA receptor antagonists inhibit characteristics of neuropathic pain such as spontaneous ongoing pain, mechanical allodynia and abnormal temporal summation. The results of long-term treatment with NMDA receptor antagonists in controlled trials have been less promising. The clinical use of these drugs in patients with neuropathic pain is restricted by their side-effect profile. Strategies to reduce the frequency and severity of side effects are highly needed. Nevertheless, the clinical trials of NMDA receptor antagonists represent a step towards mechanism based treatment of neuropathic pain. At present, it seems as NMDA receptor antagonists may become useful in a multi-drug regimen to treat certain pain characteristics (secondary hyperalgesia, mechanical allodynia, abnormal summation and spread of pain).

References

1 Parsons CG, Danysz W, Quack G (1998) Glutamate in CNS disorders as a target for drug development: An update. *Drug News Perspect* 11: 523–569

2 Dubner R (1997) Neural basis of persistent pain: Sensory specialization, sensory modulation, and neuronal plasticity. In:TS Jensen, JA Turner, Z Wiesenfeld-Hallin (eds): *Proceedings of the 8th World Congress on Pain. Progress in pain research and management*, Vol. 8, IASP Press, Seattle, 243–257

3 Dickenson A (1997) Mechanisms of central hypersensitivity: Excitatory amino acid mechanisms and their control. In: A Dickenson, JM Besson (eds): *The pharmacology of pain.* Springer Verlag, Berlin, 167–210.
4 Bennett GJ (1994) Neuropathic pain. In: PD Wall, RM Melzack (eds): *Textbook of pain.* Churchill Livingstone, Edinburgh, 201–224
5 Anis NA, Berry SC, Burton NR, Lodge D (1983) The dissociative anaesthetics, ketamine and phencyclidine, selectively reduce excitation of central mammalian neurones by N-methyl-aspartate. *Br J Pharmacology* 79: 565–575
6 Eide PK, Stubhaug A, Breivik H, Øye I (1997) Reply to ST Meller: Ketamine: Relief from chronic pain through actions at the NMDA receptor. *Pain* 72: 289–291
7 Choi DW, Peter S, Viseskul V (1987) Dextrorphan and levorphanol selectively block N-methyl-D-aspartate receptor-mediated neurotoxicity on cortical neurons. *J Pharmacol Exp Ther* 242: 713–720.
8 Parsons CG, Gruner R, Rozental J, Millar J, Lodge D (1993) Patch clamp studies on the kinetics and selectivity of N-Methyl-D-Aspartate receptor antagonism by memantine (1-amino-3,5-dimethyladamantan). *Neuropharmacology* 32: 1337–1350
9 Yaksh TL (1989) Behavioral and autonomic correlates of the tactile evoked allodynia produced by spinal glycine inhibition: Effects of modulatory receptor systems and excitatory amino acid antagonists. *Pain* 37: 111–123
10 Davar G, Hama A, Deykin A, Vos B, Maciewics R (1991) MK-801 blocks the development of thermal hyperalgesia in a rat model of experimental painful neuropathy. *Brain Res* 553: 327–330
11 Mao J, Price DD, Mayer DJ, Lu J, Hayes RL (1992) Intrathecal MK-801 and local nerve anesthesia synergistically reduce nociceptive behaviors in rats with experimental peripheral mononeuropathy. *Brain Res* 576: 254–262
12 Mao J, Price DD, Hayes RL, Lu J, Mayer DJ, Frenk H (1993) Intrathecal treatment with dextrorphan or ketamine potently reduces pain-related behaviors in a rat model of peripheral mononeuropathy. *Brain Res* 605: 164–168
13 Tal M, Bennett GJ (1993) Dextrorphan relieves neuropathic heat-evoked hyperalgesia in the rat. *Neurosci Lett* 151: 107–110
14 Yamamoto T, Yaksh TL (1992) Spinal pharmacology of thermal hyperesthesia induced by constriction injury of sciatic nerve. Excitatory amino acid antagonists. *Pain* 49: 121–128
15 Smith GD, Wiseman J, Harrison SM, Elliott PJ, Birch PJ (1994) Pretreatment with MK-801, a non-competitive NMDA antagonist, prevents development of mechanical hyperalgesia in a rat model of chronic neuropathy, but not in a model of chronic inflammation. *Neurosci Lett* 165: 79–83
16 Qian J, Brown SD, Carlton SM (1996) Systemic ketamine attenuates nociceptive behaviors in a rat model of peripheral neuropathy. *Brain Res* 715: 51–62
17 Chaplan SR, Malmberg AB, Yaksh TL (19997) Efficacy of spinal NMDA receptor antagonism in formalin hyperalgesia and nerve injury evoked allodynia in the rat. *J Pharmacol Exp Ther* 280: 829–838

18 Xu X-J, Hao J-X, Seiger Å, Wiesenfeld-Hallin Z (1993) Systemic excitatory amino acid receptor antagonists of the a-amino-3-hydroxy-5-methyl-4-isoxazolepropionic acid (AMPA) receptor and of the N-methyl-D-aspartate (NMDA) receptor relieve mechanical hypersenitivity after transient spinal cord ischemia in rats. *J Pharmacol Exp Ther* 267: 140–144
19 Hao J-X, Xu X-J (1996) Treatment of a chronic allodynia-like response in spinally injured rats: Effects of systemically administered excitatory amino acid receptor antagonists. *Pain* 66: 279–285
20 Christensen D, Gautron M, Guilbaud G, Kayser V (1999) Combined systemic administration of the glycine/NMDA receptor antagonist, (+)-HA966 and morphine attenuates pain-related behaviour in a rat model of trigeminal neuropathic pain. *Pain* 83: 433–440.
21 Yamamoto T, Yaksh TL (1992) Studies on the spinal interaction of morphine and the NMDA antagonist MK-801 on the hyperesthesia observed in a rat model of sciatic mononeuropathy. *Neurosci Lett* 135: 67–70
22 Backonja M, Arndt G, Gombar KA, Check B, Zimmermann M (1994) Response of chronic neuropathic pain syndromes to ketamine: A preliminary study. *Pain* 56: 51–57
23 Eide PK, Stubhaug A, Øye I, Breivik H (1995) Continuous subcutaneous administration of the N-methyl-D-aspartic acid (NMDA) receptor antagonist ketamine in the treatment of post-herpetic neuralgia. *Pain* 61: 221–228
24 Eide PK, Stubhaug A (1997) Relief of glossopharyngeal neuralgia by ketamine-induced N-methyl- aspartate receptor blockade. *Neurosurgery* 41: 505–508
25 Hoffmann V, Coppejans H, Vercauteren M, Adriaensen H (1994) Successful treatment of postherpetic neuralgia with oral ketamine. *Clin J Pain* 10: 240–242
26 Klepstad P, Borchgrevink PC (1997) Four years' treatment with ketamine and a trial of dextrometorphan in a patient with severe post-herpetic neuralgia. *Acta Anaesthesiol Scand* 41: 422–426
27 Mathisen LC, Skjelbred P, Skoglund LA, Øye I (1995) Effect of ketamine, an NMDA receptor inhibitor, in acute and chronic orofacial pain. *Pain* 61: 215–220
28 Nikolajsen L, Hansen PO, Jensen TS (1997) Oral ketamine therapy in the treatment of postamputation stump pain. *Acta Anaesthesiol Scand* 41: 427–429
29 Stannard CF, Porter GE (1993) Ketamine hydrochloride in the treatment of phantom limb pain. *Pain* 54: 227–230
30 Kristensen JD, Svensson B, Gordh T (1992) The NMDA-receptor antagonist CPP abolishes neurogenic "wind-up pain" after intrathecal administration in humans. *Pain* 51: 249–253
31 Takahashi H, Miyazaki M, Nanbu T, Yanagida H, Morita S (1998) The NMDA-receptor antagonist ketamine abolishes neuropathic pain after epidural administration in a clinical case. *Pain* 75: 391–394
32 Eisenberg E, Pud D (1998) Can patients with chronic neuropathic pain be cured by acute administration of the NMDA receptor antagonist amantadine? *Pain* 74: 337–339
33 Rabben T, Skjelbred P, Øye I (1999) Prolonged analgesic effect of ketamine, an N-

methyl-D-aspartate receptor inhibitor, in patients with chronic pain. *J Pharmacol Exp Ther* 289: 1060–1066
34 Eide PK, Jørum E, Stubhaug A, Bremnes J, Breivik H (1994) Relief of post-herpetic neuralgia with the N-methyl-D-aspartic acid receptor antagonist ketamine. A double-blind, cross-over comparison with morphine and placebo. *Pain* 58: 347–354
35 Felsby S, Nielsen J, Arendt-Nielsen L, Jensen TS (1995) NMDA receptor blockade in chronic neuropathic pain: A comparison of ketamine and magnesium chloride. *Pain* 64: 283–291
36 Max MB, Byas-Smith MG, Gracely RH, Bennett GJ (1995) Intravenous infusion of the NMDA antagonist, ketamine, in chronic posttraumatic pain with allodynia: A double-blind comparison to alfentanil and placebo. *Clin Neuropharmacol* 18: 360–368
37 Nikolajsen L, Hansen CL, Nielsen J, Keller J, Arendt-Nielsen L, Jensen TS (1996) The effect of ketamine on phantom pain: A central neuropathic disorder maintained by peripheral input. *Pain* 67: 69–77
38 Eide PK, Stubhaug A, Stenehjem AE (1995) Central dysesthesia pain after traumatic spinal cord injury is dependent on N-methyl-D-aspartate receptor activation. *Neurosurgery* 37: 1080–1087.
39 Pud D, Eisenberg E, Spitzer A, Adler R, Fried G, Yarnitsky D (1998) The NMDA receptor antagonist amantadine reduces surgical neuroapthic pain in cancer patients: A double blind, randomized, placebo controlled trial. *Pain* 75: 349–354
40 McQuay HJ, Carroll D, Jadad AR, Glynn CJ, Jack T, Moore RA, Wiffen PJ (1994) Dextromethorphan for the treatment of neuropathic pain: A double-blind randomised controlled crossover trial with integral n-of-1 design. *Pain* 59: 127–133
41 Nelson KA, Park KM, Robinovitz E, Tsigos C, Max MB (1997) High-dose oral dextromethorphan *versus* placebo in painful diabetic neuropathy and postherpetic neuralgia. *Neurology* 48: 1212–1218
42 Eisenberg E, Kleiser A, Dortort A, Haim T, Yarnitsky D (1998) The NMDA (N-methyl-D-aspartate) receptor antagonist memantine in the treatment of postherpetic neuralgia: A double-blind, placebo-controlled study. *Eur J Pain* 2: 321–327
43 Stubhaug A, Eide PK, Øye I, Breivik H (2002) The combination of oral morphine and ketamine improves pain relief in postherpetic neuralgia compared with morphine or ketamine alone. A randomised, double-blind, placebo-controlled crossover comparison. *Pain; in press*
44 Persson J, Axelsson G, Hallin RG, Gustafsson LL (1995) Beneficial effects of ketamine in a chronic pain state with allodynia, possibly due to central sensitization. *Pain* 60: 217–222
45 Olney JW, Farber NB (1995) NMDA antagonists as neurotherapeutic drugs, psychotogens, neurotoxins, and research tools for studying schizophrenia. *Neuropsychopharmacology* 13: 335–345

Antiepileptic drugs for treatment of neuropathic pain

Charles P. Taylor

CNS Pharmacology, Pfizer Global Research & Development, 2800 Plymouth Rd., Ann Arbor, MI 48105, USA

Introduction – neuropathic pain

Neuropathic pain arises from pathology of sensory function in the peripheral nervous system or central nervous system. Neuropathic pain usually results from damage to sensory nerves, but also may occur as an abnormal reaction to nociceptive input. Neuropathic pain is frequently associated with sensory impairment and with the painful sensory phenomena of allodynia (pain from normally innocuous stimuli), hyperalgesia (exaggerated pain from normally painful stimuli) and hyperpathia (lack of sensation that can be accompanied by severe pain).

Although neuropathic pain syndromes have diverse etiologies, the associated pain usually responds only poorly to treatment with nonsteroidal anti-inflammatory drugs, and mixed results have been reported with opiate analgesic treatments. The main drugs used to treat neuropathic pain traditionally have been the tricyclic antidepressants related to amitriptyline [1–3]. However, more recently, significant success has been achieved using medicines that were originally designed or were primarily used to prevent seizures of epilepsy, that is, antiepileptic drugs or anticonvulsant drugs.

Anticonvulsant medical treatments for neuropathic pain

Several recent reviews provide a systematic account of clinical data concerning the use of antiepileptic drugs for neuropathic pain [3–5]. This chapter will summarize the clinical literature briefly but critically, with additional information provided about drug mechanisms that may contribute to analgesic efficacy, and experimental data with new anticonvulsant drugs or mechanisms that are still under development. The chemical structures of established and novel anticonvulsant drugs discussed in this chapter are shown in Figure 1. Note that most of these compounds have very diverse chemical structures.

Mechanisms and Mediators of Neuropathic Pain, edited by Annika B. Malmberg and Sandra R. Chaplan

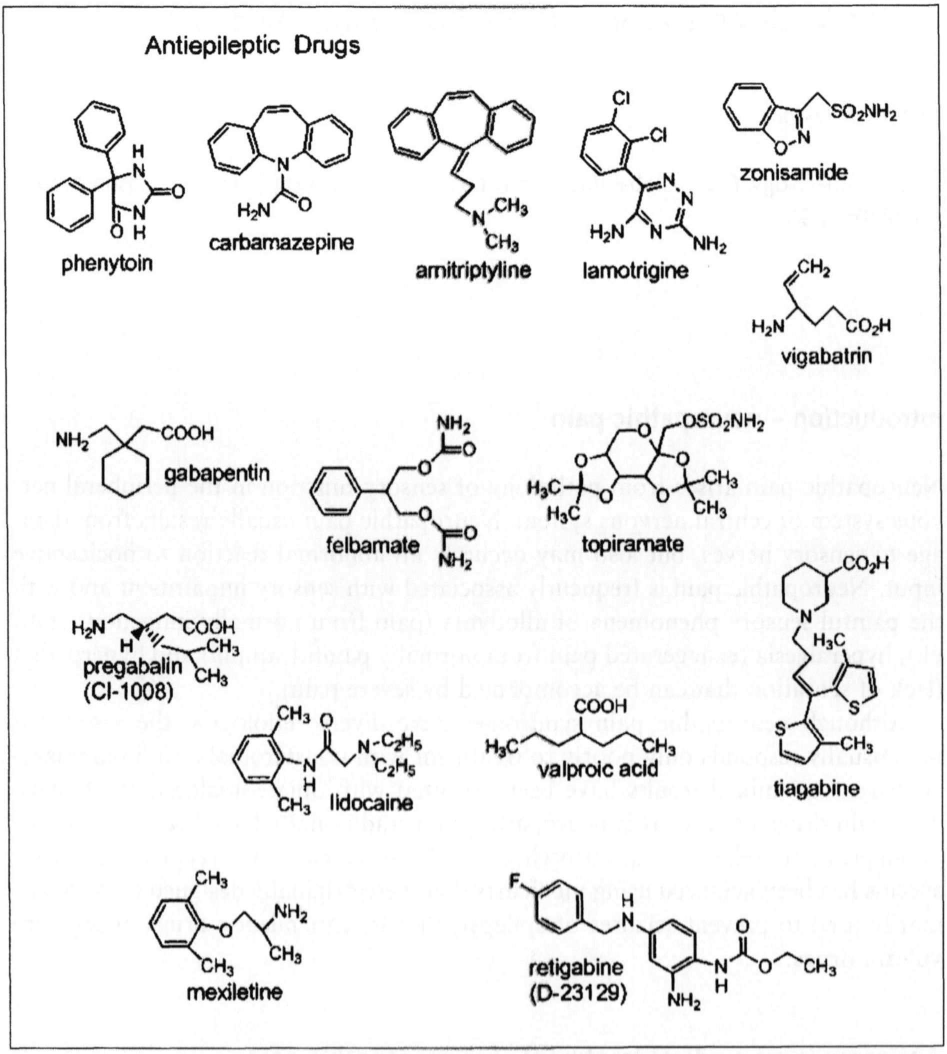

Figure 1
The chemical structures of various anticonvulsant medications are shown. Phenytoin, carbamazepine, lamotrigine, zonisamide, lidocaine and mexilitine are well-characterized voltage-dependent blockers of sodium channels. Felbamate and topiramate have multiple mechanisms that may be relevant for their pharmacolgy. Tiagabine is a well-characterized specific inhibitor of synaptic GABA uptake. Valproate is a widely-used anticonvulsant medication with mechansims of action that are controversial. Vigabatrin is an irreversible inhibitor of the main GABA degradative enzyme, GABA transaminase. Retigabine is an experimental antiepileptic drug that activates certain potassium channels. Amitriptyline is a tricyclic antidepressant that is often used for neuropathic pain treatment; note the structural similarity to carbamazepine.

Carbamazepine

Beginning in 1962, clinical trials demonstrated that treatment with anticonvulsant dosages of carbamazepine could significantly decrease pain from trigeminal neuralgia [6, 7]. Since the early clinical reports, numerous other placebo-controlled studies have confirmed that carbamazepine is effective for several different types of neuropathic pain. Carbamazepine remains the best characterized of the sodium channel compounds for neuropathic pain, with controlled clinical studies supporting use for trigeminal neuralgia, postherpetic neuralgia, painful diabetic neuropathy, lancinating tabetic pain, and central pain [5].

It is likely that the primary pharmacological effect of carbamazepine for analgesia is voltage-dependent blockade of neuronal sodium channels [8] (see Tab. 1 and section on sodium channel mechanisms, below).

Lidocaine and related compounds

Several clinical studies indicate that the sodium channel blocking drug, lidocaine, given systemically at low dosages, reduces postoperative pain and reduces pain in several human experimental models [9, 10]. Support for the use of lidocaine and structurally-related local anesthetic/antiarrhythmic drugs for neuropathic pain comes from extensive studies of low-dose intravenous lidocaine, mexiletine, tocainide, and other cardiac antiarrhythmic drugs with sodium channel blocking properties (Tab. 1, see also [10]). In addition, low dose systemic lidocaine is effective in reducing pain-related behaviors in animal models of neuropathic pain from peripheral nerve damage [11] and diabetic neuropathy with rats [12]. Clinical efficacy of lidocaine treatment [9, 13, 14] appears to be about as well documented as for carbamazepine, but lidocaine is less practical as a chronic pain treatment because of limited oral bioavailability, short plasma half-life and toxicity (ataxia, cognitive dysfunction and cardiac effects) at elevated plasma drug concentrations. Some of the same problems are encountered with the related compound, mexiletine [15, 16], which has improved oral bioavailability, allowing more routine use. A recent study of intravenous low-dose lidocaine suggests that lidocaine may alter pain originating from the central nervous system after damage to the brain or spinal cord [17].

Phenytoin

Phenytoin is the prototype anticonvulsant agent known to block sodium channels [18–20]. However, there is much less clinical evidence than for carbamazepine that phenytoin is effective for treating neuropathic pain [5]. Nevertheless, phenytoin

Table 1 - References to mechanism studies and analgesic efficacy of anticonvulsant drugs

Drug	Mechanism	Animal model	Clinical study
Phenytoin	Sodium channel block: [18–20, 24, 48, 128, 129] Calcium channel block: [133]	[21]	[23, 130–132]
Carbamazepine	Sodium channel block: [18, 24, 134, 135]	[16, 56, 136, 137]	[5–7, 138, 139]
Gabapentin	GABA-related: [37, 38, 109–113] Calcium channel related: [114–120]	[56, 73–80, 82–85, 89, 90, 140–145]	[70–72, 146–148]
Lamotrigine	Sodium and calcium channel block: [48, 149–154]	[25, 26, 56, 154, 155]	[27–29, 131, 156–158] 158] but see also [159]
Lidocaine (and derivatives mexiletine and QX-314)	Sodium channel block: [48, 160–162]	[11, 16, 74, 91, 163–167]	[9, 10, 12–17, 168, 169]
Topiramate	[35–40]	[59]	[60] but see also [170]
Felbamate	[33, 34, 41–46]	[55, 56]	[58]
Valproate	[52]	None found	Little positive data except migraine prophylaxis: [3–5]
Zonisamide	Sodium channel block: [31, 32]	None found	None found
Retigabine	Potassium channel opening: [67]	Rostock, Rundfeld, Bartsch et al., pers. comm.	None found

reduced ectopic nerve firing in animal models [21]. In a placebo-controlled clinical study, it reduced pain from diabetic neuropathy [22] and with intravenous infusion it reduced pain from a variety of neuropathic syndromes [23]. However, in another relatively small placebo-controlled study for diabetic neuropathy, phenytoin was not different from placebo. This apparently reduced efficacy of phenytoin in comparison to carbamazepine might be explained by differences in the kinetics of sodium channel block between carbamazepine and phenytoin [24]. However, it is also likely that phenytoin has not been studied as extensively as carbamazepine in well-controlled clinical trials.

Lamotrigine

Lamotrigine reduces pain responses in both animal models [25, 26] and clinical trials [27–29]. It appears likely that these effects are also due to blockade of sodium channels with additional actions on calcium channels (see below).

Sodium channel mechanisms

Anticonvulsant medications that have sodium channel blocking properties include carbamazepine [20, 30], zonisamide [31, 32] and potentially felbamate [33, 34] and topiramate [35, 36]. However, both topiramate [36–40] and felbamate [41–46] appear to have pharmacological mechanisms in addition to sodium channel modulation. Although the new antiepileptic drugs have been examined less systematically than carbamazepine in clinical studies, their apparent blockade of sodium channels at therapeutically relevant concentrations suggest that they might eventually be shown useful for treating neuropathic pain.

Some have argued that the effectiveness of tricyclic antidepressant drugs such as amitriptyline or nortriptyline derive in part from their blockade of sodium channels, and these compounds share a strong chemical resemblance to carbamazepine. Furthermore, tricyclics block sodium channels *in vitro* at concentrations similar to plasma drug concentrations observed with their use in pain management [47–51]. Finally, selective serotonin reuptake inhibitors have been tried for pain treatment, but mostly have not been found effective [1], suggesting that block of serotonin reuptake is not involved in analgesic effects of tricyclics.

Valproic acid

Valproic acid is an antiepileptic drug that has been used effectively for many years to treat primary generalized and partial seizures, but its cellular mechanisms of action remains a subject of debate [52]. Valproic acid has proven effective in placebo-controlled trials for treatment of acute mania and for prevention of migraine headaches [53]. However, its use for treatment of neuropathic pain is not clearly supported by controlled trials [4, 5, 54].

Felbamate

Felbamate is an anticonvulsant drug approved for use in epilepsy in 1993, but since associated with rare but life-threatening aplastic anemia. Because of this safety liability, felbamate is used only infrequently. However, in an animal model of neuropathic pain, felbamate is highly effective [55, 56]. The mechanisms of felbamate pharmacology are not completely understood, but may include decreases in glutamate excitatory responses [41, 42, 44, 57], increases in GABA-mediated inhibitory responses [43, 44], and possibly also modulation of sodium channels [34, 46]. Felbamate has not been studied in large placebo-controlled trials for clinical neuropathic pain, but a preliminary study [58] indicates that there may be promise for clinical treatment.

Topiramate

Topiramate is a recently approved anticonvulsant drug that has sodium channel blocking actions [35, 36], but also additional potential mechanisms of action. It is active in an animal model of neuropathic pain from chronic sciatic nerve constriction [59] and also was superior to placebo in a controlled clinical trial against painful diabetic neuropathy [60]. Note: Three placebo-controlled studies failed to confirm efficacy of topiramate for diabetic neuropathic pain [170].

Tiagabine

Tiagabine is a well-characterized selective blocker of GABA uptake at the GAT1 type of GABA transporter [61] and its pharmacology is all presumed to relate to blockade of synaptic GABA uptake in neural tissues [62]. It is effective for the prevention of partial seizures in animal models, and recently, it has been shown to prevent nociception-related behavior [63] in the hotplate, late-phase formalin test, and against chronic allodynia (measured with von Frey hairs applied to the foot) produced by intrathecal dynorphin [64, 65]. The analgesic actions of tiagabine in animal models appear to be shared with other experimental GABA uptake blockers, such as SKF 89976A, which was never developed clinically [66]. Despite these clear-cut analgesic effects in animal models, it could be argued that analgesic effects are somewhat less potent (on a mg/kg dose basis) than anticonvulsant actions [62], and that dose-related side effects (ataxia, sedation, confusion) could prevent clear analgesic effects from being observed in humans. In any case, placebo-controlled clinical trials will be needed to assess its usefulness for clinical treatment.

Retigabine

Retigabine is a drug that is now in phase 3 clinical studies for epilepsy, but is not yet available commercially. Its mechanism of action appears to be quite novel, a selective activation of one class of neuronal voltage-gated potassium channel. *In vitro*, it enhances the activity of the KCNQ2/3 potassium channel [67], a novel drug target. Retigabine is said to dose-dependently reduce behavioral pain responses in two rat models of neuropathic pain (Rostock, Rundfeldt, Bartsch, personal communication). These novel findings suggest that activation (rather than blockade) of certain voltage-gated potassium channels may be a target for both anticonvulsant and neuropathic pain therapy.

NMDA glutamate antagonists

A number of experimental anticonvulsant drugs are specific blockers of glutamate receptors, particularly of the NMDA subtype. These drugs are active in several models of epilepsy [68] but have relatively narrow margins of safety until behavioral side effects are encountered. Several clinical studies have suggested that low-affinity NMDA antagonists are useful for treating various types of pain (memantine, dextromethorphan) [1]. However, even subtype-selective NMDA antagonists that selectively block receptors containing the NR2B subunit [69] have not yet proven safe enough for routine use as analgesic drugs.

Gabapentin

Recently, gabapentin has become frequently used to treat neuropathic pain. Although gabapentin has not received approval from the US Food and Drug Administration for treatment of neuropathic pain, it has been approved for neuropathic pain treatment in Germany, France, England and several other European countries. In double-blind clinical studies, high doses of gabapentin have been effective for treatment of diabetic neuropathic pain [70] and post-herpetic neuralgia [71]. These two studies were large and well controlled and had clearly defined clinical endpoints and statistical parameters. In addition, both of these studies provided evidence for improvement of secondary measures of pain including decreased sleep disturbance and improved quality of life [72].

In animal models, gabapentin is effective against allodynia or hyperalgesia caused by chronic sciatic nerve constriction [73], ligation of L5 and L6 dorsal roots [74] or from partial ligation of the sciatic nerve [75]. It also reduces pain responses in rats from diabetes caused by the pancreatic toxin, streptozocin [76] and from systemic treatment with antibody that causes allodynia [77] or spinal cord hemisection [78]. It is noteworthy that in models of neuropathic pain in rats, gabapentin treatment reduced cold allodynia [79] and also both static tactile allodynia (measured by withdrawal to weak punctate pressure with von Frey monfilaments) and dynamic allodynia (measured by withdrawal in response to light stroking of the skin) [76]. The latter form of allodynia was not responsive to acute treatment with either morphine or amitriptyline. In addition, spinal administration of gabapentin to rats reduces the activation of immediate early genes in spinal sensory neurons, in response to intraplantar formalin injection [80]. These latter findings suggest that either as a cause or an effect of analgesia, gabapentin may alter gene expression in the spinal cord.

The chemically related compound, pregabalin [81] shares many of the analgesic-like actions of gabapentin, but pregabalin is more linearly bioavailable than gabapentin. Pregabalin is effective for reducing pain and allodynia at lower dosages than gabapentin in animal models [76, 82–85]. Recently, pregabalin has been found effective in placebo-controlled clinical studies of diabetic neuropathy [86] and post-herpetic neuralgia [87], and also with marginal positive results for spontaneous pain from arthritis [88].

Although the molecular site(s) of analgesic action of gabapentin and pregabalin are not clear, it is obvious from animal studies that it is effective if given systemically or locally into the spinal space, suggesting the spinal cord as a major site of action. However other studies have shown that local injection into an injured rat footpad [89] reduced pain responses, indicating that sites of action in the periphery cannot be ruled out. Furthermore, the pressure threshold for vocalization in a rat model of neuropathic pain [90] is more sensitive to gabapentin than foot withdrawal in the same model. These results suggest that the brain (and the cortical neocortical functions involved in integrating the response of vocalization) may be as important as the spinal cord as a site of analgesic action.

Mechanisms of antiepileptic drugs relevant for analgesia

Na^+ channel modulation

It is not yet clear exactly how voltage-dependent blockade of sodium channels reduces neuropathic pain. Focal application of local anesthetic drugs to neuromas or sensitive trigger points has prevented pain, and several sodium channel modulating drugs (phenytoin, carbamazepine, lidocaine, tocainide, mexiletine) each reduce the ectopic firing from neuromas after low systemic dosages that do not alter nerve conduction [21, 91–93]. These observations suggest that the generation of abnormal action potentials in peripheral nerves (ectopic firing) may be a major mode of sodium channel drug action (rather than blockade of nerve conduction). In addition, sensory cell bodies within dorsal root ganglia provide a site for cross-talk between sensory neurons that may excite undamaged neighboring cells [94]. Recent studies suggest that much of ectopic firing in one animal model of neuropathic pain originates from large rapidly-conducting myelinated afferents rather than pain-specific unmyelinated fibers [95].

Because of the results cited in the previous paragraphs, it is tempting to speculate that ectopic firing in large myelinated fibers in the periphery is a major target for sodium blockers such as carbamazepine, lidocaine or lamotrigine. Voltage-clamp studies show that tetrodotoxin-sensitive sodium channels (the predominant type in large sensory fibers) are particularly sensitive to modulation by voltage-dependent blockers like phenytoin. On the other hand, tetrodotoxin-resistant sodium channels (found almost exclusively in high-threshold pain-specific sensory neurons) require higher concentrations of phenytoin [96]. In addition, it appears that very low concentrations of lidocaine or the sodium channel blocker, tetrodotoxin can prevent subthreshold oscillations of membrane voltage in primary sensory neurons that are enhanced during neuropathic states [97]. Thus, anticonvulsant drugs may treat neuropathic pain at concentrations much lower than those required to block action potential propagation in nerve fibers.

However, other studies suggest that neuropathic pain may originate at least in part from increased excitability in a unique type of sodium channel (PN3 or SNS) that is relatively insensitive to the specific sodium channel toxin, tetrodotoxin [98]. Unlike tetrodotoxin, voltage-dependent sodium blockers such as lidocaine modulate all types of sodium channels (albeit, usually at slightly higher concentrations). Therefore, it is possible that future medications may specifically target tetrodotoxin insensitive sodium channels that are anatomically found only in pain-selective sensory neurons in the periphery. This type of medication might be uniquely suited to treating pain, but it is still controversial as to whether such an agent would be useful for neuropathic pain [99, 100]. There is more general agreement that PN3 selective drugs would be useful for treating inflammation related pain or visceral pain [99, 101].

Some studies suggest that low dosages of sodium channel drugs may act at second or third-order synapses in the spinal cord rather than in the periphery [102, 103]. A recent clinical study indicates that low-dose lidocaine is effective for neuropathic pain from damage to the spinal cord or brain [17], suggesting an analgesic site of action located completely within the central nervous system. Sodium channels not only underlie repetitive firing of action potentials, but also amplify synaptic potentials [104]. It is possible that voltage-dependent sodium channel blockers like lidocaine, phenytoin or carbamazepine preferentially block sustained sodium currents (rather than transient currents as during action potentials), as suggested by *in vitro* studies [105, 106]. An alternative possibility is provided by evidence that phenytoin, lamotrigine, carbamazepine and low-dose tetrodotoxin each reduce glutamate-mediated synaptic currents preferentially to GABA-mediated synaptic currents [107]. This effect occurs because sodium blockers preferentially alter the presynaptic action potential waveform in excitatory terminals, reducing the release of glutamate more than that of GABA. These effects of carbamazepine, lamotrigine and phenytoin occur at drug concentrations relevant for clinical use for neuropathic pain (as well as for antiepileptic use). Thus, it appears that voltage-dependent sodium blockers act as analgesic drugs by modulating sodium channels and reducing excitability at concentrations far less than those needed to block action potentials.

$GABA_A$ mechanisms

A number of drugs sometimes used for epilepsy have well-documented modulation of neuronal $GABA_A$ receptors (such as benzodiazepines and barbiturates). In general, this class of antiepileptic drugs has shown poor effectiveness for neuropathic pain [108], but additional well-controlled clinical trials are probably needed to establish this point. The novel antiepileptic drug vigabatrin (an inhibitor of the GABA degradative enzyme, GABA transaminase) as yet has little published support for use in neuropathic pain syndromes.

Gabapentin and pregabalin

Although the mechanism of action of gabapentin is still a topic of active research [109], it is becoming clear that the drug has more than one effect in brain or nerve tissues. It elevates whole-brain concentrations of the inhibitory neurotransmitter, γ-aminobutyric acid (GABA) in human brain through an unknown mechanism [110]. Gabapentin also increases the rate of GABA efflux from neuronal tissues or cultures [111, 112], probably by modulating the cellular location of the GABA transporter, GAT1 [113].

In actions that do not appear to relate to changes in GABA systems, gabapentin binds to an auxiliary protein of voltage-gated calcium channels, the α2δ protein [114]. Apparently in response to this binding, gabapentin reduces the release of noradrenaline, dopamine [115, 116] and (in trigeminal nucleus slices), the release of glutamate [117]. The effects of gabapentin on neurotransmitter release appear to require sustained depolarization or, in the case of glutamate release, modulation of synapses by pro-allodynic compounds like substance P or CGRP [117]. These *in vitro* findings may relate to *in vivo* results with a rat model, showing that gabapentin reduces allodynia and hyperalgesia caused by intrathecal administration of substance P [83]. However, it is also clear that synaptic transmission mediated by single action potentials in the hippocampus are not altered by gabapentin [118], and monosynaptic glutamate synaptic potentials in rat hippocampal brain slices *in vitro* are not changed (S. Donevan et al., unpublished observations). However, in the dorsal horn of spinal cord, three independent laboratories have reported that gabapentin alters synaptic potentials, particularly under conditions of neuropathic pain [119–121]. Therefore, it appears likely that gabapentin alters synaptic action in a way that is specific for prolonged depolarization or abnormally elevated neurotransmitter release, with relatively little effect on physiologic synaptic transmission.

Recent results suggest gabapentin is an agonist at a subtype of $GABA_B$ receptor [122]. However, other results refute these findings [123]. It is not yet clear whether this mechanism accounts for analgesic efficacy, and it is possible that several different pharmacological actions each contribute in part.

Several additional points can be made about analgesia with gabapentin. Firstly, analgesic actions of gabapentin are not reversed by naloxone [124], and gabapentin fails to alter the metabolism of arachidonic acid to prostanoids (Richard Gilbertsen, unpublished data). Furthermore, gabapentin does not directly block either sodium or voltage-gated calcium channels [109, 125, 126]. Therefore, gabapentin and pregabalin differ from conventional sodium channel blocking drugs (however, see also [127]). Unlike opiate analgesic drugs or NSAIDs, gabapentin appears to be highly effective only in disorders or animal models that include significant allodynia (pain in response to normally innocuous stimuli) or hyperalgesia (more severe pain than normally encountered in response to a painful stimulus). These findings suggest that gabapentin may reduce central sensitization within the spinal cord or that it acts upon one or more of the receptors or ion channels involved in central sensitization.

Conclusions

Much has been learned in recent years about the analgesic use of drugs originally designed to treat epilepsy. It is now clear that voltage-dependent drug-induced blockade of sodium channels provides a useful means of reducing neuropathic pain, and this mechanism is likely to be shared also by the tricyclic antidepressants used for pain therapy. The novel anticonvulsant drug, gabapentin has shown clear efficacy for treating several kinds of neuropathic pain. Both of these types of drugs are active in a wide range of animal models, as well as in placebo-controlled clinical trials. Other mechanisms of anticonvulsant action, such as modulation of GABA receptors or blockade of NMDA-type glutamate receptors have not shown as much promise for treatment of neuropathic pain to date, even though these mechanisms are shared with several other clinical or preclinical anticonvulsant drugs. Therefore, it appears that only certain anticonvulsant mechanisms of action will be broadly useful for treating neuropathic pain. Other authors have suggested that neuropathic pain represents a "seizure" of sensory neurons. However, it appears more likely that both conditions benefit from elevated thresholds for action potential generation (sodium channel blockade) or reduced neurotransmitter release (gabapentin).

Acknowledgements

Sincere thanks to many colleagues, including Michael Poole, Michael Quick, Lakhbir Singh, David Dooley, and Sandy McKnight.

References

1 Sindrup SH, Jensen TS (2000) Pharmacologic treatment of pain in polyneuropathy. *Neurology* 55: 915–920

2 Watson CPN (2000) The treatment of neuropathic pain: Antidepressants and opioids. *Clin J Pain* 16: S49–S55

3 Collins SL, Moore RA, McQuay HJ, Wiffen P (2000) Antidepressants and anticonvulsants for diabetic neuropathy and postherpetic neuralgia: A quantitative systematic review. *J Pain Symptom Manag* 20: 449–458

4 Collins SL, Moore RA, McQuay H, Wiffen P (2000) Anticonvulsant drugs for acute and chronic pain. *Cochrane Database System Rev* 2: CD001133

5 Backonja MM (2000) Anticonvulsants (antineuropathics) for neuropathic pain syndromes. *Clin J Pain* 16: S67–S72

6 Blom S (1962) Trigeminal neuralgia: its treatment with a new anticonvulsant drug. *Lancet* 1: 837–840

7 Blom S (1963) Tic doloureux treated with a new anticonvulsant drug. *Arch Neurol* 9: 285–290

8 Taylor CP (1997) Sodium channels and therapy of central nervous system diseases. *Adv Pharmacol* 39: 47–98

9 Petersen KL, Rowbotham MC (2000) Will ion-channel blockers be useful for management of non-neuropathic pain? *J Pain* 1: 26–34

10 Kalso E, Tramer MR, McQuay HJ (1998) Systemic local-anaesthetic type drugs in chronic pain: a systematic review. *Eur J Pain* 2: 3–14

11 Chaplan SR, Bach FW, Shafer SL, Yaksh TL (1995) Prolonged alleviation of tactile allodynia by intravenous lidocaine in neuropathic rats. *Anesthesiol* 83: 775–785

12 Bach FW, Jensen TS, Kastrup J, Stigsby B, Dejgard A (1990) The effect of intravenous lidocaine on nociceptive processing in diabetic neuropathy. *Pain* 40: 29–34

13 Galer BS, Harle J, Rowbotham MC (1996) Response to intravenous lidocaine infusion predicts subsequent response to oral mexiletine: a prospective study. *J Pain Symptom Manage* 12: 161–167

14 Wallace MS, Dych JB, Rossi SS (1996) Computer-controlled lidocaine infusion for the evaluation of neuropathic pain after peripheral nerve injury. *Pain* 66: 69–77

15 Dejgard A, Petersen P, Kastrup J (1988) Mexiletine for treatment of chronic painful diabetic neuropathy. *Lancet* 1: 9–11

16 Tanelian DL, Brose WG (1991) Neuropathic pain can be relieved by drugs that are use-dependent sodium channel blockers: lidocaine, carbamazepine and mexiletine. *Anesthesiol* 74: 949–951

17 Attal N, Gaude V, Brasseur L, Dupuy M, Guirimand F, Parker F, Bouhassira D (2000) Intravenous lidocaine in central pain: a double-blind, placebo-controlled, psychophysical study [see comments]. *Neurology* 54: 564–574

18 Willow M, Gonoi T, Catterall WA (1985) Voltage clamp analysis of the inhibitory

actions of diphenylhydantoin and carbamazepine on voltage-sensitive sodium channels in neuroblastoma cells. *Mol Pharmacol* 27: 549–558
19 McLean MJ, Macdonald RL (1983) Multiple actions of phenytoin on mouse spinal cord neurons in cell culture. *J Pharmacol Exp Ther* 227: 779–789
20 Ragsdale DS, Avoli M (1998) Sodium channels as molecular targets for antiepileptic drugs. *Brain Res Rev* 26: 16–28
21 Yaari Y, Devor M (1985) Phenytoin suppresses spontaneous ectopic discharge in rat sciatic nerve neuromas. *Neurosci Lett* 58: 117–122
22 Chadda VS, Matghaur MS (1978) Double-blind study of the effects of diphenylhydantoin sodium on diabetic neuropathy. *J Assoc Physicians India* 26: 403–406
23 McCleane GJ (1999) Intravenous infusion of phenytoin relieves neuropathic pain: a randomized, double-blinded, placebo-controlled, crossover study. *Anesth Analg* 89: 985–988
24 Kuo CC, Chen R-S, Lu L, Chen R-C (1997) Carbamazepine inhibition of neuronal Na^+ currents: quantitative distinction from phenytoin and possible therapeutic implications. *Mol Pharmacol* 51: 1077–1083
25 Bennett GJ (1998) Neuropathic pain: new insights, new interventions. *Hosp Practice* 95–114
26 Lefebvre C, Fisher K, Cahill CM, Coderre TJ (2000) Evidence that DHPG-induced nociception depends on glutamate release from primary afferent C-fibres. *Neuroreport* 11: 1631–1635
27 Simpson DM, Olney R, McArthur JC, Khan A, Godbold J, Ebel FK (2000) A placebo-controlled trial of lamotrigine for painful HIV-associated neuropathy [see comments]. *Neurol* 54: 2115–2119
28 Zakrzewska JM, Chaudhry Z, Nurmikko TJ, Patton DW, Mullens EL (1997) Lamotrigine (lamictal) in refractory trigeminal neuralgia: results from a double-blind placebo controlled crossover trial [see comments]. *Pain* 73: 223–230
29 Klamt JG, Posner J (1999) Effects of lamotrigine on pain-induced chemo-somatosensory evoked potentials. *Anaesthesia* 54: 774–777
30 Taylor CP, Narasimhan LS (1997) Sodium channels and therapy of central nervous system diseases. *Adv Pharmacol* 39: 47–98
31 Rock DM, Macdonald RL, Taylor CP (1989) Blockade of sustained repetitive action potentials in cultured spinal cord neurons by zonisamide (AD-810, CI-912), a novel anticonvulsant. *Epilepsy Res* 3: 138–143
32 Schauf CL (1987) Zonisamide enhances slow sodium inactivation in Myxicola. *Brain Res* 413: 185–189
33 Srinivasan J, Richens A, Davies JA (1996) Effects of felbamate on veratridine- and K^+-stimulated release of glutamate from mouse cortex. *Eur J Pharmacol* 315: 285–288
34 Taglialeta M, Ongini E, Brown AM, Di Renzo G, Annunziato L (1996) Felbamate inhibits cloned voltage-dependent Na^+ channels from human and rat brain. *Eur J Pharmacol* 316: 373–377
35 Coulter DA, Sombati S, DeLorenzo RJ (1993) Selective effects of topiramate on sus-

tained repetitive firing and spontaneous bursting in cultured hippocampal neurons. *Epilepsia* 34: 123 (Abstract)
36 Zona C, Ciotti MT, Avoli M (1997) Topiramate attenuates voltage-gated sodium currents in rat cerebellar granule cells. *Neurosci Lett* 231: 123–126
37 Kuzniecky RI, Hetherington HP, Ho S, Pan JW, Martin R, Gilliam F, Hugg J, Faught E (1998) Topiramate increases cerebral GABA in healthy humans. *Neurol* 51: 627–629
38 Petroff OA, Hyder F, Mattson RH, Rothman D (1999) Topiramate increases brain GABA,homocarnosine and pyrrolidinone in patients with epilepsy. *Neurol* 52: 473–478
39 Shank RP (2000) An overview of the preclinical aspects of topiramate: pharmacology, pharmacokinetics and mechanism of action. *Epilepsia* 41 (1): 3–9
40 White HS, Brown SD, Woodhead JH, Skeen GA, Wolf HH (1997) Topiramate enhances GABA-mediated chloride flux and GABA-evoked chloride currents in murine brain neurons and increases seizure threshold. *Epilepsy Res* 28: 167–179
41 Coffin V, Cohen-Williams M, Barnett A (1994) Selective antagonism of the anticonvulsant effects of felbamate by glycine. *Eur J Pharmacol* 256: R9–R10
42 Kleckner NW, Glazewski JC, Chen CC, Moscrip TD (1999) Subtype-selective antagonism of N-methyl-D-aspartate receptors by felbamate: insights into the mechanism of action. *J Pharmacol Exp Ther* 289: 886–894
43 Kume A, Greenfield LJJr, Macdonald RL (1996) Felbamate inhibits [^{3}H]t-butylbicycloorthobenzoate (TBOB) binding and enhances Cl^- current at the gamma-aminobutyric $acid_A$ ($GABA_A$) receptor. *J Pharmacol Exp Ther* 277: 1784–1792
44 Rho JM, Donevan SD, Rogawski MA (1994) Mechanism of action of the anticonvulsant felbamate: opposing effects on N-methyl-D-aspartate and gamma-aminobutyric acidA receptors. *Ann Neurol* 35: 229–234
45 Stefani A, Calabresi P, Pisani A, Mercuri NB, Siniscalchi A, Bernardi G (1996) Felbamate inhibits the dihydropyridine-sensitive calcium channels in central neurons. *J Pharmacol Exp Ther* 277: 121–127
46 Sofia RD (1995) Felbamate: Mechanisms of action. In: RH Levy, RH Mattson, BS Meldrum (eds): *Antiepileptic Drugs* (4th ed.), Raven Press, New York, NY, 791–797
47 Pancrazio JJ, Kamatchi GL, Roscoe AK, Lynch C (1998) Inhibition of neuronal Na^+ channels by antidepressant drugs. *J Pharmacol Exp Ther* 284: 208–214
48 Deffois A, Fage D, Carter C (1996) Inhibition of synaptosomal veratridine-induced sodium influx by antidepressants and neuroleptics used in chronic pain. *Neurosci Lett* 220: 117–120
49 Ishii Y, Sumi T (1992) Amitriptyline inhibits striatal efflux of neurotransmitters *via* blockade of voltage-dependent Na^+ channels. *Eur J Pharmacol* 221: 377–380
50 Lavoie PA, Beauchamp G, Elie R (1990) Tricyclic antidepressants inhibit voltage-dependent calcium channels and Na(+)-Ca2+ exchange in rat brain cortex synaptosomes. *Can J Physiol Pharmacol* 68: 1414–1418
51 Song JH, Ham SS, Shin YK, Lee CS (2000) Amitriptyline modulation of Na(+) channels in rat dorsal root ganglion neurons. *Eur J Pharmacol* 401: 297–305

52 Fariello RG (1995) Valproic acid: mechanisms of action. In: RH Levy, RH Mattson, BS Meldrum (eds): *Antiepileptic Drugs* (4th ed.), Raven Press, New York, NY, 581–588

53 Hering R, Kuritzky A (1992) Sodium valproate in hte prophylactic tratment of migraine: a double-blind study *versus* placebo. *Cephalalgia* 12: 81–84

54 Abram SE, Yaksh TL (1994) Systemic lidocaine blocks nerve injury-induced hyperalgesia and nociceptor-driven spinal sensitization in the rat. *Anesthesiol* 80: 383–391

55 Imamura Y, Bennett GJ (1995) Felbamate relieves several abnormal pain sensations in rats with an experimental peripheral neuropathy. *J Pharmacol Exp Ther* 275: 177–182

56 Hunter JC, Gogas KR, Hedley LR, Jacobson LO, Kassotakis L, Thompson J, Fontana DJ (1997) The effect of novel anti-epileptic drugs in rat experimental models of acute and chronic pain. *Eur J Pharmacol* 324: 153–160

57 McCabe RT, Wasterlain CG, Kucharczyk N, Sofia RD, Vogel JR (1993) Evidence for anticonvulsant and neuroprotective action of felbamate mediated by strychnine-insensitive glycine receptors. *J Pharmacol Exp Ther* 264: 1248–1252

58 Cheshire WP (1995) Felbamate relieved trigeminal neuralgia. *Clin J Pain* 11: 139–142

59 Shadiack AM, Molino LJ, Yagel SK, Shank RP, Rogers KE, Codd EE, Wild KD (1999) The novel anticonvulsant topiramate is anti-allodynic in a rat model of neuropathic pain. *Analgesia* 4: 173–179

60 Edwards KR, Glantz MJ, Button J, Norton J (1999) Efficacy and safety of topiramate *versus* placebo in the prophylactic treatment of painful diabetic neuropathy: a double-blind, placebo-controlled study. *Neurology* 54: A81

61 Clark JA, Amara SG (1994) Stable expression of a neuronal gamma-aminobutyric acid transporter, GAT-3, in mammalian cells demonstrates unique pharmacological properties and ion dependence. *Mol Pharmacol* 46: 550–557

62 Suzdak PD, Jansen JA (1995) A review of the preclinical pharmacology of tiagabine: A potent and selective anticonvulsant GABA uptake inhibitor. *Epilepsia* 36: 612–626

63 Laughlin TM, Tram KV, Wilcox GL, Birnbaum AK (2001) Characterization of the antiepileptic drug, tiagabine in acute, prolonged and chronic nociception. *J Pharmacol Exp Ther; in press*

64 Laughlin TM, Vanderah TW, Lashbrook J, Nichols ML, Ossipov M, Porreca F, Wilcox GL (1997) Spinally administered dynorphin A produces long-lasting allodynia: involvement of NMDA but not opioid receptors. *Pain* 72: 253–260

65 Laughlin TM, Bethea JR, Yezierski RP, Wilcox GL (2000) Cytokine involvement in dynorphin-induced allodynia. *Pain* 84: 159–167

66 Zorn SH, Enna SJ (1985) GABA uptake inhibitors produce a greater antinociceptive response in the mouse tail-immersion assay than other types of GABAergic drugs [published erratum appears in *Life Sci* 1986 Feb 3;38(5):483]. *Life Sci* 37: 1901–1912

67 Rundfeldt C, Netzer R (2000) The novel anticonvulsant retigabine activates M-currents in Chinese hamster ovary-cells tranfected with human KCNQ2/3 subunits. *Neurosci Lett* 282: 73–76

68 Rogawski MA, Porter RJ (1990) Antiepileptic drugs: pharmacological mechanisms and

clinical efficacy with consideration of promising developmental stage compounds. *Pharmacol Rev* 42: 223–286
69 Boyce S, Wyatt A, Webb JK, O'Donnell R, Mason G, Rigby M, Sirinathsinghji D, Hill RG, Rupniak NMJ (1999) Selective NMDA NR2B antagonists induce antinociception without motor dysfunction: correlation with restricted localization of NR2B subunit in dorsal horn. *Neuropharmacology* 38: 611–623
70 Backonja M, Beydoun A, Edwards KR, Schwartz SL, Fonseca V, Hes MS, Lamoreaux L, Garofalo E (1998) Gabapentin for the symptomatic treatment of painful neuropathy in patients with diabetes mellitus. *J Am Med Assoc* 280: 1831–1836
71 Rowbotham MC, Harden N, Stacey B, Bernstein P, Magnus-Miller L (1998) Gabapentin for the treatment of postherpetic neuralgia: A randomized controlled trial. *J Am Med Assoc* 280: 1837–1842
72 Vinik A, Fonseca V, Lamoreaux L, Hes M, Koto EM (1998) Neurontin (Gabapentin, GBP) improves quality of life in patients with painful peripheral diabetic neuropathy. *Diabetes* 47: A374–A374
73 Xiao W-H, Bennett GJ (1996) Gabapentin relieves abnormal pain sensations *via* a spinal site of action in a rat model of painful peripheral neuropathy. *Analges* 2: 267–273
74 Abdi S, Lee DH, Chung JM (1998) The anti-allodynic effects of amitriptyline, gabapentin, and lidocaine in a rat model of neuropathic pain. *Anesth Analg* 87: 1360–1366
75 Pan H-L, Chen S-R, Eisenach JC (1998) Gabapentin suppresses ectopic nerve discharges and reverses allodynia in neuropathic rats. *J Pharmacol Exp Ther* 288: 1026–1030
76 Field MJ, McCleary S, Hughes J, Singh L (1999) Gabapentin and pregabalin, but not morphine and amitriptyline, block both static and dynamic components of mechanical allodynia induced by streptozocin in the rat. *Pain* 80: 391–398
77 Gillin S, Sorkin LS (1998) Gabapentin reverses the allodynia produced by the administration of anti-GD2 ganglioside, an immunotherapeutic drug. *Anesth Analg* 86: 111–116
78 Hulsebosch CE, Xu G-Y, Perez-Polo JR, Westlund KN, Taylor CP, McAdoo DJ (2000) Rodent model of chronic central pain after spinal cord contusion injury and effects of gabapentin. *J Neurotrauma* 12: 1205–1217
79 Hunter JC, Gogas KR, Hedley LR, Jacobson LO, Kassotakis L, Thompson J, Fontana DJ (1997) The effect of novel anti-epileptic drugs in rat experimental models of acute and chronic pain. *Eur J Pharmacol* 324: 153–160
80 Kaneko M, Mestre C, Sanchez EH, Hammond DL (2000) Intrathecally administered gabapentin inhibits formalin-evoked nociception and the expression of Fos-like immunoreactivity in the spinal cord of the rat. *J Pharmacol Exp Ther* 292: 743–751
81 Taylor CP, Vartanian MG, Yuen PW, Bigge C, Suman-Chauhan N, Hill DR (1993) Potent and stereospecific anticonvulsant activity of 3-isobutyl GABA relates to *in vitro* binding at a novel site labeled by tritiated gabapentin. *Epilepsy Res* 14: 11–15
82 Houghton AK, Westlund KN (1998) S-(+)-3-Isobutylgaba and its stereoisomer reduces

the amount of inflammation and hyperalgesia in an acute arthritis model in the rat. *J Pharmacol Exp Ther* 285: 533–538
83 Partridge BJ, Chaplan SR, Sakamoto E, Yaksh TL (1998) Characterization of the effects of gabapentin and 3-isobutyl-gamma-aminobutyric acid on substance P-induced thermal hyperalgesia. *Anesthesiol* 88: 196–205
84 Field MJ, Holloman EF, McCleary S, Hughes J, Singh L (1997) Evaluation of gabapentin and S-(+)-3-isobutylGABA in a rat model of postoperative pain. *J Pharmacol Exp Ther* 282: 1242–1246
85 Jun JH, Yaksh TL (1998) The effect of intrathecal gabapentin and 3-isobutyl gamma-aminobutyric acid on the hyperalgesia observed after thermal injury in the rat. *Anesth Analg* 86: 348–354
86 Sharma U, Iacobellis D, Glessner C, Moore J, Young JP, LaMoreaux L, Poole RM (2000) Pregabalin effectively relieves pain in two studies of patients with painful diabetic neuropathy (studies 1008-14/-29). American Pain Society, 19th Annual Scientific Meeting (Abstract)
87 Corbin AE, Remmers AE, Lamoreaux L, Young JP, Poole RM (2001) Pregabalin reduces pain in patients with postherpetic neuralgia (PHN). American Pain Society, 20th Annual Scientific Meeting (Abstract)
88 Jaffe M, Iocobellis D, Young JP, Spiegel K, Martin S, Lamoreaux L, Glessner C, Hes M, Poole RM (2000) Post-hoc results show beneficial effects of pregabalin in patients with osteoarthritis of the hip. *Arthritis Rheum* 43 (Suppl): S337 (Abstract)
89 Carlton SM, Zhou S (1998) Attenuation of formalin-induced nociceptive behaviors following local peripheral injection of gabapentin. *Pain* 76: 201–207
90 Kayser V, Christensen D (2000) Antinociceptive effect of systemic gabapentin in mononeuropathic rats, depends on stimulus characteristics and level of test integration. *Pain* 88: 53–60
91 Devor M, Wall PD, Catalan N (1992) Systemic lidocaine silences ectopic neuroma and DRG discharge without blocking nerve conduction. *Pain* 48: 261–268
92 Burchiel KJ (1988) Carbamazepine inhibits spontaneous activity in experimental neuromas. *Exp Neurol* 102: 249–253
93 Chabal C, Russell LC, Burchiel KJ (1989) The effect of intravenous lidocaine, tocainide, and mexiletine on spontaneously active fibers originating in rat sciatic neuromas. *Pain* 38: 333–338
94 Shinder V, Amir R, Devor M (1998) Cross-excitation in dorsal root ganglia does not depend on close cell-to-cell apposition. *Neuroreport* 9: 3997–4000
95 Liu CN, Wall PD, Ben Dor E, Michaelis M, Amir R, Devor M (2000) Tactile allodynia in the absence of C-fiber activation: altered firing properties of DRG neurons following spinal nerve injury. *Pain* 85: 503–521
96 Song J-H, Nagata K, Huang C-S, Yeh JZ, Narahashi T (1996) Differential block of two types of sodium channels by anticonvulsants. *Neuroreport* 7: 3031–3036
97 Amir R, Michaelis M, Devor M (1999) Membrane potential oscillations in dorsal root

ganglion neurons: role in normal electrogenesis and neuropathic pain. *J Neurosci* 19: 8589–8596

98 Novakovic SD, Tzoumaka E, McGivern JG, Haraguchi M, Sangameswaran L, Gogas KR, Eglen RM, Hunter JC (1998) Distribution of the tetrodotoxin-resistant sodium channel PN3 in rat sensory neurons in normal and neuropathic conditions. *J Neurosci* 18: 2174–2187

99 Porreca F, Lai J, Bian D, Wegert S, Ossipov MH, Eglen RM, Kassotakis L, Novakovic S, Rabert DK, Sangameswaran L et al (1999) A comparison of the potential role of the tetrodotoxin-insensitive sodium channels, PN3/SNS and NaN/SNS2, in rat models of chronic pain. *Proc Natl Acad Sci USA* 96: 7640–7644

100 Baker MD, Wood JN (2001) Involvement of Na^+ channels in pain pathways. *Trends Pharmacol Sci* 22: 27–31

101 Khasar SG, Gold MS, Levine JD (1998) A tetrodotoxin-resistant sodium current mediates inflammatory pain in the rat. *Neurosci Lett* 256: 17–20

102 Woolf CJ, Wiesenfeld-Hallin Z (1985) The systemic administration of local anesthetics produces a selective depression of C-afferent fibre evoked activity in the spinal cord. *Pain* 23: 361–374

103 Nagy I, Woolf CJ (1996) Lignocaine selectively reduces C fibre-evoked neuronal activity in rat spinal cord *in vitro* by decreasing N-methyl-D-aspartate and neurokinin receptor-mediated post-synaptic depolarizations; implications for the development of novel centrally acting analgesics. *Pain* 64: 59–70

104 Taylor CP (1993) Na^+ currents that fail to inactivate. *Trends Neurosci* 16: 455–460

105 Segal MM, Douglas AF (1997) Late sodium channel openings underlying epileptiform activity are preferentially diminished by the anticonvulsant phenytoin. *J Neurophysiol* 77: 3021–3034

106 Chao TI, Alzheimer C (1995) Effects of phenytoin on the persistent Na^+ current of mammalian CNS neurones. *Neuroreport* 6: 1778–1780

107 Prakriya M, Mennerick S (2000) Selective depression of low-release probability excitatory synapses by sodium channel blockers. *Neuron* 26: 671–682

108 Max MB, Schafer SC, Culnane M, Smoller B, Dubner R, Gracely RH (1998) Amitriptyline, but not lorazepam, relieves postherpetic neuralgia. *Neurology* 38: 1427–1432

109 Taylor CP, Gee NS, Su TZ, Kocsis JD, Welty DF, Brown JP, Dooley DJ, Boden P, Singh L (1998) A summary of mechanistic hypotheses of gabapentin pharmacology. *Epilepsy Res* 29: 233–249

110 Petroff OA, Rothman DL, Behar KL, Lamoureux D, Mattson RH (1996) The effect of gabapentin on brain gamma-aminobutyric acid in patients with epilepsy. *Ann Neurol* 39: 95–99

111 Honmou O, Kocsis JD, Richerson GB (1995) Gabapentin potentiates the conductance increase induced by nipecotic acid in CA1 pyramidal neurons *in vitro*. *Epilepsy Res* 20: 193–202

112 Kocsis JD, Honmou O (1994) Gabapentin increases GABA-induced depolarization in rat neonatal optic nerve. *Neurosci Lett* 169: 181–184

113 Whitworth TL, Quick MW (2001) Upregulation of gamma-aminobutyric acid transporter expression: role of alkylated gamma-aminobutyric acid derivatives. *Biochem Soc Transactions* 29: 736–741

114 Gee NS, Brown JP, Dissanayake VU, Offord J, Thurlow R, Woodruff GN (1996) The novel anticonvulsant drug, gabapentin (Neurontin), binds to the alpha2delta subunit of a calcium channel. *J Biol Chem* 271: 5768–5776

115 Schlicker E, Reimann W, Gothert M (1985) Gabapentin decreases monoamine release without affecting acetylcholine release in the brain. *Drug Res* 35: 1347–1349

116 Dooley DJ, Donovan CM, Pugsley TA (2000) Stimulus-dependent modulation of [^{3}H]norepinephrine release from rat neocortical slices by gabapentin and pregabalin. *J Pharmacol Exp Ther* 296: 1086–1098

117 Maneuf YP, McKnight AT (2001) Gabapentin inhibits substance P-facilitated K^+-evoked release of [^{3}H]-glutamate from rat caudal trigeminal nucleus slices. *Pain* 93: 191–196

118 Xiong Z-Q, Stringer JL (1997) Effects of felbamate, gabapentin and lamotrigine on seizure parameters and excitability in the rat hippocampus. *Epilepsy Res* 27: 187–194

119 Shimoyama M, Shimoyama N, Hori Y (2000) Gabapentin affects glutamatergic excitatory neurotransmission in the rat dorsal horn. *Pain* 85: 405–414

120 Patel MK, Gonzalez MI, Bramwell S, Pinnock RD, Lee K (2000) Gabapentin inhibits excitatory synaptic transmission in the hyperalgesic spinal cord. *Br J Pharmacol* 130: 1731–1734

121 Moore KA, Baba H, Ji RR, Woolf CJ (1999) Effects of gabapentin on synaptic transmission and signal transduction in lamina II neurons of adult rat spinal cord. *Soc Neurosci Abstr* 25: 1943

122 Ng GYK, Bertrand S, Sullivan R, Ethier N, Wang J, Yergey J, Belley M, Trimble L, Bateman K, Alder L et al (2001) Gamma-aminobutyric acid type B receptors with specific heterodimer composition and postsynaptic actions in hippocampal neurons are targets of anticonvulsant gabapentin action. *Mol Pharmacol* 59: 144–152

123 Lanneau C, Green A, Hirst WD, Wise A, Brown JT, Donnier E, Charles KJ, Wood M, Davies CH, Paugalos MN (2001) Gabapentin is not a $GABA_B$ receptor agonist. *Neuropharmacol* 41: 965–975

124 Field MJ, Oles RJ, Lewis AS, McCleary S, Hughes J, Singh L (1997) Gabapentin (Neurontin) and S-(+)-3-isobutyl GABA represent a novel class of selective antihyperalgesic agents. *Br J Pharmacol* 121: 1513–1522

125 Rock DM, Kelly KM, Macdonald RL (1993) Gabapentin actions on ligand- and voltage-gated responses in cultured rodent neurons. *Epilepsy Res* 16: 89–98

126 Schumacher TB, Beck H, Steinhauser C, Schramm J, Elger CE (1998) Effects of gabapentin, phenytoin and carbamazepine on calcium currents in hippocampal granule cells from patients with temporal lobe epilepsy. *Epilepsia* 39: 355–363

127 Stefani A, Spadoni F, Bernardi G (1998) Gabapentin inhibits calcium currents in isolated rat brain neurons. *Neuropharmacol* 37: 83–91

128 Kuo CC, Bean BP (1994) Slow binding of phenytoin to inactivated sodium channels in rat hippocampal neurons. *J Pharmacol Exp Ther* 46: 716–725

129 Yaari Y, Selzer ME, Pincus JH (1986) Phenytoin: mechanisms of its anticonvulsant action. *Ann Neurol* 20: 171–184
130 MacFarlane BV, Wright A, O'Callaghan J, Benson HAE (1997) Chronic neuropathic pain and its control by drugs. *Pharmacol Ther* 75: 1–19
131 Webb J, Kamali F (1998) Analgesic effects of lamotrigine and phenytoin on cold-induced pain: a crossover placebo-controlled study in healthy volunteers. *Pain* 76: 357–363
132 McQuay H, Carroll D, Jadad AR, Wiffen P, Moore A (1995) Anticonvulsant drugs for management of pain: a systematic review. *Brit Med J* 311: 1047–1052
133 Twombly DA, Yoshii M, Narahashi T (1988) Mechanisms of calcium channel block by phenytoin. *J Pharmacol Exp Ther* 246: 189–195
134 McLean MJ, Macdonald RL (1986) Carbamazepine and 10,11-epoxycarbabamazepine produce use- and voltage-dependent limitation of rapidly firing action potentials of mouse central neurons in cell culture. *J Pharmacol Exp Ther* 238: 727–738
135 Zimanyi I, Weiss SRB, Lajtha A, Post RM, Reith MEA (1989) Evidence for a common site of action of lidocaine and carbamazepine in voltage-dependent sodium channels. *Eur J Pharmacol* 167: 419–442
136 Chapman V, Suzuki R, Chamarette HL, Rygh LJ, Dickenson AH (1998) Effects of systemic carbamazepine and gabapentin on spinal neuronal responses in spinal nerve ligated rats. *Pain* 75: 261–272
137 Koch BD, Faurot GF, McGuirk JR, Clarke DE, Hunter JC (1996) Modulation of mechano-hyperalgesia by clinically effective analgesics in rats with a peripheral mononeuropathy. *Analges* 2: 157–164
138 Tremont-Lukats IW, Megeff C, Backonja M-M (2000) Anticonvulsants for neuropathic pain syndromes. Mechanisms of action and place in therapy. *Drugs* 60: 1029–1052
139 McQuay H, Carroll D, Jadad AR, Wiffen P, Moore A (1995) Anticonvulsant drugs for management of pain: a sysematic review. *Brit Med J* 311: 1047–1052
140 Stanfa LC, Singh L, Williams RG, Dickenson AH (1997) Gabapentin, ineffective in normal rats, markedly reduces C-fibre evoked responses after inflammation. *Neuroreport* 8: 587–590
141 Gillin S, Sorkin LS (1998) Gabapentin reverses the allodynia produced by the administration of anti-GD2 ganglioside, an immunotherapeutic drug. *Anesth Analg* 86: 111–116
142 Shimoyama N, Shimoyama M, Davis AM, Inturrisi CE, Elliott KJ (1997) Spinal gabapentin is antinociceptive in the rat formalin test. *Neurosci Lett* 222: 65–67
143 Hwang JH, Yaksh TL (1997) Effect of subarachnoid gabapentin on tactile-evoked allodynia in a surgically-induced neuropathic pain model in the rat. *Reg Anesthesia* 22: 249–256
144 Bennett GJ (1998) Neuropathic pain: new insights, new interventions. *Hosp Practice* 95–114
145 Jacobson LO, McGuirk J, Jett MF, Gogas KR, Hunter JC (1997) Effect of gabapentin

and carbamazepine on mechanical and thermal hyperalgesia. *Pharmacologist* 39: 112 (Abstract)

146 Backonja MM (1999) Gabapentin monotherapy for the symptomatic treatment of painful neuropathy: a multicenter, double-blind, placebo-controlled trial in patients with diabetes mellitus. *Epilepsia* 40 (Suppl 6): S57–S59

147 Mellick GA, Mellicy LB, Mellick LB (1995) Gabapentin in the management of reflex sympathetic dystrophy [letter]. *J Pain Symptom Manage* 10: 265–266

148 Rosner H, Rubin L, Kestenbaum A (1996) Gabapentin adjunctive therapy in neuropathic pain states. *Clin J Pain* 12: 56–58

149 Cheung H, Kamp D, Harris E (1992) An *in vitro* investigation of the action of lamotrigine on neuronal voltage-activated sodium channels. *Epilepsy Res* 13: 107–112

150 Lang DG, Wang CM, Cooper BR (1993) Lamotrigine, phenytoin and carbamazepine interactions on the sodium current present in N4TG1 mouse neuroblastoma cells. *J Pharmacol Exp Ther* 266: 829–835

151 Wang S-J, Huang C-C, Hsu K-S, Tsai J-J, Gean P-W (1996) Inhibition of N-type calcium currents by lamotrigine in rat amygdalar neurones. *Neuroreport* 7: 3037–3040

152 Wang S-J, Tsai J-J, Gean P-W (1998) Lamotrigine inhibits depolarization-evoked Ca^{++} influx in dissociated amygdala neurons. *Synapse* 29: 355–362

153 Xie Y, Lancaster B, Peakman T, Garthwaite J (1995) Interaction of the antiepileptic drug lamotrigine with recombinant rat brain type IIA Na^+ channels and with native Na^+ channels in rat hippocampal neurones. *Pflugers Arch* 430: 437–446

154 Blackburn MG, Fleetwood WS (1997) The effects of Na^+ channel blockers on somatosensory processing by rat dorsal horn neurones. *Neuroreport* 8: 1549–1554

155 Nakamura Craig M, Follenfant RL (1995) Effect of lamotrigine in the acute and chronic hyperalgesia induced by PGE2 and in the chronic hyperalgesia in rats with streptozotocin-induced diabetes. *Pain* 63: 33–37

156 Bonicalzi V, Canavero S, Cerutti F, Piazza M, Clemente M, Chio A (1997) Lamotrigine reduces total postoperative analgesic requirement: a randomized double-blind, placebo-controlled pilot study. *Surgery* 122: 567–570

157 Lunardi G, Leandri M, Albano C, Cultrera S, Fracassi M, Rubino V, Favale E (1997) Clinical effectiveness of lamotrigine and plasma levels in essential and symptomatic trigeminal neuralgia [see comments]. *Neurol* 48: 1714–1717

158 Canavero S, Bonicalzi V (1996) Lamotrigine control of central pain. *Pain* 68: 179–181

159 McCleane G (1999) 200 mg daily of lamotrigine has no analgesic effect in neuropathic pain: a randomised, double-blind, placebo controlled trial. *Pain* 83: 105–107

160 Kaneda M, Oyama Y, Ikemoto Y, Akaike N (1989) Blockade of voltage-dependent sodium current in isolated rat hippocampal neurons by tetrodotoxin and lidocaine. *Brain Res* 484: 348–351

161 Ragsdale DS, McPhee JC, Scheuer T, Catterall WA (1996) Common molecular determinants of local anesthetic, antiarrhythmic, and anticonvulsant block of voltage-gated Na^+ channels. *Proc Natl Acad Sci USA* 93: 9270–9275

162 Ragsdale DS, Scheuer T, Catterall WA (1991) Frequency and voltage-dependent inhibi-

tion of type IIA Na⁺ channels, expressed in a mammalian cell line, by local anesthetic, antiarrhythmic, and anticonvulsant drugs. *Mol Pharmacol* 40: 756–765

163 Abram SE, Yaksh TL (1994) Systemic lidocaine blocks nerve injury-induced hyperalgesia and nociceptor-driven spinal sensitization in the rat. *Anesthesiol* 80: 383–391

164 Koch BD, Faurot GF, McGuirk JR, Clarke DE, Hunter JC (1996) Modulation of mechano-hyperalgesia by clinically effective analgesics in rats with a peripheral mononeuropathy. *Analges* 2: 157–164

165 Omana Z, I, Khabbaz MA, Hunter JC, Bley KR (1997) QX-314 inhibits ectopic nerve activity associated with neuropathic pain. *Brain Res* 771: 228–237

166 Scholz A, Kuboyama N, Hempelmann G, Vogel W (1998) Complex blockade of TTX-resistant Na⁺ currents by lidocaine and bupivacaine reduce firing frequency in DRG neurons. *J Neurophysiol* 79: 1746–1754

167 Tanelian DL, MacIver MB (1991) Analgesic concentrations of lidocaine suppress tonic A-δ and C fiber discharges produced by injury. *Anesthesiol* 74: 934–936

168 Dirks J, Fabricius P, Petersen KL, Rowbotham MC, Dahl JB (2000) The effect of systemic lidocaine on pain and secondary hyperalgesia associated with the heat/capsaicin sensitization model in healthy volunteers. *Anesth Analg* 91: 967–972

169 Rowbotham MC, Reisner-Keller LA, Fields HL (1991) Both intravenous lidocaine and morphine reduce the pain of postherpetic neuralgia. *Neurol* 441: 1024–1028

170 Anonymous (2001) Topamox fails in diabetic neuropathy. *Scrip* 2680: 23

Calcium channel antagonists for the treatment of pain

Mark S. Wallace

Department of Anesthesiology, University of California, San Diego, School of Medicine, 9500 Gilman Drive #0924, La Jolla, CA 92093-0924, USA

Introduction

Several lines of evidence developed in preclinical models suggest that both spontaneous pain and evoked pain are mediated in part by voltage-dependent calcium channels (VDCCs) [1]. Both the central and peripheral nervous systems have an abundance of calcium channels. However, in addition to their roles in sensory processing, VDCCs play critical roles in various physiological processes including cardiac electrophysiology, vascular tone, motor control and other neural communication. The ubiquity and importance of VDCCs naturally raises concerns that drug-induced blockade could have adverse impact on multiple physiological systems. The number of subtypes of VDCC that has been recognized to date gives hope, however, that just as it has been possible to use subtype-selective VDCC antagonists as antihypertensive and antiarrhythmic drugs, it may be possible to selectively modulate subytpes involved in painful sensations without unacceptable side effects.

There are six functional classes of calcium channels expressed throughout the nervous system, designated L, N, P/Q, R and T [2–4]; each class may include alpha (pore-forming) subunits encoded by several different genes. L-type channels are mainly located on cell bodies and dendrites and N, P/Q, and R-type channels are located at synaptic sites and are involved in both excitatory and inhibitory neurotransmitter release [5–8]. The L, N, P/Q, and R-type channels are high-threshold, in that they are activated at strongly depolarized membrane thresholds. Many studies have shown that L, N, and P/Q-type channels are involved in nociception [9]. To date no selective pharmacological antagonists are available for the recently characterized R- or T-type channels. The R-type is the least understood channel; while it may play a role in nociception, the details remain to be elucidated [10, 11]. The low-threshold, or T-type channel, requires little voltage change for activation, and appears to play a role in the early phase of action potential initiation or depolarization; its role in nociception also remains to be characterized [12, 13]. Table 1 summarizes the calcium channel antagonists that have been studied to date. This review will focus on the L-, N- and P/Q-type channel antagonists.

Mechanisms and Mediators of Neuropathic Pain, edited by Annika B. Malmberg and Sandra R. Chaplan

Table 1 - Antagonists of high-threshold voltage-dependent calcium channels

L:	diltiazem, nicardipine, nifedipine, nimodipine, nitrendipine, verapamil
N:	CTxGVIA (SNX-124), Ziconotide (SNX-111), SNX-159, SNX-239
P/Q:	AgaIVA

Effects of the calcium channel antagonists on nociceptive processing

Based on preclinical studies, the processing of nociceptive information may be characterized in terms of several discrete mechanisms. These mechanisms reflect: (1) the processing of acute nociceptive input that results in short-lasting depolarizations of Aδ and C fibers; (2) facilitated states that arise from persistent small unmyelinated afferent input (as after tissue injury); and (3) altered processing that arises secondary to nerve injury, yielding anomalous states of both hyperalgesia and allodynia. There is an abundance of research on the effects of the calcium channel antagonists on these different mechanisms of nociceptive processing.

Effects on acute nociceptive processing

L-type channels

Animal studies: Hara et al. demonstrated that intrathecal verapamil and diltiazem significantly increased both the heat-induced tail flick latency and colorectal distention threshold in rats [14]. Yildiran et al. demonstrated a significant positive affect of verapamil on acetic acid induced writhing and hot plate testing in mice. The effects were comparable to salicylate analgesia [15]. In anesthetized rats, Neugebauer et al. showed that spinally administered nimodipine significantly decreased spinal dorsal horn nociceptive responses to noxious pressure applied to a normal knee [16]. Iwasaki et al. showed that the local peripheral application of verapamil, diltiazem and nicardipine had no effect on tail-flick latency in the rat [17]. However, when administered together with lidocaine, the area under the curve was significantly increased in a dose-dependent fashion. Omote et al. showed similar results with the intrathecal delivery of verapamil [18]; verapamil alone had no sensory motor effect; however, when administered together with lidocaine or tetracaine, there was a significant prolongation of the anesthesia. There is one study that found no effect of intrathecal verapamil on the hot plate test in rats [19] and one study that found no effect of intrathecal verapamil, diltiazem, and nicardipine on the tail flick model in rats. However, when administered with morphine, there was a synergistic effect of these three agents [20].

Human studies: The only human study evaluating the effect of L-type calcium channel antagonists on acute pain was performed by Taniguchi et al. They showed that the peripheral delivery of nicardipine, verapamil or diltiazem by iontophoresis elevated the pain threshold to the same degree as lidocaine. When calcium channel antagonists were administered in combination with lidocaine, the pain threshold elevation was not changed, but the duration of the analgesic action was prolonged when compared with lidocaine alone [21].

N-type channels

The most recent studies have focused on the N-type calcium channel. VDCC of the N-type exist in the superficial laminae of the dorsal horn associated with primary afferent terminals, and are thought to modulate nociceptive processing by a central mechanism [22]. Blockade of the N-type calcium channel in the superficial dorsal horn modulates membrane excitability and inhibits neurotransmitter release, resulting in pain relief [23]. Ziconotide, a 25-amino acid peptide that is a synthetic version of a naturally-occurring peptide found in the venom of the marine snail, *Conus magus*, specifically and selectively binds to N-type voltage-sensitive calcium channels [24–26]. It is the first N-type calcium channel antagonist to enter clinical development.

Animal and human studies: Although the N-type calcium channel antagonists have not been studied on acute nociceptive processing in humans, there are animal studies that have demonstrated a significant reduction of the response to the hot-plate and paw pressure in the rat, particularly after chronic intrathecal infusion of ziconotide [19, 27, 28]. Omote et al. demonstrated a significant increase in tail flick latency and paw pressure withdrawal after intrathecal CTxGVIA [29]. In addition, Neugebauer et al. showed that spinally administered CTxGVIA significantly decreased spinal dorsal horn nociceptive responses to noxious pressure applied to a normal knee [16]. However, two studies have shown no effect of spinally administered CTxGVIA [30] and ziconotide [31] on paw pressure withdrawal and one study showed no effect of intravenous ziconotide, SNX239, and SNX159 on paw pressure withdrawal [1]. A study by Wang et al. evaluated the effect of both single dosing and a 7-day continuous infusion of ziconotide on acute nociceptive processing. They demonstrated that both the single dose and continuous infusion produced analgesia in the rat hot-plate and tail flick models [28]. In comparison, morphine had a significant effect only with acute dosing, and tolerance developed to morphine during the 7-day infusion.

P/Q-type channels

Animal and human studies: There are no human studies of P/Q-type channel antagonists on acute nociceptive processing, and the animal studies have shown a general lack of effect. In the anesthetized rat, Nebe et al. demonstrated that spinally administered AgaIV produced both increases and decreases in the responses of single nociceptive neurons to noxious pressure applied to the normal joint [32]. This is consistent with studies that show that the P/Q-type channel is involved in the release of both excitatory and inhibitory neurotransmitters [33, 34]. Other studies have shown no effect of P/Q-type antagonists in the hot-plate test [19], or on mechanical or heat stimulation of the normal rat hindpaw [30, 31].

Effects on facilitated pain processing

L-type channels

Animal studies: Several animal studies have demonstrated a lack of effect of verapamil and diltiazem on the early and late phase of the formalin model [19, 35]. However, models using the inflamed knee joint have shown mixed results. Neugebauer et al. demonstrated that spinally administered nimodipine or diltiazem depressed spinal nociceptive neuron responses to noxious pressure applied to the inflamed knee joint of the rat [16]. On the other hand, Sluka found no effect of spinally administered nifedipine on central sensitization induced by the inflamed knee joint of the rat [31].

Human studies: There have been several studies on the effects of spinally and systemically delivered L-type channel antagonists on facilitated pain in humans. Choe et al. demonstrated that the combination of epidural verapamil and bupivacaine significantly decreased total opioid requirements over bupivacaine alone in patients undergoing lower abdominal surgery [36]. Lehmann et al. demonstrated a significant reduction in fentanyl requirements in hysterectomy patients who received a nimodipine infusion [37]. In addition, there are numerous reports on the efficacy of the calcium channel antagonists on the prevention and treatment of migraines and chronic daily headaches [38, 39].

N-type channels

Animal studies: N-type channel antagonists significantly affect formalin induced hyperalgesia; the spinal delivery of ziconotide [19, 27, 28, 40] and CTxGVIA [35] has been shown to be effective in both phase I and, more robustly, phase II of the formalin test. In addition, the spinal delivery of ziconotide [31] and CTxGVIA [16, 32] decreased primary and secondary hyperalgesia in the acute arthritis model in the rat. Using the intradermal capsaicin model, Sluka demonstrated that spinally administered CTxGVIA prevented the development of secondary hyperalgesia and allodynia in a dose-dependent manner [30]. Using a rat postoperative pain model, Wang et al. demonstrated that the intrathecal delivery of ziconotide resulted in a dose-dependent reduction in heat hyperalgesia [41]. Unlike μ-opioid agonists, there is no development of tolerance with chronic infusions of up to 7 days [27].

Human studies: A pilot study of 24 patients undergoing orthopedic and lower abdominal surgery showed a significant reduction in intravenous morphine consumption with the continuous infusion of 7 µg/h of epidural ziconotide. The side effects and safety profiles were acceptable [42]. There is currently a large phase III trial in progress on the efficacy of ziconotide in postoperative pain.

P/Q-type channels

Animal and human studies: Whereas AgaIVA has minimal to no effects on acute nociceptive processing, several studies have shown significant effects on facilitated pain. The spinal delivery of AgaIVA in rats has been shown to significantly affect the formalin model in the rat, with the effect being greater on the second phase [19, 35]. Sluka demonstrated an effect of intraspinal AgaIVA on the secondary mechanical hyperalgesia and allodynia induced by intradermal capsaicin [30]. There are no human studies on the effects of the P/Q channel antagonists on facilitated pain.

Effects on neuropathic pain

L-type channels

Animal studies: Using the tight spinal nerve ligation model of neuropathic pain in the rat [13], Chaplan et al. demonstrated that intrathecal diltiazem, verapamil and nimodipine had no significant effect on the pain behavior at the highest dose studied [1]. In a study using streptozotocin-induced diabetic neuropathy in the rat [43], Calcutt et al. demonstrated no effect of intrathecal diltiazem on paw withdrawal response. In a study on the systemic delivery of L-type channel antagonists, White and Cousins demonstrated no effect of subcutaneously delivered nifedipine on the mechanical hyperalgesia induced by partially ligating the sciatic nerve in the rat [44].

Human studies: Intractable cancer pain is often due to nervous system invasion of the cancer or nervous system injury secondary to the cancer treatment. Using a double-blind, placebo-controlled study, Santillan et al. demonstrated that the addition of nimodipine in cancer patients requiring a morphine dose escalation significantly reduced the morphine requirements [45]. In a study on complex regional pain syndrome type I and II, Muizelaar et al. demonstrated an effect of nifedipine on pain that was comparable to phenoxybenzamine. The effects were greater on early-stage than late-stage disease [46].

N-type channels

Animal studies: Using the spinal nerve ligation model of neuropathic pain [13], Chaplan et al. demonstrated that intrathecal ziconotide, SNX-159 and SNX-239, pro-

duced a dose-dependent decrease in mechanical allodynia [1]. Bowersox et al. demonstrated similar results with both intrathecal bolus and infusion dosing of ziconotide [40]. On the other hand, Yamamoto and Sakashita found a significant effect of N-type calcium channel antagonists on the heat hyperalgesia in the chronic constriction nerve injury model [47] but no effect on the heat hyperalgesia of the spinal nerve ligation model [48]. In a study of streptozotocin-induced diabetic neuropathy in the rat [43], Calcutt et al. demonstrated a significant effect of intrathecal SNX-239 on paw withdrawal responses. Studies on the peripheral effects of the N-type channel antagonists on neuropathic pain are mixed. Xiao and Bennett demonstrated a significant effect of ziconotide on heat and mechanical hyperalgesia when chronically applied to the injured area of the axon in the chronic constriction injury model [49]. White and Cousins demonstrated a significant effect of subcutaneously delivered ziconotide on the mechanical hyperalgesia induced by partially ligating the sciatic nerve in the rat [44]. On the other hand, Chaplan et al., using the spinal nerve ligation model, showed no effect of ziconitide or SNX-239 acutely applied to the injured portion of the nerve, or delivered by the intravenous route [1].
Human studies: The first clinical report on the use of ziconotide for neuropathic pain described successful management with intrathecal ziconotide of a patient with refractory pain due to brachial plexus avulsion injury [50]. One double-blind, placebo controlled study has been conducted on intrathecal ziconotide on refractory neuropathic pain. Two hundred and fifty-nine patients with a wide variety of neuropathic pain syndromes were enrolled; a preliminary report shows that intrathecal ziconotide resulted in a significant reduction of pain over placebo. In this study, the following side effects were also reported, all of which resolved upon dosage reduction or discontinuation: dizziness, nausea, nystagmus, gait imbalance, confusion, constipation and urinary retention [51]. Orthostatic hypotension has also been reported with intrathecal ziconotide, thought to be secondary to the sympatholytic effect of the calcium channel blockade [52]. Another study on the short-term epidural delivery of ziconotide for postoperative pain did not report these side effects [42]. The selectivity of ziconotide for the N-type VDCC is very high [53]; thus these effects are likely due to the importance of N-type VDCC in various pathways in addition to nociception, rather than nonspecific actions. It appears that spinally delivered ziconotide has a narrow therapeutic window, but provided that dosing is carefully titrated, analgesia does appear to be achievable without unacceptable side effects. Intrathecal ziconotide is currently in phase III trials for the management of neuropathic pain.

P/Q-type channels

Animal and human studies: All studies on the effects of the P/Q-type channel antagonists on neuropathic pain have been negative. Chaplan et al. demonstrated no effect of intrathecal AgaIVA on mechanical allodynia in the SNL model [1].

Table 2 - Summary of animal studies on the effect of intrathecal calcium channel antagonists on nociception

	Acute		Facilitated			Neuropathic	
			Formalin				
	Thermal	Mechanical	Phase I	Phase II	Other	Thermal	Mechanical
L	+ [15]	+ [16]	– [19]	– [19]	+ [16]		– [1]
	+ [14]						
	– [20]]		– [35]	– [35]	– [31]		
N	+ [27]						
	+ [29]	+ [27]	+ [27]	+ [27]	+ [16]	+ [48]	+ [1]
	+ [41]	+ [16]	+ [40]	+ [40]	+ [30]	(Bennett model)	+ [40]
		+ [29]	+ [35]	+ [35]	+ [31]	– [48];	+ [43]
		+ [41]	+ [41]	+ [41]	+ [32]	(Chung model)	
		– [30]			+ [41]		
		– [31]					
P/Q	– [30]	– [32]	– [19]	– [19]	+ [30]	– [48]	– [1]
	– [31]		– [32]	– [35]			– [44]

+, positive effect on nociception; –, no effect on nociception

Yamamoto and Sakashita showed no effects of intrathecal AgaIVA on the heat hyperalgesia of the CCI model and the SNL model [48]. In a study on the systemic delivery of P/Q channel antagonists, White and Cousins demonstrated no effects of subcutaneously delivered SNX-230 on the mechanical hyperalgesia induced by partially ligating the sciatic nerve in the rat [44]. There are no human studies on the effects of the P/Q channel antagonists on neuropathic pain.

Conclusions

Tables 2 and 3 summarize animal and human studies on the effects of the VDCC antagonists on nociception. It appears that the N-type antagonists are more likely to be clinically useful compared to the L- and P/Q-type antagonists. The therapeutic ratio of N-type VDCC antagonists increases with spinal delivery; systemic and local peripheral delivery are less effective. There also appear to be synergistic effects in combination with either local anesthetics (both peripheral and central delivery) and opioids. Preliminary double-blind, placebo-controlled studies in human postoperative and neuropathic pain are very promising for the N-type channel antago-

Table 3 - Summary of human studies on the effect of calcium channel antagonists on nociception

	Acute	Facilitated	Neuropathic
L	No human studies	+ [36]	+ [46]
		+ [37]	+ [45]
N		+ [42]	+ [50]
			+ [51]
P/Q		No human studies	No human studies

+, positive effect on nociception; –, no effect on nociception

nists; ziconotide is likely to be the first FDA approved VDCC agent for the treatment of neuropathic and postoperative pain.

Acknowledgement
I would like to thank Linda Sutherland for her editorial assistance.

References

1 Chaplan SR, Pogrel JW, Yaksh TL (1994) Role of voltage-dependent calcium channel subtypes in experimental tactile allodynia. *J Pharmacol Exp Ther* 269: 1117–1123
2 Bean BP (1989) Classes of calcium channels in vertebrate cells. *Annu Rev Physiol* 51: 367–384
3 Hess P (1990) Calcium channels in vertebrate cells. *Annu Rev Neurosci* 13: 337–356
4 Swandulla D, Carbone E, Lux HD (1991) Do calcium channel classifications account for neuronal calcium channel diversity? *Trends Neurosci* 14: 46–51
5 Bertolino M, Llinas RR (1992) The central role of voltage-activated and receptor-operated calcium channels in neuronal cells. *Annu Rev Pharmacol Toxicol* 32: 399–421
6 Dunlap K, Luebke JI, Turner TJ (1995) Exocytotic Ca^{2+} channels in mammalian central neurons. *Trends Neurosci* 18: 89–98
7 Miljanich GP, Ramachandran J (1995) Antagonists of neuronal calcium channels: structure, function, and therapeutic implications. *Annu Rev Pharmacol Toxicol* 35: 707–734
8 Olivera BM, Miljanich GP, Ramachandran J, Adams ME (1994) Calcium channel diversity and neurotransmitter release: the omega-conotoxins and omega-agatoxins. *Annu Rev Biochem* 63: 823–867
9 Vanegas H, Schaible H (2000) Effects of antagonists to high-threshold calcium channels upon spinal mechanisms of pain, hyperalgesia and allodynia. *Pain* 85: 9–18

10 Saegusa H, Kurihara T, Zong S, Minowa O, Kazuno A, Han W, Matsuda Y, Yamanaka H, Osanai M, Noda T et al (2000) Altered pain responses in mice lacking alpha 1E subunit of the voltage-dependent Ca2+ channel. *Proc Natl Acad Sci USA* 97: 6132–6137

11 Smith SM, Piedras-Rentera ES, Namkung Y, Shin HS, Tsien RW (1999) Neuronal voltage-activated calcium channels: on the roles of the alpha 1E and beta 3 subunits. *Ann NY Acad Sci* 868: 175–198

12 McRory JE, Santi CM, Hamming JM, Sutton KG, Baillie DL, Stea A, Snutch TP (2001) Molecular and functional characterization of a family of rat brain T-type calcium channels. *J Biol Chem* 276: 3999–4011

13 Kim SH, Chung JM (1992) An experimental model for peripheral neuropathy produced by segmental spinal nerve ligation in the rat. *Pain* 50: 355–363

14 Hara K, Saito Y, Kirihara Y, Sakura S, Kosaka Y (1998) Antinociceptive effects of intrathecal L-type calcium channel blockers on visceral and somatic stimuli in the rat. *Anesth Analg* 87: 382–387

15 Yildiran G, Ulak G, Utkan T, Goldeli E, Gacar N, Erden F (1997) Antinociception induced by verapamil and chloramphenicol in mice. *Biol Neonate* 72: 28–31

16 Neugebauer V, Vanegas H, Nebe J, Rumenapp P, Schaible HG (1996) Effects of N- and L-type calcium channel antagonists on the responses of nociceptive spinal cord neurons to mechanical stimulation of the normal and the inflamed knee joint. *J Neurophysiol* 76: 3740–3749

17 Iwasaki H, Ohmori H, Omote K, Kawamata M, Sumita S, Yamauchi M, Namiki A (1996) Potentiation of local lignocaine-induced sensory block by calcium channel blockers in rats. *Br J Anaesth* 77: 243–247

18 Omote K, Iwasaki H, Kawamata M, Satoh O, Namiki A (1995) Effects of verapamil on spinal anesthesia with local anesthetics. *Anesth Analg* 80: 444–448

19 Malmberg AB, Yaksh TL (1994) Voltage-sensitive calcium channels in spinal nociceptive processing: blockade of N- and P-type channels inhibits formalin-induced nociception. *J Neurosci* 14: 4882–4890

20 Omote K, Sonoda H, Kawamata M, Iwasaki H, Namiki A (1993) Potentiation of antinociceptive effects of morphine by calcium-channel blockers at the level of the spinal cord. *Anesthesiology* 79: 746–752

21 Taniguchi K, Miyagawa A, Mizutani A, Honda N, Oyama T (1995) The effect of calcium channel antagonist administered by iontophoresis on the pain threshold. *Acta Anaesthesiol Belg* 46: 69–73

22 Gohil K, Bell JR, Ramachandran J, Miljanich GP (1994) Neuroanatomical distribution of receptors for a novel voltage-sensitive calcium-channel antagonist, SNX-230 (omega-conopeptide MVIIC). *Brain Res* 653: 258–266

23 Augustine GJ, Charlton MP, Smith SJ (1987) Calcium action in synaptic transmitter release. *Annu Rev Neurosci* 10: 633–693

24 Olivera BM, Gray WR, Zeikus R, McIntosh JM, Varga J, Rivier J, de Santos V, Cruz LJ (1985) Peptide neurotoxins from fish-hunting cone snails. *Science* 230: 1338–1343

25 Aosaki T, Kasai H (1989) Characterization of two kinds of high-voltage-activated Ca-

channel currents in chick sensory neurons. Differential sensitivity to dihydropyridines and omega-conotoxin GVIA. *Pflugers Arch* 414: 150–156
26 Sher E, Clementi F (1991) Omega-conotoxin-sensitive voltage-operated calcium channels in vertebrate cells. *Neuroscience* 42: 301–307
27 Malmberg AB, Yaksh TL (1995) Effect of continuous intrathecal infusion of omega-conopeptides, N-type calcium-channel blockers, on behavior and antinociception in the formalin and hot-plate tests in rats. *Pain* 60: 83–90
28 Wang YX, Gao D, Pettus M, Phillips C, Bowersox SS (2000) Interactions of intrathecally administered ziconotide, a selective blocker of neuronal N-type voltage-sensitive calcium channels, with morphine on nociception in rats. *Pain* 84: 271–281
29 Omote K, Kawamata M, Satoh O, Iwasaki H, Namiki A (1996) Spinal antinociceptive action of an N-Type voltage-dependent calcium channel blocker and the synergistic interaction with morphine. *Anesthesiology* 84: 636–643
30 Sluka KA (1997) Blockade of calcium channels can prevent the onset of secondary hyperalgesia and allodynia induced by intradermal injection of capsaicin in rats. *Pain* 71: 157–164
31 Sluka KA (1998) Blockade of N- and P/Q-type calcium channels reduces the secondary heat hyperalgesia induced by acute inflammation. *J Pharmacol Exp Ther* 287: 232–237
32 Nebe J, Vanegas H, Neugebauer V, Schaible HG (1997) Omega-agatoxin IVA, a P-type calcium channel antagonist, reduces nociceptive processing in spinal cord neurons with input from the inflamed but not from the normal knee joint—an electrophysiological study in the rat *in vivo*. *Eur J Neurosci* 9: 2193–2201
33 Takahashi T, Momiyama A (1993) Different types of calcium channels mediate central synaptic transmission. *Nature* 366: 156–158
34 Turner TJ, Adams ME, Dunlap K (1993) Multiple Ca2+ channel types coexist to regulate synaptosomal neurotransmitter release. *Proc Natl Acad Sci USA* 90: 9518–9522
35 Diaz A, Dickenson AH (1997) Blockade of spinal N- and P-type, but not L-type, calcium channels inhibits the excitability of rat dorsal horn neurones produced by subcutaneous formalin inflammation. *Pain* 69: 93–100
36 Choe H, Kim JS, Ko SH, Kim DC, Han YJ, Song HS (1998) Epidural verapamil reduces analgesic consumption after lower abdominal surgery [see comments]. *Anesth Analg* 86: 786–790
37 Lehmann KA, Kriegel R, Ueki M (1989) [The clinical significance of drug interactions between opiates and calcium antagonists. A randomized double-blind study using fentanyl and nimodipine within the framework of postoperative intravenous on-demand analgesia]. *Anaesthesist* 38: 110–115
38 Micieli G, Piazza D, Sinforiani E, Cavallini A, Trucco M, Gabellini S, Mancuso A, Pacchetti C (1985) Antimigraine drugs in the management of daily chronic headaches: clinical profiles of responsive patients. *Cephalalgia* 5: 219–224
39 Schuler ME, Goldman MP, Munger MA (1988) The role of calcium channel blocking agents in the prevention of migraine. *Drug Intell Clin Pharm* 22: 187–191
40 Bowersox SS, Gadbois T, Singh T, Pettus M, Wang YX, Luther RR (1996) Selective N-

type neuronal voltage-sensitive calcium channel blocker, SNX-111, produces spinal antinociception in rat models of acute, persistent and neuropathic pain. *J Pharmacol Exp Ther* 279: 1243–1249
41 Wang YX, Pettus M, Gao D, Phillips C, Scott Bowersox S (2000) Effects of intrathecal administration of ziconotide, a selective neuronal N-type calcium channel blocker, on mechanical allodynia and heat hyperalgesia in a rat model of postoperative pain. *Pain* 84: 151–158
42 Atanasoff PG, Wallace MS, Wendel D, Edell TA, Ross M (1998) Analgesic efficacy of epidural ziconitide in acute postoperative pain. *American Pain Society 17th Annual Scientific Meeting* A894
43 Calcutt NA, Chaplan SR (1997) Spinal pharmacology of tactile allodynia in diabetic rats. *Br J Pharmacol* 122: 1478–1482
44 White DM, Cousins MJ (1998) Effect of subcutaneous administration of calcium channel blockers on nerve injury-induced hyperalgesia. *Brain Res* 801: 50–58
45 Santillan R, Hurle MA, Armijo JA, de los Mozos R, Florez J (1998) Nimodipine-enhanced opiate analgesia in cancer patients requiring morphine dose escalation: a double-blind, placebo-controlled study. *Pain* 76: 17–26
46 Muizelaar JP, Kleyer M, Hertogs IA, DeLange DC (1997) Complex regional pain syndrome (reflex sympathetic dystrophy and causalgia): management with the calcium channel blocker nifedipine and/or the alpha-sympathetic blocker phenoxybenzamine in 59 patients. *Clin Neurol Neurosurg* 99: 26–30
47 Bennett G, Xie Y-K (1988) A peripheral mononeuropathy in rat that produces disorders of pain sensation like those seen in man. *Pain* 33: 87–107
48 Yamamoto T, Sakashita Y (1998) Differential effects of intrathecally administered N- and P-type voltage-sensitive calcium channel blockers upon two models of experimental mononeuropathy in the rat. *Brain Res* 794: 329–332
49 Xiao WH, Bennett GJ (1995) Synthetic omega-conopeptides applied to the site of nerve injury suppress neuropathic pains in rats. *J Pharmacol Exp Ther* 274: 666–672
50 Brose WG, Gutlove DP, Luther RR, Bowersox SS, McGuire D (1997) Use of intrathecal SNX-111, a novel, N-type, voltage-sensitive, calcium channel blocker, in the management of intractable brachial plexus avulsion pain. *Clin J Pain* 13: 256–259
51 Presley R, Charapata S, Ferrar-Brecher T, Yearwood T, Staats P, Wallace MS, Dubois M, Gaeta R, Follett K, Gaudette M et al (1998) Chronic, opioid resistant, neuropathic pain: marked analgesic efficacy of intrathecal ziconitide. *American Pain Society 17th Annual Scientific Meeting* A697
52 McGuire D, Bowersox S, Fellmann JD, Luther RR (1997) Sympatholysis after neuron-specific, N-type, voltage-sensitive calcium channel blockade: first demonstration of N-channel function in humans. *J Cardiovasc Pharmacol* 30: 400–403
53 Wang YX, Bezprozvannaya S, Bowersox SS, Nadasdi L, Miljanich G, Mezo G, Silva D, Tarczy-Hornoch K, Luther RR (1998) Peripheral *versus* central potencies of N-type voltage-sensitive calcium channel blockers. *Naunyn Schmiedebergs Arch Pharmacol* 357: 159–168

Cholinergic, adenosinergic and α_2 adrenergic mechanisms in chronic neuropathic pain

[1]Hartmut Buerkle and [2]Stuart A. Dunbar

[1]Department of Anesthesiology and Critical Care Medicine, Westfälische Wilhelms-Universität Münster, Albert-Schweitzer-Str. 33, 48149 Münster, Germany; [2]Department of Anesthesiology, Tufts University School of Medicine, Baystate Medical Center, 759 Chestnut Street, Springfield, MA 01199, USA

Cholinergic, purinergic and α_2-adrenergic agonists in neuropathic pain

The G-protein connected receptor systems of these three respective drug classes have widespread distributions throughout the body in addition to the central nervous system. Systemic administration of all three of these types of drugs is associated with pronounced effects that are clinically useful in different therapeutic situations, but may be profoundly adverse if uncontrolled. Acetylcholinesterase inhibitors are used to treat myasthenia gravis, and, in combination with muscarinic antagonists, to reverse neuromuscular blockade; numerous muscarinic side effects include bradycardia, bronchorrhea, bronchospasm and lacrimation. Adenosine is used therapeutically as a vasodilator and for pharmacological termination of tacharrhythmias; α_2 adrenergic agonists are useful antihypertensives. Because of these systemic effects, most reports investigating their analgesic potential describe spinal or epidural administration of these various agents, although systemic administration of α_2 agonists is described in the management of spinal spasticity, and systemic delivery of adenosine has been studied for neuropathic pain relief.

Cholinergic mechanisms in neuropathic pain

The spinal delivery of cholinomimetic agents produces analgesia, which is blocked by muscarinic receptor antagonists [1, 2]. In addition to a direct mechanism, nicotinergic stimulation can result in spinal antinociception through the release of norepinephrine and acetylcholine. There is a body of evidence suggesting that cholinergic antinociception is partly due to nitric oxide release at the spinal dorsal horn level [3–5]. The cholinergic receptors that mediate nociceptive impulses are

Mechanisms and Mediators of Neuropathic Pain, edited by Annika B. Malmberg and Sandra R. Chaplan

localized in the superficial and deep dorsal horn of the spinal cord, and their action is primarily intrinsic. Currently, equivocal data exist for distinct muscarinic receptors being involved in the analgesic effects of intrathecally administered muscarinic agonists [6, 7]. Of the five known muscarinic acetylcholine receptors, M_1 receptors in sheep [8] and rats [9] and M_2 receptors are thought to be the main subtypes involved in spinal antinociception.

Although direct intrathecal delivery of muscarinergic and nicotinergic agents will produce antinociception, the dosages required also result in adverse behavioral effects [10]. Hence the focus of most studies on antinociceptive cholinergic effects has been to enhance endogenous acetylcholine concentrations by blocking its degradation by acetylcholinesterase using neostigmine. Neostigmine, a quaternary amide, does not cross the blood brain barrier, but can be administered epidurally or spinally for pain control. Recent animal experiments suggest that intrathecal neostigmine antagonizes touch- evoked allodynia by a spinal mechanism in a rat model of neuropathic pain, and that the antiallodynic action of cholinesterase inhibitors is probably mediated by a spinal muscarinic receptor, likely the M_1 subtype [9].

Based on the lack of toxicity in preclinical studies, spinal administration of neostigmine is generally considered to be acceptable, and numerous clinical studies have employed this route of administration. However, there are so far no data on the use of neostigmine in human neuropathic pain syndromes [11–19].

Purinergic mechanisms in neuropathic pain

Analgesic interactions between purinergic and adrenergic systems have been demonstrated in studies of combined administration of adenosine and clonidine. This combination resulted in reduced hypersensitivity and interacted in an additive manner in a rat model of neuropathic pain [20]. The observed effects of adenosine were blocked by the intrathecal injection of A1 but not A2 adenosine receptor antagonists, by an α_2-adrenergic antagonist, and as well by the destruction of spinal noradrenergic nerve terminals [20]. The authors of this study concluded that after nerve injury, adenosine acts to reduce hypersensitivity through spinal norepinephrine release [20]. This finding is in accord with another recent study that found no effect of adenosine on acute pain, but significant effects on a subsequent hyperalgesic state. In addition it was found that spinal adenosine results in an enhancement of clonidine antinociception in an acute nociceptive setting [21].

Intravenous administration of adenosine reduces clinical and experimental pain in humans [22, 23]. As described above, this route has limited therapeutic potential due to cardivascular side effects. Intrathecal delivery of adenosine agonists produces antinociceptive effects in animals, and its safety has been demonstrated in rats and dogs [24]. Adenosine reduces experimental pain in human volunteers by IT administration [25]. One healthy volunteer receiving IT adenosine (2000 μg) reported transient lumbar pain. Further phase I safety trials of acute intrathecal adenosine administration in humans also revealed no apparent data for toxicity [25] although adenosine can also elicit an algogenic response, reported by one volunteer in a tolerability study, limiting the spinal bolus dose to 2000 µg. Motor weakness, one of the most predominant side effects, is only observed at concentrations exceeding those needed for antinociception. The antinociceptive activity exceeds the very short biological half-life of adenosine (≈ 10 s) and it is most likely mediated through A1 receptors located in lamina I and II of the dorsal spinal horn. Activation of A1 receptors hyperpolarizes intrinsic neurones and inhibits the release of excitatory neurotransmitters such as substance P and glutamate. A study in humans showed that patients with neuropathic pain syndromes have reduced levels of blood and CSF adenosine [26]. This adenosine deficiency could explain the potential therapeutic effects of administering adenosine or its analogs [26].

α-adrenergic mechanisms in neuropathic pain

There is preclinical evidence that spinal α_2 agonists block spinal and peripheral neurophysiological adaptations that accompany chronic neuropathic pain. Animal studies show an enhancement of adrenergic sensitivity in intact sensory axons that innervate areas of skin made neuropathic by transection of neighboring nerves in animal models [27]. This shows that adrenergic hypersensitivity occurs beyond the immediate innervation of a neuropathic injury. Furthermore, in a rat model of spinal nerve ligation, injured DRG neurons giving rise to Aβ fibers become sensitive to catecholamines *via* activation of α_2-adrenoceptors on their surface [28]. Spinal α_2 agonists have been shown to reverse such nerve ligation induced allodynia in the rat [29]. It is tempting to speculate that inhibition of catecholamine release contributes to this effect. However, both intrathecal clonidine and neostigmine (which augments sympathetic outflow), but not surgical sympathectomy, were antihyperalgesic after nerve ligation in a recent study [30]. Clonidine, but not morphine, delays the development of thermal hyperesthesia induced by sciatic nerve constriction injury [31]. Other animal studies appear to show that α_2-agonists may be more selective as antihyperalgesic agents than morphine when administered spinally in neuropathic pain states [32, 33]. These studies suggest that spinal clonidine may be capable of blocking spinally mediated enhancement of nociception likely pursuant to upregulation of α_2 receptors in the dorsal horn in association with chronic pain.

Clinical studies also appear to support the use of neuraxial α_2 agonists in the management of chronic neuropathic pain. A description of the use of epidural clonidine in patients with deafferentation pain demonstrates its efficacy [34]. A subsequent study examined the effect of epidural clonidine in reflex sympathetic dystrophy (RSD). In a randomized, blinded, placebo-controlled study of 26 patients with RSD, 19 patients were found to obtain significant pain relief from epidural clonidine (mean infusion rate of 32 μg/h, range 15–50 μg/h, mean duration 43 days, range 7–225 days) [35]. Furthermore in a study of patients with severe intractable cancer pain unresponsive to maximally tolerated doses of oral or epidural opioids, epidural clonidine was successfully used as an analgesic for periods of up to five months in some patients. In the same study clonidine was found to be particularly useful in a subgroup of patients that had predominantly neuropathic pain [36]. Because these patients were terminally ill it is difficult to apply this data to the chronic non-cancer pain population. However this study formed the basis upon which epidural clonidine was accepted for use in the cancer patient population in the USA. It did not report any obvious long term side effects that might be consistent with neurotoxicity, despite the duration of some infusions, which were as long as 23 months. Patients showed higher VAS scores towards the end of the study reflecting either tolerance or a change in their underlying pain level.

The spinal antinociceptive mechanisms of α_2 agonists are likely due to effects at both presynaptic and postsynaptic sites. Presynaptic effects include inhibition of primary afferent neurotransmitter release in the spinal cord dorsal horn. These neurotransmitters include calcitonin gene-related peptide, substance P, vasoactive intestinal polypeptide, and glutamate. Postsynaptically, clonidine has an excitatory effect on spinal neurons with low threshold receptive fields, but an inhibitory effect upon wide dynamic range and high threshold neurons in the rat dorsal horn [37]. This suggests a mechanism of action whereby clonidine may inhibit wide dynamic neuron hyperexcitability, a feature of chronic pain. In keeping with this formulation, α_2 agonists may possess direct antagonist properties at NMDA receptors present on wide dynamic range neurons [38]. Xylazine, romifidine, detomidine, dexmedetomidine and clonidine selectively depress NMDA receptor-mediated spinal reflexes.

One of the problems with chronic administration of opioids in patients with neuropathic pain has been the phenomenon of tolerance. The clinical sequelae of tolerance are loss of pain control and the need for opioid dose escalation; tolerance may in some cases be extreme enough to result in complete failure of therapy. Opioid tolerance theoretically might be treatable by substituting α_2 agonist therapy. Unfortunately just as for morphine, tolerance has been reported with all α_2 agonists, at least in animal models [39]. Animals typically become rapidly tolerant to α_2 agonists such as clonidine, dexmedetomidine and ST91, so that in 4–5 days there is no remaining analgesic effect. Encouragingly, however, animal studies show a lack of significant cross tolerance between opioids and α_2 agonists, suggesting that in each case tolerance is mediated *via* separate receptor effector pathways. A lack of cross tolerance between ST91 and morphine has also been demonstrated in primate studies [40], but there are few clinical observations, and no formal studies to show a lack of significant cross tolerance between spinal clonidine and morphine in man. As an incidental observation, in Eisenach's study of opioid tolerant cancer patients discussed previously [36], most responded to epidural clonidine, which argues against clinically significant cross tolerance in man.

Analgesic synergism between spinally administered α_2 agonists and opioids occurs in animal models of nerve injury [41]. Thus lower concentrations of both agonists administered together produce equivalent analgesia with fewer side-effects. Synergism between α_2 agonists and cholinergic agents has also been described [42]. There is just one outcome study in man of the combination of local anesthetics and α_2 agonists. Glynn et al. evaluated the effects of epidural clonidine, epidural lidocaine, and the combination of epidural clonidine and lidocaine in a double-blind cross-over study in 20 patients with chronic low-back pain and nine patients with neuropathic pain. Twelve of the 17 patients who completed all three arms of the study reported superior pain control from the combination of epidural clonidine and lidocaine [43]. Whether this represents synergism or an additive effect is unclear.

Co-administration of lower doses of α_2 agonists and opioids may offer several advantages. First, improved analgesia may result from synergistic effects. Second, a lack of cross tolerance between these agents raises hope that the combination of lower concentrations of each might slow the onset of tolerance, because tolerance development is a function not only of duration of exposure, but of concentration. Third, since the side effects of each drug class are both different and, in general, concentration-dependent, combination therapy may offer fewer side-effects. A small, randomized double blinded study by Glynn and associates of 20 patients with chronic non-cancer pain, illustrates differences between the side effects of equianalgesic bolus doses of spinal clonidine (150 µg) given to one group, and spinal morphine (5 mg) given to another [43]. Both groups had an equivalent analgesic response, but only patients given morphine had pruritus, nausea or vomiting. Only patients given clonidine developed hypotension and sedation.

Clonidine is approved for chronic epidural administration in patients with cancer pain. Epidurally administered clonidine appears to partition preferentially into the CSF rather than the bloodstream: a study by Glynn and associates showed that 150 µg of epidural clonidine in man produced a mean maximum CSF concentration of 228 ng/ml, but that the mean plasma concentration at all time points was less than 0.7 ng/ml [44]. In a study by Eisenach and co-workers, epidural clonidine was demonstrated to be as effective as spinal clonidine at blocking hyperalgesia associated with capsaicin administration in human volunteers [45]. Intrathecal administration may be a preferable route in many clinical situations, however, such as patients with pre-existing fully implantable intrathecal infusion devices. However, data supporting a lack of neurotoxicity after prolonged intrathecal administration in man are still insufficient. Two cancer patients having received prolonged infusions of opioid plus clonidine were examined post mortem and found to have no histopathological evidence of spinal toxicity. Otherwise, there are no formal outcome studies that examine the efficacy or toxicity of chronic spinal infusions of clonidine in the treatment of chronic pain.

Combined mechanisms of action

The roles of the cholinergic, purinergic, adrenergic and indeed the opioidergic receptor systems modulating neuropathic pain states are intricately interconnnected, and by no means simple. Our understanding of the receptor crosstalk mechanisms at play in these systems is evidently still very limited, and much remains to be learned in this regard. Of note, spinal opioid analgesia has been suggested to result in part from spinal adenosine release; furthermore, the decreased efficacy of opioids in neuropathic pain treatment may result from a disruption of opioid-purinergic mechanisms [46]. Analgesic effects of α_2 agonists may also be substantially indirect: it has been shown that spinally administered clonidine increases concentrations of acetylcholine in cerebrospinal fluid and microdialysates from spinal cord dorsal horn [47, 48], and a recent animal study demonstrated that the analgesic effect of intrathecally administered clonidine on neuropathic pain is mediated by both spinal muscarinic and nicotinic receptors [49]. As above, the interaction of spinal muscarinic and α_2 receptors may occur through an interaction with nitric oxide [3]. Thus, in additions to the mechanisms noted above, complex actions result from administration of specific agonists, and our ultimate understanding of one receptor system calls for a complete analysis of receptor system interrelationships.

In conclusion, there is little doubt that the spinal cord is a significant site of antinociceptive action for cholinergic, purinergic and adrenergic agonists [50]. These drug classes offer potentially valuable additions to the armamentarium of clinically meaningful antihyperalgesic agents. Such breakthroughs encourage us to continue to extend preclinical studies into man and experiment with additional new drugs and drug classes. As always, however, spinal administration of any substance requires careful consideration of the potential for neurotoxicity and the ratio of risk to benefit.

References

1 Yaksh TL, Dirksen R, Harty GJ (1985) Antinociceptive effects of intrathecally injected cholinomimetic drugs in the rat and cat. *Eur J Pharmacol* 117: 81–88

2 Naguib M, Yaksh TL (1994) Antinociceptive effects of spinal cholinesterase inhibition and isobolographic analysis of the interaction with mu and alpha 2 receptor systems. *Anesthesiology* 80: 1338–1348

3 Xu Z, Chen SR, J CE, Pan HL (2000) Role of spinal muscarinic and nicotinic receptors in clonidine-induced nitric oxide release in a rat model of neuropathic pain. *Brain Res* 861: 390–398

4 Xu Z, Tong C, Eisenach JC (1996) Acetylcholine stimulates the release of nitric oxide from rat spinal cord. *Anesthesiology* 85: 107–111

5 Xu Z, Li P, Tong C, Figueroa J, Tobin JR, Eisenach JC (1996) Location and character-

istics of nitric oxide synthase in sheep spinal cord and its interaction with alpha(2)-adrenergic and cholinergic antinociception. *Anesthesiology* 84: 890–899

6 Naguib M, Yaksh TL (1997) Characterization of muscarinic receptor subtypes that mediate antinociception in the rat spinal cord. *Anesth Analg* 85: 847–853

7 Pan HL, Song HK, Eisenach JC (1998) Effects of intrathecal neostigmine, bupivacaine, and their combination on sympathetic nerve activity in rats. *Anesthesiology* 88: 481–486

8 Bouaziz H, Tong C, Eisenach JC (1995) Postoperative analgesia from intrathecal neostigmine in sheep. *Anesth Analg* 80: 1140–1144

9 Hwang JH, Hwang KS, Leem JK, Park PH, Han SM, Lee DM (1999) The antiallodynic effects of intrathecal cholinesterase inhibitors in a rat model of neuropathic pain. *Anesthesiology* 90: 492–499

10 Kesingland AC, Gentry CT, Panesar MS, Bowes MA, Vernier JM, Cube R, Walker K, Urban L (2000) Analgesic profile of the nicotinic acetylcholine receptor agonists, (+)-epibatidine and ABT-594 in models of persistent inflammatory and neuropathic pain. *Pain* 86: 113–118

11 Guan X, Liang X, Liu X (1990) Interaction between intrathecal neostigmine and epidural clonidine in human. *Chen Tzu Yen Chiu* 15: 136–139

12 Eisenach JC, Hood DD, Curry R (1997) Phase I human safety assessment of intrathecal neostigmine containing methyl- and propylparabens. *Anesth Analg* 85: 842–846

13 Hood DD, Eisenach JC, Tuttle R (1995) Phase I safety assessment of intrathecal neostigmine methylsulfate in humans [see comments]. *Anesthesiology* 82: 331–343

14 Hood DD, Mallak KA, Eisenach JC, Tong C (1996) Interaction between intrathecal neostigmine and epidural clonidine in human volunteers. *Anesthesiology* 85: 315–325

15 Lauretti GR, Reis MP, Prado WA, Klamt JG (1996) Dose-response study of intrathecal morphine *versus* intrathecal neostigmine, their combination, or placebo for postoperative analgesia in patients undergoing anterior and posterior vaginoplasty. *Anesth Analg* 82: 1182–1187

16 Krukowski JA, Hood DD, Eisenach JC, Mallak KA, Parker RL (1997) Intrathecal neostigmine for post-cesarean section analgesia: dose response. *Anesth Analg* 84: 1269–1275

17 Lauretti GR, Reis MP (1997) Postoperative analgesia and antiemetic efficacy after subarachnoid neostigmine in orthopedic surgery. *Reg Anesth* 22: 337–342

18 Lauretti GR, Hood DD, Eisenach JC, Pfeifer BL (1998) A multi-center study of intrathecal neostigmine for analgesia following vaginal hysterectomy. *Anesthesiology* 89: 913–918

19 Yaksh TL, Grafe MR, Malkmus S, Rathbun ML, Eisenach JC (1995) Studies on the safety of chronically administered intrathecal neostigmine methylsulfate in rats and dogs [see comments]. *Anesthesiology* 82: 412–427

20 Gomes JA, Li X, Pan HL, Eisenach JC (1999) Intrathecal adenosine interacts with a spinal noradrenergic system to produce antinociception in nerve-injured rats. *Anesthesiology* 91: 1072–1079

21 Chiari AI, Eisenach JC (1999) Intrathecal adenosine: interactions with spinal clonidine and neostigmine in rat models of acute nociception and postoperative hypersensitivity. *Anesthesiology* 90: 1413–1421
22 Segerdahl M, Ekblom A, Sandelin K, Wickman M, Sollevi A (1995) Peroperative adenosine infusion reduces the requirements for isoflurane and postoperative analgesics. *Anesth Analg* 80: 1145–1149
23 Segerdahl M, Ekblom A, Sjolund KF, Belfrage M, Forsberg C, Sollevi A (1995) Systemic adenosine attenuates touch evoked allodynia induced by mustard oil in humans. *Neuroreport* 6: 753–756
24 Chiari A, Yaksh TL, Myers RR, Provencher J, Moore L, Lee CS, Eisenach JC (1999) Preclinical toxicity screening of intrathecal adenosine in rats and dogs. *Anesthesiology* 91: 824–832
25 Rane K, Segerdahl M, Goiny M, Sollevi A (1998) Intrathecal adenosine administration: a phase 1 clinical safety study in healthy volunteers, with additional evaluation of its influence on sensory thresholds and experimental pain. *Anesthesiology* 89: 1108–1115; discussion 1109A
26 Guieu R, Peragut JC, Roussel P, Hassani H, Sampieri F, Bechis G, Gola R, Rochat H (1996) Adenosine and neuropathic pain. *Pain* 68: 271–274
27 Moon DE, Lee DH, Han HC, Xie J, Coggeshall RE, Chung JM (1999) Adrenergic sensitivity of the sensory receptors modulating mechanical allodynia in a rat neuropathic pain model. *Pain* 80: 589–595
28 Leem JW, Gwak YS, Nam TS, Paik KS (1997) Involvement of alpha2-adrenoceptors in mediating sympathetic excitation of injured dorsal root ganglion neurons in rats with spinal nerve ligation. *Neurosci Lett* 234: 39–42
29 Yaksh TL, Pogrel JW, Lee YW, Chaplan SR (1995) Reversal of nerve ligation-induced allodynia by spinal alpha-2 adrenoceptor agonists. *J Pharmacol Exp Ther* 272: 207–214
30 Lavand'homme P, Pan HL, Eisenach JC (1998) Intrathecal neostigmine, but not sympathectomy, relieves mechanical allodynia in a rat model of neuropathic pain. *Anesthesiology* 89: 493–499
31 Yamamoto T, Nozaki-Taguchi N (1996) Clonidine, but not morphine, delays the development of thermal hyperesthesia induced by sciatic nerve constriction injury in the rat. *Anesthesiology* 85: 835–845
32 Xu XJ, Puke MJ, Wiesenfeld-Hallin Z (1992) The depressive effect of intrathecal clonidine on the spinal flexor reflex is enhanced after sciatic nerve section in rats. *Pain* 51: 145–151
33 Leiphart JW, Dills CV, Zikel OM, Kim DL, Levy RM (1995) A comparison of intrathecally administered narcotic and nonnarcotic analgesics for experimental chronic neuropathic pain. *J Neurosurg* 82: 595–599
34 Petros A, Wright R (1987) Epidural and Oral Clonidine in domicillary control of deafferentation pain. *Lancet* 1: 1034
35 Rauck RL, Eisenach JC, Jackson K, Young LD, Southern J (1993) Epidural clonidine

treatment for refractory reflex sympathetic dystrophy [see comments]. *Anesthesiology* 79: 1163–1169; discussion 1127A
36 Eisenach JC, DuPen S, Dubois M, Miguel R, Allin D; The Epidural Clonidine Study Group (1995) Epidural clonidine analgesia for intractable cancer pain. *Pain* 61 (3): 391–399
37 Millar J, O'Brien FE, Williams GV, Wood J (1993) The effects of iontophoretic clonidine on neurones in the rat superficial dorsal horn. *Pain* 53: 137–145
38 Faber E, Chambers J, Evans R (1998) Depression of NMDA receptor-mediated synaptic transmission by four alpha2 adrenoceptor agonists on the *in vitro* rat spinal cord preparation. *Br J Pharmacol* 124: 507–512
39 Takano Y, Yaksh TL (1993) Chronic spinal infusion of dexmedetomidine, ST-91 and clonidine: spinal alpha 2 adrenoceptor subtypes and intrinsic activity. *J Pharmacol Exp Ther* 264: 327–335
40 Yaksh TL, Reddy SV (1981) Studies in the primate on the analgetic effects associated with intrathecal actions of opiates, alpha-adrenergic agonists and baclofen. *Anesthesiology* 54: 451–467
41 Ossipov MH, Lopez Y, Bian D, Nichols ML, Porreca F (1997) Synergistic antinociceptive interactions of morphine and clonidine in rats with nerve ligation injury. *Anesthesiology* 86: 196–204
42 Abram SE, Winne RP (1995) Intrathecal acetyl cholinesterase inhibitors produce analgesia that is synergistic with morphine and clonidine in rats. *Anesth Analg* 81: 501–507
43 Glynn C, O'Sullivan K (1996) A double-blind randomised comparison of the effects of epidural clonidine, lignocaine and the combination of clonidine and lignocaine in patients with chronic pain. *Pain* 64: 337–343
44 Glynn C, Jamous M, Teddy P (1992) Cerebrospinal fluid kinetics of epidural clonidine in man. *Pain* 49: 361–367
45 Eisenach J, Hood D, Curry R (2000) Relative potency of epidural to intrathecal clonidine differs between acute thermal pain and capsaicin-induced allodynia. *Pain* 84: 57–64
46 Lavand'homme PM, Eisenach JC (1999) Exogenous and endogenous adenosine enhance the spinal antiallodynic effects of morphine in a rat model of neuropathic pain. *Pain* 80: 31–36
47 Xu Z, Tong C, Pan HL, Cerda SE, Eisenach JC (1997) Intravenous morphine increases release of nitric oxide from spinal cord by an alpha-adrenergic and cholinergic mechanism. *J Neurophysiol* 78: 2072–2078
48 Detweiler DJ, Eisenach JC, Tong C, Jackson C (1993) A cholinergic interaction in alpha 2 adrenoceptor-mediated antinociception in sheep. *J Pharmacol Exp Ther* 265: 536–542
49 Pan HL, Chen SR, Eisenach JC (1999) Intrathecal clonidine alleviates allodynia in neuropathic rats: interaction with spinal muscarinic and nicotinic receptors. *Anesthesiology* 90: 509–514
50 Poon A, Sawynok J (1998) Antinociception by adenosine analogs and inhibitors of adenosine metabolism in an inflammatory thermal hyperalgesia model in the rat. *Pain* 74: 235–245

Index

The PIR-Series
Progress in Inflammation Research

Homepage: http://www.birkhauser.ch

Up-to-date information on the latest developments in the pathology, mechanisms and therapy of inflammatory disease are provided in this monograph series. Areas covered include vascular responses, skin inflammation, pain, neuroinflammation, arthritis cartilage and bone, airways inflammation and asthma, allergy, cytokines and inflammatory mediators, cell signalling, and recent advances in drug therapy. Each volume is edited by acknowledged experts providing succinct overviews on specific topics intended to inform and explain. The series is of interest to academic and industrial biomedical researchers, drug development personnel and rheumatologists, allergists, pathologists, dermatologists and other clinicians requiring regular scientific updates.

Available volumes:

T Cells in Arthritis, P. Miossec, W. van den Berg, G. Firestein (Editors), 1998
Chemokines and Skin, E. Kownatzki, J. Norgauer (Editors), 1998
Medicinal Fatty Acids, J. Kremer (Editor), 1998
Inducible Enzymes in the Inflammatory Response, D.A. Willoughby, A. Tomlinson (Editors), 1999
Cytokines in Severe Sepsis and Septic Shock, H. Redl, G. Schlag (Editors), 1999
Fatty Acids and Inflammatory Skin Diseases, J.-M. Schröder (Editor), 1999
Immunomodulatory Agents from Plants, H. Wagner (Editor), 1999
Cytokines and Pain, L. Watkins, S. Maier (Editors), 1999
In Vivo *Models of Inflammation*, D. Morgan, L. Marshall (Editors), 1999
Pain and Neurogenic Inflammation, S.D. Brain, P. Moore (Editors), 1999
Anti-Inflammatory Drugs in Asthma, A.P. Sampson, M.K. Church (Editors), 1999
Novel Inhibitors of Leukotrienes, G. Folco, B. Samuelsson, R.C. Murphy (Editors), 1999
Vascular Adhesion Molecules and Inflammation, J.D. Pearson (Editor), 1999
Metalloproteinases as Targets for Anti-Inflammatory Drugs, K.M.K. Bottomley, D. Bradshaw, J.S. Nixon (Editors), 1999
Free Radicals and Inflammation, P.G. Winyard, D.R. Blake, C.H. Evans (Editors), 1999
Gene Therapy in Inflammatory Diseases, C.H. Evans, P. Robbins (Editors), 2000
New Cytokines as Potential Drugs, S. K. Narula, R. Coffmann (Editors), 2000
High Throughput Screening for Novel Anti-inflammatories, M. Kahn (Editor), 2000
Immunology and Drug Therapy of Atopic Skin Diseases, C.A.F. Bruijnzeel-Komen, E.F. Knol (Editors), 2000
Novel Cytokine Inhibitors, G.A. Higgs, B. Henderson (Editors), 2000
Inflammatory Processes. Molecular Mechanisms and Therapeutic Opportunities, L.G. Letts, D.W. Morgan (Editors), 2000

Cellular Mechanisms in Airways Inflammation, C. Page, K. Banner, D. Spina (Editors), 2000
Inflammatory and Infectious Basis of Atherosclerosis, J.L. Mehta (Editor), 2001
Muscarinic Receptors in Airways Diseases, J. Zaagsma, H. Meurs, A.F. Roffel (Editors), 2001
TGF-β and Related Cytokines in Inflammation, S.N. Breit, S. Wahl (Editors), 2001
Nitric Oxide and Inflammation, D. Salvemini, T.R. Billiar, Y. Vodovotz (Editors), 2001
Neuroinflammatory Mechanisms in Alzheimer's Disease. Basic and Clinical Research, J. Rogers (Editor), 2001
Disease-modifying Therapy in Vasculitides, C.G.M. Kallenberg, J.W. Cohen Tervaert (Editors), 2001
Inflammation and Stroke, G.Z. Feuerstein (Editor), 2001
NMDA Antagonists as Potential Analgesic Drugs, D.J.S. Sirinathsinghji, R.G. Hill (Editors), 2002
Migraine: A neuroinflammatory disease? E.L.H. Spierings, M. Sanchez del Rio (Editors), 2002

MIX
Papier aus verantwortungsvollen Quellen
Paper from responsible sources
FSC® C105338

If you have any concerns about our products,
you can contact us on
ProductSafety@springernature.com

In case Publisher is established outside the EU,
the EU authorized representative is:
Springer Nature Customer Service Center GmbH
Europaplatz 3, 69115 Heidelberg, Germany

Printed by Libri Plureos GmbH
in Hamburg, Germany